ADVANCED SPACE PLASMA PHYSICS

Rudolf A. Treumann
Max-Planck-Institut für extraterrestrische Physik, Garching & Österreichische Akademie der Wissenschaften, Institut für Weltraumforschung, Graz

Wolfgang Baumjohann
Max-Planck-Institut für extraterrestrische Physik, Garching & Österreichische Akademie der Wissenschaften, Institut für Weltraumforschung, Graz

ADVANCED SPACE PLASMA PHYSICS

ICP

Imperial College Press

Published by

Imperial College Press
57 Shelton Street
Covent Garden
London WC2H 9HE

Distributed by

World Scientific Publishing Co. Pte. Ltd.
P O Box 128, Farrer Road, Singapore 912805
USA office: Suite 1B, 1060 Main Street, River Edge, NJ 07661
UK office: 57 Shelton Street, Covent Garden, London WC2H 9HE

British Library Cataloguing-in-Publication Data
A catalogue record for this book is available from the British Library.

First published 1997
Reprinted 2001

ADVANCED SPACE PLASMA PHYSICS

Copyright © 1997 by Imperial College Press

All rights reserved. This book, or parts thereof, may not be reproduced in any form or by any means, electronic or mechanical, including photocopying, recording or any information storage and retrieval system now known or to be invented, without written permission from the Publisher.

For photocopying of material in this volume, please pay a copying fee through the Copyright Clearance Center, Inc., 222 Rosewood Drive, Danvers, MA 01923, USA. In this case permission to photocopy is not required from the publisher.

ISBN 1-86094-026-9

Printed in Singapore.

Preface

This book is the second volume of our introductory text on Space Plasma Physics. The first volume is published under the title *Basic Space Plasma Physics* and covers the more fundamental aspects, i.e., single particle dynamics, fluid equilibria, and waves in space plasmas. This second volume extends the material to the more advanced fields of plasma instabilities and nonlinear effects.

Actually, there are already a number of monographs, where the general nonlinear plasma methods are described in considerable detail. But many of these books are quite specialized. The present book selects those methods, which are applied in space plasma physics, and, on the expense of detailedness, tries to make them accessible to the more practically oriented student and researcher by putting the new achievements and methods into the context of general space physics.

The first part of the book is concerned with the evolution of linear instabilities in plasmas. Instabilities have turned out to be the most interesting and important phenomena in physics. They arise when free energy has accumulated in a system which the system wants to get rid of. In plasma physics there is a multitude of reasons for the excitation of instabilities. Inhomogeneities may evolve both in real space and in velocity space. These inhomogeneities lead to the generation of instabilities as a first linear and straightforward reaction of the plasma to such deviations from thermal equilibrium. The first chapters cover a representative selection of the many possible macro- and microinstabilities in space plasmas, from the Rayleigh-Taylor and Kelvin-Helmholtz to electrostatic and electromagnetic kinetic instabilities. Their quasilinear stabilization and nonlinear evolution and their application to space physics problems is treated.

As a natural extension of the linear evolution, nonlinear effects do inevitably evolve in an unstable plasma, simply because an instability cannot persist forever but will exhaust the available free energy. Therefore all instabilities are followed by nonlinear evolution. The second part of the book, the chapters on nonlinear effects, can only give an overview about the vast field of nonlinearities. These chapters include the nonlinear evolution of single waves, weak turbulence, and strong turbulence, all presented from the view-point of their relevance for space plasma physics. Special topics include soliton formation, caviton collapse, anomalous transport, auroral particle acceleration, and elements of the theory of collisionless shocks.

Linear theory occupies about half of the book. The second half reviews nonlinear theory as systematically as possible, given the restricted space. The last chapter presents a number of applications. The reader may find our selection a bit unsystematic, but we have chosen to select only those which, in our opinion, demonstrate the currently more important aspects of space physics. There are many other small effects which need to be treated using nonlinear theory, but have been neglected here, since we did not find them fundamental enough to be included in a textbook like the present one. Nevertheless, we hope that the reader will find the book useful as a guide to unstable and nonlinear space plasma physics, giving him a taste of the complexity of the problems.

Since space plasma physics has in the past served as a reservoir of ideas and tools also for astrophysics, the present volume will certainly be useful for the needs of a course in non-relativistic plasma astrophysics and for scientists working in this field. With a slight extension to the parameter ranges of astrophysical objects most of the instabilities and nonlinear effects do also apply to astrophysics, as long as high-energy effects and relativistic temperatures are not important.

It is a pleasure to thank Rosmarie Mayr-Ihbe for turning our often rough sketches into the figures contained in this book and Thomas Bauer, Anja Czaykowska, Thomas Leutschacher, Reiner Lottermoser, and especially Joachim Vogt for carefully reading the manuscript. We gratefully acknowledge the continuous support of Gerhard Haerendel, Gregor Morfill and Heinrich Soffel.

Last not least, we would like to mention that we have profited from many books and reviews on plasma and space physics. References to most of them have been included into the suggestions for further reading at the end of each chapter. These suggestions, however, do not include the large number of original papers, which we made use of and are indebted to.

We have made every effort to make the text error-free in this revised edition; unfortunately this is a never ending task. We hope that the readers will kindly inform us about misprints and errors, preferentially by electronic mail to *baumjohann@oeaw.ac.at*.

Contents

1. Introduction

Space physics is to a large part plasma physics. This was realized already in the first half of this century, when plasma physics started as an own field of research and when one began to understand geomagnetic phenomena as effects caused by processes in the uppermost atmosphere, the ionosphere and the interplanetary space. Magnetic storms, bay disturbances, substorms, pulsations and so on were found to have their sources in the ionized matter surrounding the Earth.

In our companion volume, *Basic Space Plasma Physics*, we have presented its concepts, the basic processes and the basic observations. The present volume builds on the level achieved therein and proceeds into the domain of instabilities and nonlinear effects in collisionless space plasmas. In this introduction we review some of the very basics from the companion volume.

1.1. Plasma Properties

Classical non-relativistic plasmas are defined as quasineutral, i.e., in a global sense non-charged mixtures of gases of negatively charged electrons and positive ions, containing very large numbers of particles such that it is possible to define quantities like number densities, n_s, thermal velocities, $v_{\mathrm{th}s}$, bulk velocities, $\mathbf{v}_s$, pressures, p_s, temperatures, T_s, and so on. Viewed from kinetic theory, it must be possible to define a distribution function, $f_s(\mathbf{x}, \mathbf{v}, t)$, for each species $s = e, i$ (electrons, ions) in the plasma such that it gives the probability of finding a certain number of particles in the phase space interval $[\mathbf{x}, \mathbf{v}; \mathbf{x} + d\mathbf{x}, \mathbf{v} + d\mathbf{v}]$. If this is the case, any microscopic electric fields of a test charge in the plasma, i.e., of every point charge or every particle in the plasma, will be screened out by the Coulomb fields of the many other charges over the distance of a *Debye length* given in Eq. (I.1.3) of our companion book (equation numbers from that volume are prefixed by the roman numeral). Here it is written for the particle species s

$$\lambda_{Ds} = \left(\frac{\epsilon_0 k_B T_s}{n_e e^2} \right)^{1/2} \tag{1.1}$$

where k_B is the Boltzmann constant. The Debye length of electrons is abbreviated as $\Lambda_D = \Lambda_{De}$ throughout this book. The condition for considering a group of particles to constitute a plasma is then that the number of particles in the Debye sphere is large, or after Eq. (I.1.5) that the *plasma parameter*

$$\Lambda = n_e \lambda_D^3 \gg 1 \tag{1.2}$$

In this book we deal mainly with collisionless plasmas. These are plasmas where the Coulomb collision time, $\tau_c = 1/\nu_c$, is much longer than any other characteristic time of variation in the plasma. The quantity ν_c is the collision frequency between the particles. For Coulomb collisions between electrons and ions it has been derived in Eq. (I.4.9) of our companion volume, *Basic Space Plasma Physics*. Plasmas are collisional if

$$\omega \ll \nu_c \tag{1.3}$$

where ω is the frequency of the variation under consideration.

Plasmas, in general, have a number of such characteristic frequencies. The most fundamental one is the *plasma frequency* of a species s

$$\omega_{ps} = \left(\frac{n_s q_s^2}{m_s \epsilon_0}\right)^{1/2} \tag{1.4}$$

It increases with charge, q_s, and density, n_s, but decreases with increasing mass, m_s, of the particle species. It gives the frequency of oscillation of a column of particles of species s against the background plasma consisting of all other plasma populations. Thus it is the characteristic frequency by which quasineutrality in a plasma can be violated if no external electric field is applied to the plasma. The *electron plasma frequency* is the highest plasma frequency, since the electron mass is small and, furthermore, quasi-neutrality requires $n_e = \sum_i n_i$. Between the plasma frequency and the thermal velocity of a species there is the simple relation

$$v_{\mathrm{ths}} = \omega_{ps} \lambda_{Ds} \tag{1.5}$$

Magnetized plasmas have another fundamental frequency, the *cyclotron frequency* given in Eq. (I.2.12). For a magnetic field of strength B this frequency is

$$\omega_{gs} = \frac{q_s B}{m_s} \tag{1.6}$$

The cyclotron frequency increases with magnetic field and charge, but, as in the case of the plasma frequency, heavier particles have a lower cyclotron frequency. Physically the cyclotron frequency counts the rotations of the charge around a magnetic field line in its gyromotion (see Sec. 2.2 of *Basic Space Plasma Physics*). A given plasma particle population can be considered to be magnetized if its cyclotron frequency is larger than the frequency of any variation applied to the plasma, $\omega_{gs} \gg \omega$. In the opposite case, when its cyclotron frequency is low, this particular species behaves as if the plasma would not contain a magnetic field. Because of the different particle masses, different plasma components may have a different magnetization behavior for a given variation frequency, ω.

As with the plasma frequency, there is a relation between the thermal velocity of a species and the cyclotron frequency of its particles

$$v_{\mathrm{th}s} = \omega_{gs} r_{gs} \tag{1.7}$$

This equation defines the *gyroradius*, r_{gs}, of species s.

The gyroradius given above is actually the thermal gyroradius, because it is defined through the thermal velocity of the species. It is the average gyroradius of the particles of the particular species. Of course, each particle has its own gyroradius, depending on its velocity component perpendicular to the magnetic field. The gyroradius increases with velocity and also with mass or, better, it increases with particle energy. Energetic particles thus have large gyroradii.

Finally, we introduce one particular important quantity used in plasma physics, i.e., the ratio of thermal-to-magnetic energy density, the so-called *plasma beta*

$$\beta = \frac{nk_BT}{B^2/2\mu_0} \tag{1.8}$$

This ratio tells us whether the plasma is dominated by the thermal pressure or if the magnetic field dominates the dynamics of the plasma. Clearly, for $\beta > 1$ the former case is realized, and the magnetic field plays a relatively subordinate role, while in the opposite case, when $\beta < 1$, the magnetic field governs the dynamics of the plasma.

1.2. Particle Motions

Single particle motion in a plasma is naturally strongly distorted by the presence of all the other particles, the propagation of disturbances across a plasma, and a number of other effects. However, due to the Debye screening, the particles move approximately freely in a dilute collisionless and hot plasma for distances larger than one Debye length. One can assume that the small distortions of the particles caused by their participation in the *Debye screening* of the Coulomb fields of the other particles they pass along in their motion will in the average be small and will constitute only negligible wiggles around their collisionless orbits. This kind of wiggling in a more precise theory can be described by the *thermal fluctuations* of the particle density and velocity.

Within these assumptions it is possible to calculate the particle orbits. The particle orbits satisfy the single particle equation of motion in which all the collisional interactions with other particles and fields are neglected. Given external magnetic, $\mathbf{B}$, and electric fields, $\mathbf{E}$, this equation of motion reads

$$m_s \frac{d\mathbf{v}_s}{dt} = q_s(\mathbf{E} + \mathbf{v}_s \times \mathbf{B}) \tag{1.9}$$

The motion of the particles along the field lines is independent of the magnetic field and, in the absence of a parallel electric field component, $E_\parallel = 0$, the parallel particle velocity remains constant, $v_\parallel = \text{const}$.

The transverse particle motion can be split into a number of independent velocities if it is assumed that the gyromotion is sufficiently fast with respect to a bulk speed perpendicular to the magnetic field (see Chap. 2 of *Basic Space Plasma Physics*). Averaging over the circular *gyromotion*, the particle itself can be replaced by its *guiding center*, i.e., the center of its gyrocircle.

The velocity of the guiding center may be decomposed into a number of *particle drifts*. In a stationary perpendicular electric field the *Lorentz force* term in the above equation of motion tells that a simple transformation of the whole plasma into a coordinate system moving with the *convection* or *$E \times B$ drift* given in Eq. (I.2.19)

$$\mathbf{v}_E = \frac{\mathbf{E} \times \mathbf{B}}{B^2} \tag{1.10}$$

cancels the electric field. In this co-moving system the particle motion is independent of $\mathbf{E}_\perp$, the perpendicular component of the electric field. It is force-free. Obviously, all particles independent of their mass or charge experience this drift motion, which is a mere result of the Lorentz transformation.

For time varying electric fields another drift arises, the so-called *polarization drift* given in Eq. (I.2.24)

$$\mathbf{v}_P = \frac{1}{\omega_{gs} B} \frac{d\mathbf{E}_\perp}{dt} \tag{1.11}$$

where the time derivative is understood as the total convective derivative. This drift depends on the mass and charge state of the species under consideration. Heavy particle drift faster than light particles. In addition, the directions of the drifts are opposite for opposite charges, leading to current generation. This drift is important for all low-frequency transverse plasma waves.

These drifts follow from a consideration of single-particle motions in electric and magnetic fields. As pointed out, plasmas do usually not behave like single particles. Only in rare cases, of which the *ring current* in the inner magnetosphere is an example (see Chap. 3 of the companion volume, *Basic Space Plasma Physics*), the motion of a single energetic particle mimics the motion of the entire energetic plasma component, and the single particle drifts are useful tools for the description of the plasma dynamics. In all other cases one must refer to a *collective behavior* of the plasma which arises from the internal correlations between particles and fields even in the collisionless case. The plasma may then be considered not to consist of single particles but of particle fluids species. Each fluid can have its own density, bulk speed, pressure and temperature.

Such fluids when immersed into a magnetic field experience a *diamagnetic drift* which has been derived in Eq. (I.7.72). Obviously, this drift is a *collective effect* insofar as the collective particle pressure comes into play

$$\mathbf{v}_{dia,s} = \frac{\mathbf{B} \times \nabla_{\perp} p}{q_s n_s B^2} \tag{1.12}$$

Like the polarization drift, this bulk *pressure gradient drift* motion leads to currents, drift waves, may cause instability and nonlinear effects.

1.3. Basic Kinetic Equations

Single particle effects, like the particle motion reviewed in the previous section, are often hidden in a plasma. In general, plasma dynamics cannot be described in such a simple way, but is determined by complicated correlations between particles and fields. The full set of basic equations of a plasma consists of the two *Maxwell equations*

$$\nabla \times \mathbf{B} = \mu_0 \mathbf{j} + \frac{1}{c^2}\frac{\partial \mathbf{E}}{\partial t} \tag{1.13}$$

$$\nabla \times \mathbf{E} = -\frac{\partial \mathbf{B}}{\partial t} \tag{1.14}$$

which must be completed by the two additional conditions, the absence of magnetic charges and Poisson's equation for the electric charge density, ρ

$$\nabla \cdot \mathbf{B} = 0 \tag{1.15}$$

$$\nabla \cdot \mathbf{E} = \rho/\epsilon_0 \tag{1.16}$$

The current and charge densities are defined as the sums over the current and charge densities of all species

$$\mathbf{j} = \sum_s q_s n_s \mathbf{v}_s \tag{1.17}$$

$$\rho = \sum_s q_s n_s \tag{1.18}$$

The bulk velocities and densities must be calculated from the basic equations determining the dynamics of the plasma. In a purely collisionless state the most fundamental equation describing the plasma dynamics is the *Vlasov equation*, taken separately for each species

$$\left[\frac{\partial}{\partial t} + \mathbf{v}\cdot\nabla + \frac{q_s}{m_s}(\mathbf{E} + \mathbf{v}\times\mathbf{B})\cdot\frac{\partial}{\partial \mathbf{v}}\right] f_s(\mathbf{x}, \mathbf{v}, t) = 0 \tag{1.19}$$

which is a scalar equation for the particle *distribution function*. For its justification and derivation see Chap. 6 of the companion volume, *Basic Space Plasma Physics*. The densities and bulk velocities entering the current and charges are determined as the *moments* of the distribution function, f_s, as solution of the Vlasov equation

$$n_s = \int d^3v\, f_s(\mathbf{x}, \mathbf{v}, t) \tag{1.20}$$

$$n_s\mathbf{v}_s = \int d^3v\, \mathbf{v} f_s(\mathbf{x}, \mathbf{v}, t) \tag{1.21}$$

The Vlasov equation together with the system of field equations and definitions of densities and currents turns out to be a highly nonlinear system of equations, in which the fields determine the behavior of the distribution function and the fields themselves are determined by the distribution function through the charges and currents.

This self-consistent system of equations forms the basis for collisionless plasma physics. In our companion volume we present a number of solutions of this system of equations for equilibrium and linear deviations from equilibrium. In the following we extend this approach to a number of unstable solutions and into the domain where nonlinearities become important.

The Vlasov equation (1.19) may be used to derive *fluid equations* for the different particle components. The methods of constructing fluid equations is given in Chap. 7 of the companion volume, *Basic Space Plasma Physics*. It is based on a moment integration technique of the Vlasov equation which is well known from general kinetic theory. One multiplies the Vlasov equation successively by rising powers of the velocity **v** and integrates the resulting equation over the entire velocity space. The system of hydrodynamic equations obtained consists of an infinite set for the infinitely many possible moments of the one-particle distribution function, f_s. The first two moment equations are the continuity equation for the particle density and the momentum conservation equation

$$\frac{\partial n_s}{\partial t} + \nabla\cdot(n_s\mathbf{v}_s) = 0 \tag{1.22}$$

$$\frac{\partial n_s\mathbf{v}_s}{\partial t} + \nabla\cdot(n_s\mathbf{v}_s\mathbf{v}_s) = n_s\frac{q_s}{m_s}(\mathbf{E} + \mathbf{v}_s\times\mathbf{B}) - \frac{1}{m_s}\nabla p_s \tag{1.23}$$

where, for simplicity, the pressure has been assumed to be isotropic. These equations have to be completed by another equation for the pressure or by an energy law.

1.4. Plasma Waves

The system of Vlasov-Maxwell equations or its hydrodynamic simplifications allow for the propagation of disturbances on the background of the plasma. Generally, these disturbances are nonlinear time-varying states the plasma can assume. But as long as their amplitudes are small when compared with the undisturbed field and particle variables, they can be treated in a linear approximation as small disturbances. This condition can be written as $|\delta A(\mathbf{x},t)| \ll |A_0(\mathbf{x},t)|$, where δA is the amplitude of the variation of some quantity $A(\mathbf{x},t)$, and A_0 is its equilibrium undisturbed value which may also vary in time and space. In the linear approximation such disturbances of the plasma state represent propagating waves of frequency, $\omega(\mathbf{k})$, and wavenumber, $\mathbf{k}$. As usual, the phase and group velocities of these waves are defined as

$$\mathbf{v}_{ph} = \frac{\omega(\mathbf{k})}{k^2}\mathbf{k} \tag{1.24}$$

$$\mathbf{v}_{gr} = \frac{\partial\omega(\mathbf{k})}{\partial\mathbf{k}} = \nabla_{\mathbf{k}}\omega(\mathbf{k}) \tag{1.25}$$

The *phase velocity* is directed parallel to $\mathbf{k}$ and gives direction and speed of the propagation of the wave front or phase

$$\phi(\mathbf{x},t) = \mathbf{k}\cdot\mathbf{x} - \omega(\mathbf{k})t \tag{1.26}$$

while the *group velocity* can point into a direction different from the phase velocity. It gives the direction of the flow of energy and information contained in the wave. Both can be calculated from knowledge of the frequency. The latter is the solution of the wave dispersion relation in both the linear approximation and the full nonlinear theory.

In the linear approximation the dispersion relation is particularly simple to derive. Because of the linear approximation, the full set of Maxwell-Vlasov or Maxwell-hydrodynamic equations contains only linear disturbances. Thus the system can be reduced to a set of linear algebraic equations with vanishing determinant

$$D(\omega,\mathbf{k}) = 0 \tag{1.27}$$

the *dispersion relation*. The analytical form of the dispersion relation is obtained from the linearized wave equation (I.9.45)

$$\nabla^2\delta\mathbf{E} - \nabla(\nabla\cdot\delta\mathbf{E}) - \epsilon_0\mu_0\frac{\partial^2\delta\mathbf{E}}{\partial t^2} = \mu_0\frac{\partial\delta\mathbf{j}}{\partial t} \tag{1.28}$$

The linear current density, $\delta\mathbf{j}$, on the right-hand side is expressed by the linear *Ohm's law* given in Eq. (I.9.46)

$$\delta\mathbf{j}(\mathbf{x},t) = \int d^3x' \int_{-\infty}^{t} dt'\boldsymbol{\sigma}(\mathbf{x}-\mathbf{x}',t-t')\cdot\delta\mathbf{E} \tag{1.29}$$

with $\boldsymbol{\sigma}(\mathbf{x}-\mathbf{x}', t-t')$ the linear conductivity tensor. Fourier transformation of Eqs. (1.28) and (1.29) with respect to time and space gives as equation for the Fourier amplitude of the wave field

$$\left[\left(k^2-\frac{\omega^2}{c^2}\right)\mathsf{I}-\mathbf{k}\mathbf{k}-i\omega\mu_0\boldsymbol{\sigma}(\omega,\mathbf{k})\right]\cdot\delta\mathbf{E}(\omega,\mathbf{k})=0 \tag{1.30}$$

The linear conductivity, $\boldsymbol{\sigma}(\omega, \mathbf{k})$, is a function of frequency, ω, and wavenumber, $\mathbf{k}$. The fields and the conductivity satisfy the following symmetry relations

$$\begin{aligned}\delta\mathbf{E}(-\mathbf{k},-\omega) &= \delta\mathbf{E}^*(\mathbf{k},\omega)\\ \boldsymbol{\sigma}(-\mathbf{k},-\omega) &= \boldsymbol{\sigma}^*(\omega,\mathbf{k})\end{aligned} \tag{1.31}$$

The dispersion relation follows from the condition that Eq. (1.30) should have nontrivial solutions

$$D(\omega,\mathbf{k})=\mathrm{Det}\left[\left(k^2-\frac{\omega^2}{c^2}\right)\mathsf{I}-\mathbf{k}\mathbf{k}-i\omega\mu_0\boldsymbol{\sigma}(\omega,\mathbf{k})\right]=0 \tag{1.32}$$

It is convenient to introduce the *dielectric tensor* of the plasma

$$\boldsymbol{\epsilon}(\omega,\mathbf{k})=\mathsf{I}+\frac{i}{\omega\epsilon_0}\boldsymbol{\sigma}(\omega,\mathbf{k}) \tag{1.33}$$

and to rewrite the dispersion relation into the shorter version

$$D(\omega,\mathbf{k})=\mathrm{Det}\left[\frac{k^2c^2}{\omega^2}\left(\frac{\mathbf{k}\mathbf{k}}{k^2}-\mathsf{I}\right)+\boldsymbol{\epsilon}(\omega,\mathbf{k})\right]=0 \tag{1.34}$$

This dispersion relation is the basis of all linear plasma theory and is also used in nonlinear plasma theory. The dielectric tensor which appears in this relation must be calculated from the dynamical model of the plasma. Its most general analytical form derived from the linearized set of the Maxwell-Vlasov equations has been given in Eq. (I.10.94) of Chap. 10 of the companion volume, *Basic Space Plasma Physics*. For further reference we repeat this equation here

$$\begin{aligned}\boldsymbol{\epsilon}(\omega,\mathbf{k}) \;=\; & \left(1-\sum_s\frac{\omega_{ps}^2}{\omega^2}\right)\mathsf{I}-\sum_s\sum_{l=-\infty}^{l=\infty}\frac{2\pi\omega_{ps}^2}{n_{0s}\omega^2}\\ & \int_0^\infty\int_{-\infty}^{\infty}v_\perp dv_\perp dv_\parallel\left(k_\parallel\frac{\partial f_{0s}}{\partial v_\parallel}+\frac{l\omega_{gs}}{v_\perp}\frac{\partial f_{0s}}{\partial v_\perp}\right)\frac{\mathsf{S}_{ls}(v_\parallel,v_\perp)}{k_\parallel v_\parallel+l\omega_{gs}-\omega}\end{aligned} \tag{1.35}$$

The tensor appearing in the integrand, S_{ls}, is of the form

$$\mathsf{S}_{ls}(v_\|, v_\perp) = \begin{bmatrix} \dfrac{l^2\omega_{gs}^2}{k_\perp^2} J_l^2 & \dfrac{ilv_\perp\omega_{gs}}{k_\perp} J_l J_l' & \dfrac{lv_\|\omega_{gs}}{k_\perp} J_l^2 \\ -\dfrac{ilv_\perp\omega_{gs}}{k_\perp} J_l J_l' & v_\perp^2 J_l'^2 & -iv_\| v_\perp J_l J_l' \\ \dfrac{lv_\|\omega_{gs}}{k_\perp} J_l^2 & iv_\| v_\perp J_l J_l' & v_\|^2 J_l^2 \end{bmatrix} \tag{1.36}$$

and the Bessel functions, $J_l, J_l' = dJ_l/d\xi_s$, depend on the argument $\xi_s = k_\perp v_\perp/\omega_{gs}$.

The determinant of the dispersion relation, $D(\omega, \mathbf{k})$, is a function of frequency, wavenumber and a set of plasma parameters. Its solution yields the frequency relation $\omega = \omega(\mathbf{k})$. Different versions of the dispersion determinant are derived in the companion volume, *Basic Space Plasma Physics*, and solved in several approximations. Generally spoken, in contrast to the vacuum where a continuum of electromagnetic waves can propagate, there is no continuum of plasma waves. Even in the linear approximation, neglecting all couplings, correlations and nonlinear interactions, plasmas are highly complicated dielectrics which possess only a few narrow windows where linear disturbances are allowed. These disturbances are the eigenmodes of the plasma. They appear as the *discrete spectrum* of eigenvalues of the basic linear system of equations governing the dynamics of a plasma as solutions of Eq. (1.27).

A further difference between wave propagation in vacuum and in a plasma is that the plasma allows for two types of waves, *transverse electromagnetic waves* and *longitudinal electrostatic waves*. The latter are nothing else but oscillations of the electrostatic potential and are not accompanied by magnetic fluctuations. Somehow they resemble sound waves in ordinary hydrodynamics, but there is a large zoo of electrostatic waves in a plasma most of which are not known in simple hydrodynamics.

The *electrostatic modes* are confined to the plasma, because oscillations of the electrostatic potential can be maintained only inside the plasma boundaries. Only two of the *electromagnetic modes* smoothly connect to the free-space electromagnetic wave and can leave the plasma, the *O-mode* and the high-frequency branch of the *X-mode*. The other low-frequency electromagnetic waves, the *Z-mode*, *whistlers* and *electromagnetic ion-cyclotron modes*, and the three magnetohydrodynamic wave modes, the *Alfvén wave*, and the *fast mode* and the *slow mode*, are all confined to the plasma. We have discussed the properties and propagation characteristics of these modes in Chaps. 9 and 10 of the companion volume, *Basic Space Plasma Physics*.

Waves propagating in a plasma can experience *reflection* and *resonance*. Reflection occurs when the wavenumber vanishes for finite frequency, $k \to 0$. Here the direction of the wave turns by an angle π, indicating that the wave is reflected from the particular point where its wavenumber vanishes. Resonance occurs where the wavenumber diverges at finite frequency, $k \to \infty$. At such a point the wavelength becomes very short, and the interaction between the plasma particles becomes very strong. Here the wave may either dissipate its energy or extract energy from the plasma in order to grow.

As long as one looks only into the real solutions of the dispersion relation, no information can be obtained about the possible growth of a wave or its damping at the resonant point. However, as the possibility of resonances in a plasma shows, plasmas are active media. This is also realized when remembering that the charges and their motions themselves are sources of the fields. In order to investigate these processes one must include the possibility of complex solutions of the dispersion relation. The fluctuations of the fields can be excited or amplified or, in the opposite case, they can be absorbed in the plasma. The frequency becomes complex under these conditions

$$\omega(\mathbf{k}) \rightarrow \omega(\mathbf{k}) + i\gamma(\omega, \mathbf{k}) \tag{1.37}$$

Here $\gamma(\omega, \mathbf{k})$ is the growth or damping rate of the wave, which depends on the real part of the frequency and on the wavenumber. The wave grows for $\gamma > 0$, and it becomes damped for $\gamma < 0$.

In the companion volume we treated the damping rate, i.e., solutions with $\gamma < 0$. In the present volume, we will start with $\gamma > 0$ solutions. The next chapters are devoted to the discussion of these still linear effects leading to instability, before turning to nonlinear effects which arise when the amplitudes of the waves become so large that the linear assumption must be abandoned.

Introductory Texts

The literature listed below is a selection of introductory texts into plasma physics and space plasma physics which should be consulted before attempting to read this book.

[1] W. Baumjohann and R. A. Treumann, *Basic Space Plasma Physics* (Imperial College Press, London, 1996).

[2] F. F. Chen, *Introduction to Plasma Physics and Controlled Fusion, Vol. 1* (Plenum Press, New York, 1984).

[3] N. A. Krall and A. M. Trivelpiece, *Principles of Plasma Physics* (McGraw-Hill, New York, 1973).

[4] E. M. Lifshitz and L. P. Pitaevskii, *Physical Kinetics* (Pergamon Press, Oxford, 1981).

[5] D. C. Montgomery and D. A. Tidman, *Plama Kinetic Theory* (McGraw-Hill, New York, 1964).

[6] D. R. Nicholson, *Introduction to Plasma Theory* (Wiley, New York, 1983).

2. Concept of Instability

Generation of instability is the general way of redistributing energy which has accumulated in a *non-equilibrium* state. Figure 2.1 demonstrates in a simple mechanical analogue how a heavy sphere situated in an external potential field can find itself in several different situations which may be either stable or unstable. The first of these situations is the *stable equilibrium*, where the sphere lies on the lowest point of an infinitely high potential trough. In this position the sphere can only perform oscillations around its equilibrium position, which will damp out due to friction until the sphere comes to rest at the bottom of the potential trough. In the contrary situation the sphere finds itself on top of a potential hill. The slightest linear distortion of its position will let it roll down the hill. This is an unstable case, a *linear instability*, which sets in spontaneously. In the *metastable state* the sphere lies on a plateau on top of a hill and can wander around until it reaches the crest and rolls down. In the last example of a *nonlinear instability* the sphere is stable against small-amplitude disturbances, but becomes unstable for larger amplitudes.

In plasma physics the potential troughs and wells are replaced by sources of *free energy*, and the heavy sphere corresponds to a certain wave mode, in most cases an eigenmode of the plasma. There is a multitude of free energy sources in the Earth's environment. Neither the ionosphere nor the magnetosphere are closed systems in thermal equilibrium, but are driven by energy, momentum and mass input from outside, e.g., from the solar wind. On the macroscopic scale this input produces spatial gradients and inhomogeneities. On the microscopic scale it leads to deformation and distortions of the local distribution functions. The former free energy sources are the causes of the large-scale macroinstabilities, while the latter cause small-scale microinstabilities.

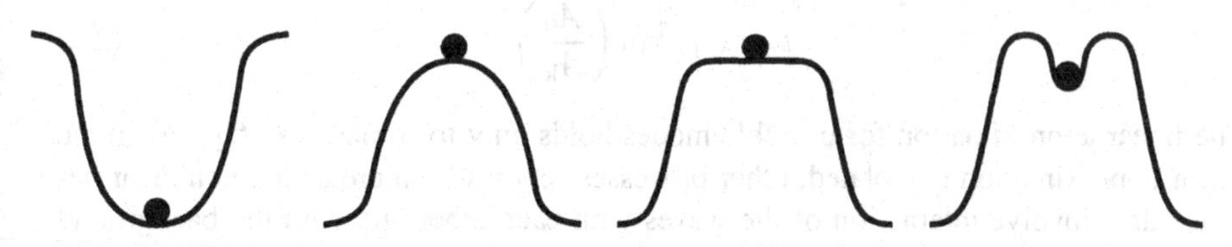

Fig. 2.1. Different non-equilibrium configurations leading to instability.

2.1. Linear Instability

The concept of instability arises from a formal consideration of the wave function. In linear wave theory the amplitude of the waves is much less than the stationary state vector, so that the wave can be considered a small disturbance. For instance, if the wave is a disturbance δn of the density n, then $\delta n(\mathbf{x}, t) \ll n_0$, where n_0 can still be a function of space and time, but it is assumed that its variation is much slower than that of the disturbance. If this is the case, the wave function can be represented by a superposition of plane waves oscillating at frequency $\omega(\mathbf{k})$, where ω is the solution of the linear dispersion relation $D(\omega, \mathbf{k}) = 0$. Any wave field component $\delta A(\mathbf{x}, t)$ can then be Fourier decomposed as

$$\delta A = \sum_{\mathbf{k}} A_{\mathbf{k}} \exp(i\mathbf{k} \cdot \mathbf{x} - i\omega t) \tag{2.1}$$

In general the dispersion relation is a complex equation and has a number of frequency solutions which are also complex $\omega = \omega_r + i\gamma$. From Eq. (2.1) it is clear that, for real ω, the disturbances are oscillating waves. On the other hand, for complex solutions the behaviour of the wave amplitude depends heavily on the sign of the imaginary part of the frequency $\gamma(\omega_r, \mathbf{k})$. If $\gamma < 0$ the real part of the amplitude becomes an exponentially decreasing function of time, and the wave is damped. On the other hand, for $\gamma > 0$ the wave amplitude grows exponentially in time, and we encounter a linear instability. In this case the decrement γ is called the *growth rate* of the corresponding eigenmode. Note, however, that instability can only arise if there are free energy sources in the plasma which feed the growing waves. If this is not the case, then a solution with a positive γ is a fake solution which violates energy conservation and causality.

Growth Rate

The amplitude of an unstable wave increases as

$$A_{\mathbf{k}}(t) = A_{\mathbf{k}} \exp[\gamma(\omega_r, \mathbf{k})t] \tag{2.2}$$

Thus the linear approximation breaks down when the amplitude becomes comparable to the background value of the field, i.e., $A_{\mathbf{k}}(t)/A_0 \approx 1$, or at the nonlinear time

$$t_{\mathrm{nl}} \approx \gamma^{-1} \ln\left(\frac{A_0}{A_{\mathbf{k}}}\right) \tag{2.3}$$

The linear approximation for unstable modes holds only for times $t \ll t_{\mathrm{nl}}$. When the linear approximation is violated, other processes set on which are called nonlinear because they involve interaction of the waves with each other and with the background plasma, which cannot be treated by linear methods. The time t_{nl} is reached the earlier the larger the growth rate is. When the growth rate becomes larger than the wave frequency, $\gamma > \omega$, the wave amplitude explodes and the wave has no time to perform even one single oscillation during one wave period. The wave concept becomes obsolete in this case and it is reasonable to consider in the first place only instabilities of comparably small growth rates which satisfy the conditions of linearity during many wave periods

$$\gamma/\omega \ll 1 \tag{2.4}$$

This remark does not preclude the existence of instabilities with growth rates larger than the wave frequency. In fact, one of the first examples of an instability will deal with this case below. One then speaks of *purely growing* or *non-oscillating* instabilities with about zero real frequency. Such instabilities appear only in the lowest frequency range of a magnetohydrodynamic plasma model.

Weak Instability

Instabilities can be either strong or weak. Strong instabilities have growth rates which violate the condition (2.4) and thus coincide in many cases with non-oscillating instabilities. For weakly growing instabilities with a growth rate satisfying Eq. (2.4), one can design a general procedure to deduce the growth rate from the general dispersion relation given in Eq. (I.9.55) of the companion book, *Basic Space Plasma Physics*. The dispersion relation, $D(\omega, \mathbf{k}) = 0$, is an implicit relation between the wave frequency and the wavenumber. Given the wavenumber, it is possible to determine the wave frequency. In general, $D(\omega, \mathbf{k})$ is a complex function

$$D(\omega, \mathbf{k}) = D_r(\omega, \mathbf{k}) + iD_i(\omega, \mathbf{k}) \tag{2.5}$$

It therefore provides two equations which can be used to determine either the frequency, ω, in dependence on the wavenumber, $\mathbf{k}$, or vice versa. It is convenient to assume that $\mathbf{k}$ is real. Then the frequency is complex

$$\omega(\mathbf{k}) = \omega_r(\mathbf{k}) + i\gamma(\omega_r, \mathbf{k}) \tag{2.6}$$

For growth rates satisfying Eq. (2.4) these expressions can be simplified by expanding $D(\omega, \mathbf{k})$ around the real part of the frequency, ω_r, up to first order. This procedure, which is the same as used for calculating the damping rate in Sec. 10.6 of our companion book, *Basic Space Physics*, yields

$$D(\omega, \mathbf{k}) = D_r(\omega_r, \mathbf{k}) + (\omega - \omega_r) \left.\frac{\partial D_r(\omega, \mathbf{k})}{\partial \omega}\right|_{\gamma=0} + iD_i(\omega_r, \mathbf{k}) = 0 \tag{2.7}$$

Since $\omega - \omega_r = i\gamma$, this equation enables us to obtain a dispersion relation for the real part of the frequency and at the same time an expression for the growth rate as

$$D_r(\omega_r, \mathbf{k}) = 0 \tag{2.8}$$

$$\gamma(\omega_r, \mathbf{k}) = -\frac{D_i(\omega_r, \mathbf{k})}{\partial D_r(\omega, \mathbf{k})/\partial\omega|_{\gamma=0}} \tag{2.9}$$

The first of these equations is the dispersion relation for real frequencies, while the second equation determines the linear growth rate, γ. In the following, as long as no confusion is caused, we will drop the index r, taking ω as the real frequency of the wave.

Spontaneous Cherenkov Emission

An illustrative example of wave amplification is the Cherenkov emission. We already know from Sec. 10.2 of the companion book, *Basic Space Plasma Physics*, that inverse Landau damping leads to amplification of plasma waves. This process depends on the shape of the equilibrium distribution function. It is a process in which plasma wave emission is induced by the overpopulation of a higher level, very similar to Laser emission, and is called *induced emission*. Before we come to consider some of the most important plasma instabilities subject to this kind of induced emission, we point out that there is also a different direct or *spontaneous emission* mechanism, which is independent of the overpopulation of the distribution function and also independent of the wave amplitude.

This kind of spontaneous emission is closely related to spontaneous emission of electromagnetic waves in a medium of large refraction index and therefore reduced light velocity, c', from particles moving faster than the speed of light in the medium, $v_e > c'$, the *Cherenkov effect*. In a plasma the role of the light velocity is taken over by the phase velocity of the plasma waves. There are many more than one possible wave modes in a plasma. Hence, spontaneous emission can appear in any of the plasma modes if some test particles exceed a certain critical velocity. For instance, if this velocity is the reduced speed of light, the emission will be in the high-frequency electromagnetic modes. This requires high relativistic velocities of the particles. Since in a Maxwellian distribution of a thermal plasma there are only very few such particles, this emission is negligible. But emission in one of the electrostatic plasma modes is still possible. Such a spontaneous emission occurs if fast test particles are in resonance with the plasma wave. In other words, the fast test particles, typically electrons, must have a velocity which is close to being equal to the phase velocity of the plasma wave

$$v_e = \omega(\mathbf{k})/k \tag{2.10}$$

If we take the Langmuir wave, we have $\omega \approx \omega_{pe}$. This is very accurate because emission of the waves by test particles can take place only for small

$$k^2\lambda_D^2 < 1 \tag{2.11}$$

corresponding to long wavelengths. Shorter wavelengths only contribute to the screening of the particle in the Debye sphere. Both conditions together give the condition for spontaneous emission of Langmuir waves

$$v_e = \omega_{pe}/k > \omega_{pe}\lambda_D = v_{\mathrm{th},e} \tag{2.12}$$

from fast electrons in a plasma. Hence, electrons with velocities faster than the electron thermal velocity contribute to spontaneous emission of Langmuir waves. This effect can be understood from the expression for the total charge density variation

$$\delta\rho_e(\omega, \mathbf{k}) = \frac{\delta\rho_{\mathrm{ex}}(\omega, \mathbf{k})}{\epsilon(\omega, \mathbf{k})} \tag{2.13}$$

through the external charge fluctuation and the dielectric response function. Because in each of the eigenmodes of the plasma the dielectric function vanishes, $\epsilon(\omega, \mathbf{k}) = 0$, the plasma can still excite finite amplitude density fluctuations in the absence of external fluctuations. In this case both the numerator and denominator of the above expression vanish at the eigenmode yielding a finite charge fluctuation. This is the necessary condition for spontaneous emission. If the plasma in addition contains particles which satisfy Eq. (2.12), it will spontaneously emit Langmuir waves. The rate of spontaneous emission is equal to the energy loss of the fast particles in the 'collision' with the long-wavelength eigenmodes. This energy loss of one single particle is

$$\frac{dW_e}{dt} = -\frac{\pi m_e \omega_{pe}^4}{2n_0} \sum_{\mathbf{k}} k^{-2} \left[\delta(\omega - \mathbf{k}\cdot\mathbf{v}_e) + \delta(\omega + \mathbf{k}\cdot\mathbf{v}_e)\right] \tag{2.14}$$

The sum is over all wavenumbers satisfying $k\lambda_D < 1$ and the two (not dimensionless!) delta functions account for the parallel and antiparallel resonant wave modes. Changing the sign and integrating over the velocity distribution yields as a final result

$$\left(\frac{dW_l}{dt}\right)_{\mathrm{Ch}} = \frac{\pi m_e \omega_{pe}^4}{2n_0 k^2} \int d^3v\, f_{0e}(\mathbf{v})\delta(\omega_{pe} - \mathbf{k}\cdot\mathbf{v}) \tag{2.15}$$

as spontaneous emission rate of Langmuir waves. Inserting a Maxwellian distribution of temperature T_e and carrying out the integration, one obtains for the emission rate

$$\boxed{\left(\frac{dW_l}{dt}\right)_{\mathrm{Ch}} = \frac{4\sqrt{\pi}\omega_{pe} k_B T_e}{k^5 \lambda_D^5}} \tag{2.16}$$

in a thermal plasma in equilibrium. This emission is weak. In fact, it is proportional to $(k\lambda_D)^{-5}$. Comparing this dependence with the $(k\lambda_D)^{-3}$ dependence of Landau damping, one recognizes that, for short wavelengths, Landau damping dominates over spontaneous Cherenkov emission. This is the reason for the weakness of thermal Langmuir fluctuations in the short-wavelength range. But for long wavelengths spontaneous emission becomes stronger, and the Langmuir fluctuation level is relatively high. Clearly, if one adds a nonthermal component of higher velocity, for instance a beam of fast electrons, the spontaneous emission rate is drastically enhanced.

2.2. Electron Stream Modes

Let us construct the simplest electrostatic dispersion relation leading to instability. We consider a cold plasma in order to dismiss any complications due to thermal effects. And we assume sufficiently high frequencies so that ion effects can be neglected. To provide a free-energy source we assume that a cold electron beam of density n_b and velocity $\mathbf{v}_b$ streams across the electron background of density n_0 and velocity $v_0 = 0$. It is clear that this system is not at equilibrium and that the electrostatic interaction between the two plasmas should ultimately lead to dissipation of the extra energy stored in the streaming motion of the beam. The beam will be decelerated and the beam electrons will mix into the background plasma. During this process the plasma will be heated. The ignition of this complicated process leading to thermodynamic equilibrium will be caused by an instability.

Beam-Plasma Dispersion Relation

The dispersion relation of a beam-plasma system can be constructed by remembering that the plasma response function, $\epsilon(\omega, \mathbf{k})$, is the sum of the contributions of the plasma components. Since both components are cold and the plasma is isotropic, we get

$$\epsilon(\omega, \mathbf{k}) = 1 - \frac{\omega_{p0}^2}{\omega^2} - \frac{\omega_{pb}^2}{(\omega - \mathbf{k} \cdot \mathbf{v}_b)^2} = 0 \tag{2.17}$$

The first term on the right-hand side is the background plasma contribution which, in the absence of the beam, would yield Langmuir oscillations. The second term is of exactly the same structure, but with the background plasma frequency replaced by the beam plasma frequency, $\omega_{pb}^2 = n_b e^2/\epsilon_0 m_e$, and the frequency being Doppler-shifted by the beam velocity. Setting $v_b = 0$, it is easily seen that the above dispersion relation reproduces Langmuir oscillations at the total plasma frequency, $\omega_{pe}^2 = \omega_{p0}^2 + \omega_{pb}^2$. For $v_b \neq 0$, Eq. (2.17) is a fourth-order equation in frequency, ω, which can have conjugate complex solutions, one of them with a positive imaginary part leading to instability, the other having a negative imaginary part and thus being damped and fading away in the long-time limit.

If we neglect the background plasma by setting $\omega_{p0}^2 = 0$, the dispersion relation can be solved for the streaming part, yielding

$$\omega = \mathbf{k} \cdot \mathbf{v}_b \pm \omega_{pb} \tag{2.18}$$

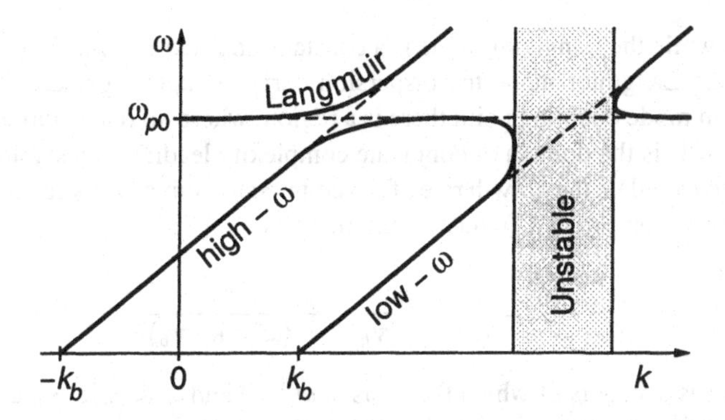

Fig. 2.2. Coupling of Langmuir and beam modes in the two-stream instability.

The two waves described by this relation are the *beam modes*. They exist only in the presence of the beam and have purely linear dispersion. Moreover, for the negative sign in front of the beam plasma frequency the beam mode frequency becomes $\omega < \mathbf{k} \cdot \mathbf{v}_b$, while for the positive sign the frequency is higher, $\omega > \mathbf{k} \cdot \mathbf{v}_b$. The former wave carries negative energy and is a *negative energy wave*. Using Eqs. (I.9.84)

$$W_w(\omega, \mathbf{k}) = \frac{\epsilon_0}{2} \langle |\delta\mathbf{E}(\omega, \mathbf{k})|^2 \rangle \frac{\partial[\omega\, \epsilon(\omega, \mathbf{k})]}{\partial\omega} \tag{2.19}$$

and (I.9.86), $W_{\rm E} = \epsilon_0 |\delta E|^2/2$, both from the companion book, this can be verified by calculating the wave energy in the two modes [make use of Eq. (2.17)]

$$\frac{W_w}{W_{\rm E}} = \frac{\partial[\omega\epsilon(\omega, \mathbf{k})]}{\partial\omega} = \omega\frac{\partial\epsilon(\omega, \mathbf{k})}{\partial\omega} = \frac{2\omega_{p0}^2}{\omega^2} + \frac{2\omega\omega_{pb}^2}{(\omega - \mathbf{k} \cdot \mathbf{v}_b)^3} \tag{2.20}$$

The first part is the Langmuir wave energy, while the second part is the energy which the beam contributes to the waves. The interesting point about this contribution is that it depends on the third power of the Doppler-shifted frequency in the denominator. Hence, when the Doppler-shifted frequency is negative, the energy of the beam mode becomes negative, which is the case for the low-frequency beam mode. Extracting energy from this mode will thus lead to instability, which in the present case means accumulation of 'negative energy' in the wave and growth of its amplitude.

Two-Stream Instability

The instability resulting from coupling between the negative energy beam mode and the Langmuir plasma mode is the *two-stream instability*. The coupling of the modes is shown in Fig. 2.2. In the (ω, k) diagram the beam modes are centered around $k = 0$

with slope v_b, while the Langmuir mode is a constant line at $\omega = \omega_{p0}$. Where the beam modes cross the Langmuir mode the dispersion curves couple together. At the low-frequency beam mode coupling point there is a region, where no real solution exists for either k or ω. This is the domain of conjugate complexity leading to instability.

In order to calculate the growth rate of the counterstreaming two-stream instability, the dispersion relation Eq. (2.17) must be rewritten as

$$1 - \frac{\omega_{p0}^2}{\omega^2} = \frac{\omega_{pb}^2}{(\omega - \mathbf{k} \cdot \mathbf{v}_b)^2} + \frac{\omega_{pb}^2}{(\omega + \mathbf{k} \cdot \mathbf{v}_b)^2} \tag{2.21}$$

This equation has six roots of which the roots at $\omega \approx 0$ and $\omega \approx \pm \mathbf{k} \cdot \mathbf{v}_b$ are the most interesting. Putting $\omega = 0$ on the right-hand side, the solution is

$$\omega = \pm \omega_{p0} \mathbf{k} \cdot \mathbf{v}_b / (\mathbf{k} \cdot \mathbf{v}_b^2 - 2\omega_{pb}^2)^{1/2} \tag{2.22}$$

At short wavelengths $\mathbf{k} \cdot \mathbf{v}_b > 2\omega_{pb}$ this yields a real-frequency oscillation near $\omega < \omega_{p0}$, valid for ω_{p0} small. At large wavelengths the dispersion relation has two purely imaginary roots

$$\omega = \pm i \sqrt{2n_0 / n_b} \tag{2.23}$$

one of them being unstable. The solution near $\omega \sim \mathbf{k} \cdot \mathbf{v}_b$ satisfies the simplified dispersion relation

$$1 - \left(\frac{\omega_{p0}}{\mathbf{k} \cdot \mathbf{v}} \right)^2 = \frac{\omega_{pb}^2}{(\omega - \mathbf{k} \cdot \mathbf{v}_b)^2} + \frac{\omega_{pb}^2}{4\mathbf{k} \cdot \mathbf{v}^2} \tag{2.24}$$

Solving for ω yields

$$\omega = \mathbf{k} \cdot \mathbf{v}_b \pm \frac{\omega_{pb} \mathbf{k} \cdot \mathbf{v}_b}{\left[\mathbf{k} \cdot \mathbf{v}_b^2 - (\omega_{p0}^2 + \omega_{pb}^2/4) \right]^{1/2}} \tag{2.25}$$

which for large values of $\mathbf{k} \cdot \mathbf{v}_b > \omega_{p0}^2 + \omega_{pb}^2/4$ is a real-frequency oscillation. At long wavelengths one, however, finds a conjugate complex solution

$$\boxed{\omega_{\mathrm{ts}} = \mathbf{k} \cdot \mathbf{v}_b \left\{ 1 \pm i\omega_{pb} (n_0 + n_b/4)^{-1/2} \right\}} \tag{2.26}$$

that yields one damped and one unstable mode. The latter has the growth rate

$$\boxed{\gamma_{\mathrm{ts}} = \omega_{pb} \mathbf{k} \cdot \mathbf{v}_b / \sqrt{n_0}} \tag{2.27}$$

Because the situation is symmetric, a similar instability is obtained for negative frequency $\omega \sim -\mathbf{k} \cdot \mathbf{v}_b$.

The graphical representation of the solution is shown in Figure 2.3 plotting the two sides of the two-stream dispersion relation Eq. (2.21) as two separate functions, ϵ_l and ϵ_b, of frequency, ω. shows the principal shape of these functions. ϵ_l has a negative pole at $\omega = 0$ and approaches the horizontal line at 1 for large $|\omega|$, while ϵ_b is always positive, has poles at $\omega = \pm\mathbf{k}\cdot\mathbf{v}_b$, vanishes for $|\omega| \rightarrow \infty$ and has a minimum at $\omega = 0$. The two crossing points outside the poles are the real high-frequency solutions of the dispersion relation. These are two of the six solutions of the dispersion relation. The remaining solutions are conjugate-complex and correspond to low-frequency imaginary crossings at frequency $\omega < kv_b$. One of these is the above unstable solution.

The two-stream instability is the simplest instability known. It is a cold electron fluid instability which in practice rarely occurs, because other kinetic instabilities set in before it can develop.

For a single cold beam in cold plasma the right-hand side of Eq. (2.21) contains only one beam term. Near $\omega \approx kv_b$ this term is much larger than 1. In this case we get

$$\omega_{p0}^2(\omega - kv_b)^2 + \omega_{pb}^2\omega^2 = 0 \tag{2.28}$$

It is easily shown that it has the solutions

$$\omega = \frac{kv_b}{2 + n_b/n_0}\left[1 \pm i\left(\frac{n_b}{n_0}\right)^{1/2}\right] \tag{2.29}$$

which yields an oscillation in the negative energy mode with a frequency just below kv_b

$$\boxed{\omega_{\mathrm{sb}} = \frac{kv_b}{2 + n_b/n_0}} \tag{2.30}$$

and instability of this mode with growth rate

$$\boxed{\gamma_{\mathrm{sb}} = \omega_{\mathrm{ts}}\left(\frac{n_b}{n_0}\right)^{1/2}} \tag{2.31}$$

Weak Beam Instability

The two-stream instability is considerably modified when the beam density is much less than the density of the ambient plasma, $n_b \ll n_0$. When this happens the Langmuir mode at $\omega = -\omega_{p0}$ decouples from the other solutions of the two-stream dispersion relation. At this frequency the plasma behaves as if no beam exists. Decoupling of

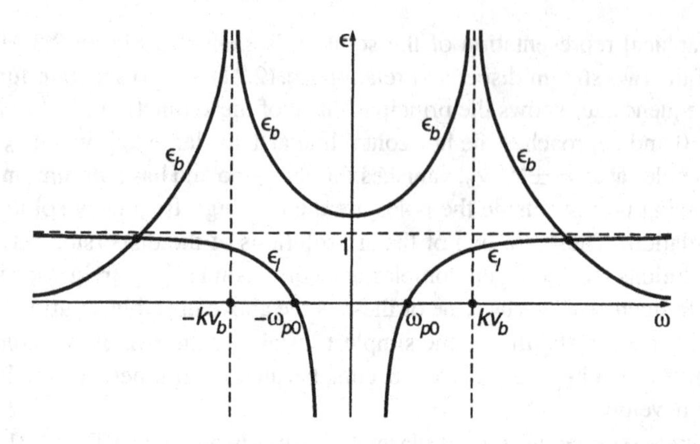

Fig. 2.3. Solution of the dispersion relation of the two-stream instability.

this mode implies that the remaining dispersion relation will be of third order in the frequency only. It is convenient to introduce a new variable as

$$\Omega = \omega - \omega_{p0} \tag{2.32}$$

and to define $\Delta = \omega_{p0} - \mathbf{k} \cdot \mathbf{v}_b$. Equation (2.17) then reduces to

$$\Omega(\Omega + \Delta)^2 - \frac{\omega_{p0}\omega_{pb}^2}{2} = 0 \tag{2.33}$$

Here we used the approximation that $1 - \omega_{p0}^2/\omega^2 \approx 2\Omega/\omega_{p0}$, for $\omega \approx \omega_{p0}$. With the help of the new dimensionless variable X defined through

$$\Omega = \omega_{p0} \left(\frac{n_b}{n_0} \right)^{1/3} X \tag{2.34}$$

and using the abbreviation $\delta = \Delta\omega_{p0}^{-1}(n_0/n_b)^{1/3}$, the above equation is brought into the dimensionless form

$$2X(X + \delta)^2 - 1 = 0 \tag{2.35}$$

For $\Delta = \delta = 0$ this equation has a threefold degenerate real solution $X = 2^{-1/3}$ which can be used to approximate the frequency of the weak beam mode

$$\boxed{\omega_{\text{wb}} = \omega_{p0} \left[1 + \left(\frac{n_b}{2n_0} \right)^{1/3} \right]} \tag{2.36}$$

The growth rate of the unstable solution is found by inserting $X = 2^{-1/3} + i\gamma$ into the third-order equation for X. Solving for the imaginary part one finds that $\gamma^2 = 3\Omega^2$. This yields the growth rate of the weak beam instability

$$\boxed{\gamma_{\rm wb} = \sqrt{3}\omega_{p0}\left(\frac{n_b}{2n_0}\right)^{1/3}} \tag{2.37}$$

This growth rate is much lower than that of the two-stream instability, because the free energy supplied by the weak beam is small. On the other hand, the instability is a high-frequency instability close to the background plasma frequency. Weak beams excite Langmuir waves at small growth rates.

Stabilization and Quenching

The weak beam instability is the zero temperature limit of the more general hot beam instability. One can show that a finite temperature will stabilize the weak beam instability. The condition for instability was $|\omega - \mathbf{k}\cdot\mathbf{v}_b| \approx \omega_{p0}(n_b/2n_0)^{1/3}$. If the beam is Maxwellian and has a thermal spread of $1.4kv_{\rm th\it b}$, Landau damping can be neglected as long as $|\omega - \mathbf{k}\cdot\mathbf{v}_b| \gg 2v_{\rm th\it b}$. On the other hand, the excited waves have $k \approx \omega_{p0}/v_b$. Combining these expressions, Landau damping can be neglected if

$$\frac{v_{\rm th\it b}}{v_b} \ll \left(\frac{n_b}{n_0}\right)^{1/3} \ll 1 \tag{2.38}$$

Since the beam densities must be small, only relatively fast beams will cause weak beam instabilities to grow. Otherwise the instability will make the transition to the two-stream instability. This is the case more relevant to space plasma physics, where most beams have sufficient time to relax and to become warm. But the initial stages of beam injection when narrow nearly monoenergetic beams leave from an acceleration source as for instance auroral electric potential drops or electron beam reflection from perpendicular shocks will lead to the weak beam instability.

One can estimate when the weak beam instability quenches itself. On p. 224 of our companion book, *Basic Space Physics*, it was shown that for Langmuir waves about half the energy is contained in the wave electric fluctuation while the other half is contained in the irregular thermal motion of the electrons. Hence, equating $W_w/2$ with the thermal energy of the beam electrons, $W_{\rm th\it b} = m_e n_b v_{\rm th\it b}^2/2$, and using the threshold condition in Eq. (2.38), one finds

$$W_w \approx 2W_b\left(\frac{n_b}{n_0}\right)^{2/3} \tag{2.39}$$

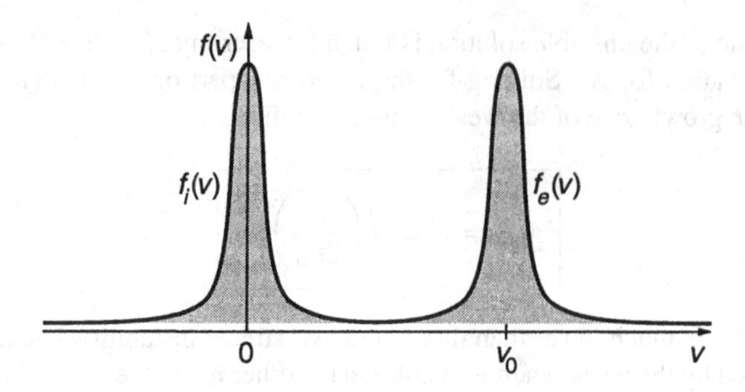

Fig. 2.4. Buneman-unstable velocity distribution.

Thus, when the wave energy reaches the fraction $(n_b/n_0)^{2/3}$ of the beam kinetic energy, $W_b = m_e n_b v_b^2/2$, and the weak beam instability ceases.

In the foreshock region of the Earth's bow shock the kinetic energy of the electron beam is about 10 eV (assuming specular reflection of the electrons). Measured Langmuir wave energies suggest a ratio of wave to solar wind thermal energy of about 10^{-4}. The solar wind electrons have thermal energy densities of about 10^8 eV/m^3. The wave energy density is thus $W_w \approx 10^{-15}$ J/m^3. This yields a beam density of $n_b/n_0 \approx 10^{-4}$. Such densities require that the thermal spread of the beam must be less than $v_{\mathrm{th}b} \ll 0.05\, v_b \approx 50$ km/s corresponding to a beam of temperature $k_B T_b \approx 0.03$ eV for weak beam instability. The beams are very cold and will readily spread out in velocity space.

2.3. Buneman-Instability

Another instability which is closely related to the two-stream instability is the *electron-ion two-stream* or *Buneman instability*. It arises from current flow across an unmagnetized plasma and can also be treated in the fluid picture. Currents are associated with the relative flow of electrons and ions. Fig. 2.4 shows a typical Buneman-unstable electron-ion distribution.

Growth Rate and Frequency

For the Buneman instability one considers the contribution of the motionless ions to the two-stream instability. Assuming that all plasma components are cold, the dispersion relation can be written as

$$\epsilon(\omega, \mathbf{k}) = 1 - \frac{\omega_{pi}^2}{\omega^2} - \frac{\omega_{pe}^2}{(\omega - kv_0)^2} = 0 \tag{2.40}$$

Here the ions take over the position of the motionless component, while all electrons are assumed to stream across the ion fluid at their bulk velocity, v_0. Clearly this will cause a current to flow in the plasma. Because the ion plasma frequency is much smaller than the electron plasma frequency the dominating term is the electron term. Instability will arise at the slow negative energy mode

$$\omega_{\rm n} \approx kv_0 - \omega_{pe} \tag{2.41}$$

while the positive energy wave $\omega_{\rm p} = kv_0 + \omega_{pe}$ does not couple to the instability. One can thus rewrite the above relation as

$$(\omega - \omega_{\rm n})\omega^2 = \frac{\omega_{pi}^2(\omega - kv_0)^2}{\omega - \omega_{\rm p}} \tag{2.42}$$

The wavenumber of interest is $k \approx \omega_{pe}/v_0$, because for a two-stream instability this wavenumber couples to the negative energy wave. In contrast to the electron two-stream instability, the frequency is small compared to the electron plasma frequency, $\omega \ll \omega_{pe}$. With these approximations the dispersion relation becomes

$$\omega^3 \approx -\frac{m_e}{2m_i}\omega_{pe}^3 \tag{2.43}$$

Of the three roots of this equation one is a real negative frequency wave

$$\omega = -\left(\frac{m_e}{2m_i}\right)^{1/3} \omega_{pe} \tag{2.44}$$

The other two are complex conjugate, and one of them has positive imaginary part. To find these two solutions we put $\omega \to \omega + i\gamma$ to obtain the following two equations

$$\begin{aligned} \omega(\omega^2 - 3\gamma^2) &= -\frac{m_e\omega_{pe}^3}{2m_i} \\ \gamma^2 &= 3\omega^2 \end{aligned} \tag{2.45}$$

The second equation gives $\gamma = \pm\sqrt{3}\omega$. Inserting into the first equation yields for the frequency of the maximum unstable Buneman mode

$$\boxed{\omega_{\rm bun} = \left(\frac{m_e}{16m_i}\right)^{1/3} \omega_{pe} \approx 0.03\,\omega_{pe}} \tag{2.46}$$

from which the growth rate is found to be

$$\boxed{\gamma_{\rm bun} = \left(\frac{3m_e}{16m_i}\right)^{1/3} \omega_{pe} \approx 0.05\,\omega_{pe}} \tag{2.47}$$

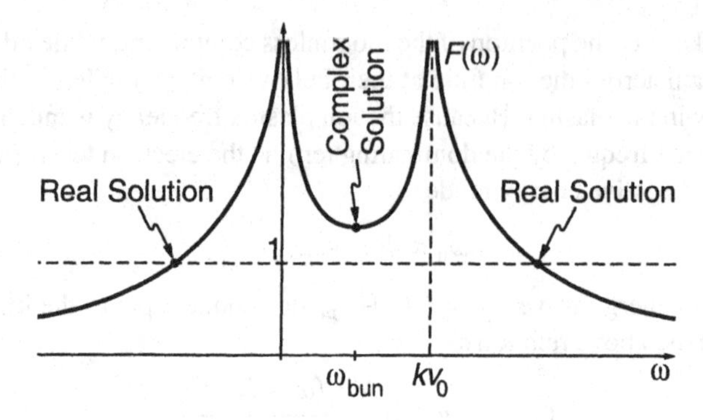

Fig. 2.5. Graphical solution of the Buneman instability.

This growth rate is very large, of the order of the frequency itself. Hence, the Buneman instability is a strong instability driven by the fast bulk motion of all the electrons moving across the plasma. One can expect that this instability will cause violent effects on the current flow, retarding the current and feeding its energy into heating of the plasma. It is interesting to note that the Buneman two-stream waves propagate parallel to the current flow but otherwise are electrostatic waves. As we will show later, they are the fast part of an ion-acoustic wave, which becomes unstable in weak current flow across a plasma.

Mechanism

To obtain an idea of the mechanism of the Buneman instability, we again use a graphical representation of the dispersion relation in the form

$$1 = \frac{\omega_{pi}^2}{\omega^2} + \frac{\omega_{pe}^2}{(\omega - kv_0)^2} = F(\omega) \tag{2.48}$$

The function $F(\omega)$ is shown in Fig. 2.5. It has two poles at $\omega = 0$ and $\omega = kv_0$. In between it has a minimum, whose position is found by calculating $\partial F(\omega)/\partial\omega = 0$

$$\omega_{\rm bun} = \frac{kv_0}{1 + (m_i/m_e)^{1/3}} \tag{2.49}$$

Inserting this value into $F(\omega)$ and demanding that the minimum of $F(\omega_{\rm bun}) > 1$, the condition for instability is found to be

$$k^2v_0^2 < \omega_{pe}^2 \left[1 + \left(\frac{m_e}{m_i}\right)^{1/3}\right]^3 \tag{2.50}$$

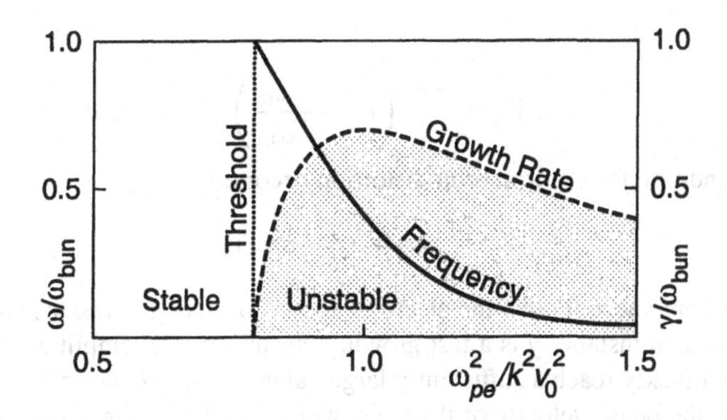

Fig. 2.6. Growth rate and frequency of Buneman instability.

which justifies our first choice of the unstable wavenumber showing in addition that any sufficiently long wave will become unstable against the electron flow. The threshold value for which the instability becomes marginal with vanishing growth rate, $\gamma \rightarrow 0$, is found by replacing the smaller sign with the equal sign. The marginally stable waves have the Buneman wavelength

$$\lambda_{\text{bun}} = \frac{2\pi v_0}{\omega_{pe}} \left[1 + \left(\frac{m_e}{m_i} \right)^{1/3} \right]^{-3/2} \tag{2.51}$$

The numerical solution of the Buneman dispersion relation, in dependence of the instability condition in Eq. (2.50), is shown in Fig. 2.6. The instability exists only above the threshold where its frequency is highest. The maximum growth rate in Eq. (2.47), normalized to the Buneman frequency, ω_{bun}, is found near the position where this value is close to 1. For larger speeds the growth rate decreases. More interesting is the observation that the normalized Buneman-unstable frequency, $\omega/\omega_{\text{bun}}$, decreases steeply to values far below the Buneman frequency of Eq. (2.46), when the conditions for instability are far above the threshold. This implies that Buneman-unstable long-wavelength modes have low frequencies and that low electron speeds excite low-frequency waves.

Quenching

We conclude this section with an outlook on the further evolution of the Buneman instability. Remember that the plasma, electrons as well as ions, have been assumed very cold. Hence, even small electrostatic potentials arising in the plasma will be able to distort the electron particle motion and thus modify the electric current which is the driving source of the Buneman instability. In the wave frame the streaming electron energy is

$$W_e = \frac{m_e}{2}\left(v_0 - \frac{\omega_{\rm bun}}{k_{\rm bun}}\right)^2 \tag{2.52}$$

so that the condition for electron orbit distortion becomes

$$W_w(t) > nW_e = \tfrac{1}{2} n m_e v_0^2 \tag{2.53}$$

where we took into account that the electrons move considerably faster than the wave. Now, the Buneman instability is a fast growing instability. The amplitude of the wave will therefore quickly reach a sufficiently large value to trap the electrons and slowing them down to the phase velocity of the wave which lies below threshold. The current will be disrupted in this case and the instability quenches itself. From the above condition we can estimate the time until quenching will happen, assuming that the wave amplitude grows from thermal level

$$W_{\rm tf} \approx k_B T_e / \lambda_D^3 \tag{2.54}$$

which was given in Eq. (I.9.27), as

$$W_{\rm E}(t) = W_{\rm tf} \exp(2\gamma_{\rm bun} t) \tag{2.55}$$

For the energy density we find from the Buneman dielectric response function

$$W_w(t) \approx \left(\frac{16 m_i}{m_e}\right)^{2/3} W_{\rm E}(t) \tag{2.56}$$

For the thermal level we can assume that it is well approximated by the thermal level of high-frequency Langmuir waves given in Eq. (2.54). Inserting all this into Eq. (2.53) and solving for the current disruption time, $t_{\rm cd}$, gives

$$\omega_{pe} t_{\rm cd} \approx \left(\frac{2m_i}{3m_e}\right)^{1/3} \ln\left[\left(\frac{m_e}{16 m_i}\right)^{2/3} \frac{v_0^2}{v_{\rm the}^2} n\lambda_D^3\right] \tag{2.57}$$

This expression depends only weakly on the electron current speed above threshold. The dominating number in the argument of the logarithm is the Debye number, $N_D = n\lambda_D^3$. Its logarithm is typically of the order of 15–30. Hence, in terms of the electron plasma frequency the self-disruption time of the current due to Buneman instability in an electron-proton plasma takes about 200 electron plasma periods or about 10 Buneman oscillations. Thus the Buneman instability will manifest itself in spiky oscillations of the current and in bursty emission of electrostatic waves below and up to the Buneman frequency, $\omega < \omega_{\rm bun} \approx 0.03\,\omega_{pe}$.

2.4. Ion Beam Instability

As for a last introductory example we discuss an instability generated by two counter-streaming plasma flows, the *electrostatic streaming instability* or *counterstreaming ion beam instability*. It is important in many kinds of plasma flows as, for instance, the solar wind. Its electromagnetic counterpart plays a significant role in the foreshock region of the Earth's bow shock.

Cold Electron Background

The dispersion relation of the counterstreaming ion beam instability including hot electrons consists of the two cold beam terms and a general hot electron term including the electron plasma dispersion function

$$1-\frac{1}{2k^2\lambda_D^2}Z'\left(\frac{\omega}{kv_{\mathrm{the}}}\right)-\frac{1}{2}\left[\frac{\omega_{pi}^2}{(\omega-kv_b)^2}+\frac{\omega_{pi}^2}{(\omega+kv_b)^2}\right]=0 \tag{2.58}$$

If the electrons are cold the distribution functions of the three components are well separated as shown in the left part of Fig. 2.7. The electron dispersion function reduces to ω_{pe}^2/ω^2, and the dispersion relation simplifies and can be written in a form similar to the cold beam instability

$$1-\frac{\omega_{pe}^2}{\omega^2}=\frac{1}{2}\left[\frac{\omega_{pi}^2}{(\omega-kv_b)^2}+\frac{\omega_{pi}^2}{(\omega+kv_b)^2}\right] \tag{2.59}$$

Its graphical representation is given in Fig. 2.8. The dispersion relation has three poles at $\omega = 0, \pm kv_b$, and the function $F(\omega)$ has two minima at low frequencies which both can be unstable. For sufficiently large beam velocities these minima are separated far enough to let the instability split into two Buneman-like instabilities, with growth rates given in the previous subsection, and one with positive, the other with negative frequency.

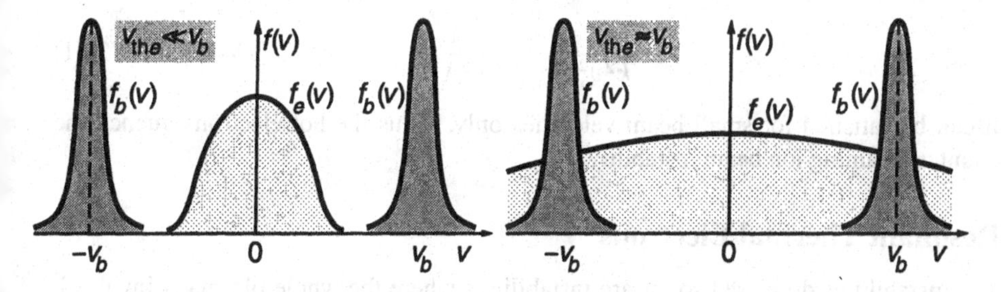

Fig. 2.7. Configuration of counterstreaming ion beam distributions.

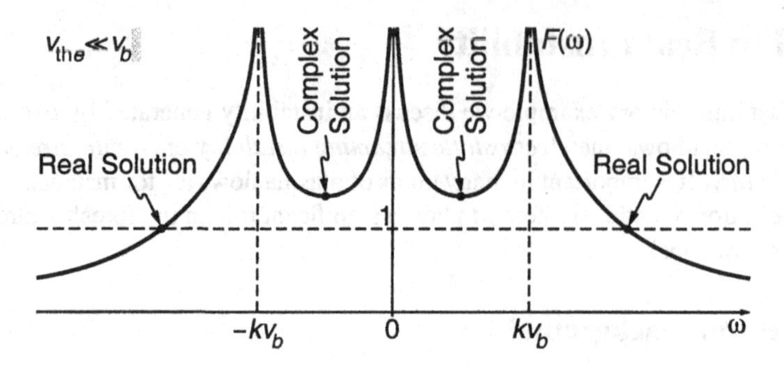

Fig. 2.8. Cold counterstreaming ion dispersion relation.

Hot Electron Background

On the other hand, when the electrons are hot (right-hand part of Fig. 2.7), we can use the small argument expansion for the plasma dispersion function because for small frequencies $\omega/kv_{\rm the} \ll 1$. Introducing the expansion given in App. A.7 of our companion book, we write the dispersion relation as

$$1 + \frac{1}{k^2\lambda_D^2} = \frac{1}{2}\left[\frac{\omega_{pi}^2}{(\omega - kv_b)^2} + \frac{\omega_{pi}^2}{(\omega + kv_b)^2}\right] \tag{2.60}$$

Instead of Fig. 2.8 we now have Fig. 2.9. The horizontal line at $1 + (k\lambda_D)^{-2}$ is the electron contribution. The poles of the combined ion terms are at $\omega = \pm kv_b$. The solutions are the cross-overs of the ion function with the horizontal. There exist two real solutions at frequencies well outside the two poles. But at frequencies $|\omega| < kv_b$ real solutions are possible only for low ratios ω_{pi}/kv_b. Here the possibility for instability arises. The instability is a low-frequency instability with frequency $\omega \approx 0$. In a non-symmetric configuration with differing beam densities and beam velocities the symmetry of the curves in Fig. 2.9 will be distorted, and the frequency will differ from zero. The minimum at $\omega = 0$ has the value $\omega_{pi}^2/k^2v_b^2$. Hence, instability sets in for

$$\frac{\omega_{pi}^2}{k^2v_b^2} > 1 + \frac{1}{k^2\lambda_D^2} \tag{2.61}$$

It can be satisfied for small beam velocities only. Thus the hot electrons quench the counterstreaming ion beam instability.

Resonant Thermal Electrons

The instabilities discussed so far are instabilities where the whole plasma is involved. They are bulk or *non-resonant* instabilities. In the case of the counterstreaming ion

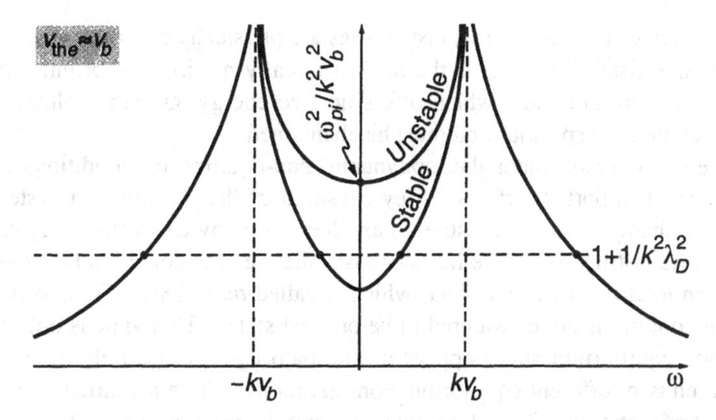

Fig. 2.9. Solution of the counterstreaming ion beam dispersion relation.

beam instability the quenching of the instability by the hot electron component applies only to the non-resonant instability. There is another range of frequencies, where the frequency is of the order of the electron thermal velocity, implying resonant contribution of electrons with $v \approx v_{\text{the}} \approx \omega/k$. The frequency of this wave is still of the order of the ion plasma frequency. Hence, their wavelength is very large compared with the Debye length. In this case the plasma dispersion function cannot be expanded and solutions are found only by numerical methods. The maximum growth rate of this resonant counterstreaming ion beam instability in hot electron plasmas is considerably smaller than the maximum Buneman growth rate

$$\boxed{\gamma_{\text{ib,max}} \approx 0.1\,\gamma_{\text{bun,max}}} \tag{2.62}$$

The small value of $\gamma_{\text{ib,max}}$ is easily understood, because in contrast to the Buneman instability, where the whole plasma contributes, the counterstreaming ion beam instability is fed by the small number of hot resonant electrons only. As a consequence the waves cannot gain much energy. They are weakly growing waves and will cause much less violent effects on the plasma than the ordinary Buneman instability.

Concluding Remarks

The present chapter introduces the reader to the concept of instability and explains this concept with a few illustrative examples. Although the concept of instability is mathematically relatively simple, it presents a number of fundamental physical difficulties. Obtaining a positive imaginary part from a given dispersion relation does not necessarily imply that one really encounters an instability. Instabilities do arise only if free energy is available. In other words, instabilities are physically real only when the state from which the instability starts is thermodynamically not in equilibrium. In a thermodynamic equilibrium state, which offers no free energy, growing solutions of the dispersion relation are spurious and must be abandoned.

On the other hand, under thermodynamic non-equilibrium conditions instabilities are the most important effects. They cause all of the transitions a system experiences when changing from one state to another. In many cases they may cause the formation of new structure, while in other cases they lead to some kind of transitional state between total disorder and order, which is called *turbulence*. Traditionally, non-equilibrium conditions are considered to be ordered states. This view is only partially correct. Non-equilibrium states appear in the majority of cases only as transitional states between two different equilibrium configurations, where the structure is formed via the onset of instability. Therefore, though instabilities act primarily to re-distribute the available free energy, they cause structure and order which may end up as another long-living ordered equilibrium, which is very different from the most probable thermodynamic equilibrium state.

Further Reading

Only a small selection of the many books on instabilities is given here. The general theory of instabilities is found in [5]. A useful introduction into a number of instabilities is given in [2]. Reference [3] contains a more or less systematic but not complete compilation of many instabilities which are to some extent relevant for space and astrophysics. Ion beam instabilities are completely reviewed in [4], electron beam instabilities in [1].

[1] R. J. Briggs, *Electron-Stream Interaction with Plasmas* (MIT Press, Cambridge, 1964).

[2] A. Hasegawa, *Plasma Instabilities and Nonlinear Effects* (Springer Verlag, Heidelberg, 1975).

[3] D. B. Melrose, *Instabilities in Space and Laboratory Plasmas*, (Cambridge University Press, Cambridge, 1991).

[4] M. V. Nezlin, *Physics of Intense Beams in Plasmas* (Institute of Physics Publ., Bristol, 1993).

[5] T. H. Stix, *Waves in Plasma* (American Institute of Physics, New York, 1992).

3. Macroinstabilities

Because of the multitude of free energy sources, a very large number of instabilities can develop in a plasma. It is sometimes convenient to divide them into two large groups according to the spatial scale involved in the instability. If this scale is of macroscopic size, comparable to the bulk scales of the plasma, the instabilities are called macroinstabilities. On the other hand, if the characteristic size of the instabilities is microscopic, of the scale size of the particle inertial lengths and gyroradii, the instabilities are called microinstabilities. In the latter case it is natural to assume that kinetic effects will become of greater importance than in the former case. Thus microinstabilities are typically also kinetic instabilities while macroinstabilities can be treated in the framework of fluid plasma theory. In some cases, however, it is useful to account for kinetic effects in macroinstabilities as well. The present section will cover the most important macroinstabilities appearing in space plasmas.

3.1. Rayleigh-Taylor Instability

On global scales plasma inhomogeneities cannot be neglected and several macroinstabilities are caused by plasma gradients. The simplest such instability is the *Rayleigh-Taylor instability* or *interchange instability*. It is the instability of a plasma boundary under the influence of a gravitational field. Because of this reason it is also called *gravitational instability*. Since the centrifugal force acts on a particle moving along curved magnetic field lines in a similar way as a gravitational force (see Sec. 2.4 of our companion book, *Basic Space Plasma Physics*), this can lead to similar effects. This instability is called *flute instability*.

Mechanism

Consider a heavy plasma supported against the gravitational force by a magnetic field as shown in Fig. 3.1. The boundary between plasma and magnetic field is the horizontal (x, y) plane, and the magnetic field points in the direction of x, so that $\mathbf{B}_0 = B_0\hat{\mathbf{e}}_x$. The gravitational acceleration $\mathbf{g} = -g\hat{\mathbf{e}}_z$ acts downward, while the plasma density gradient,

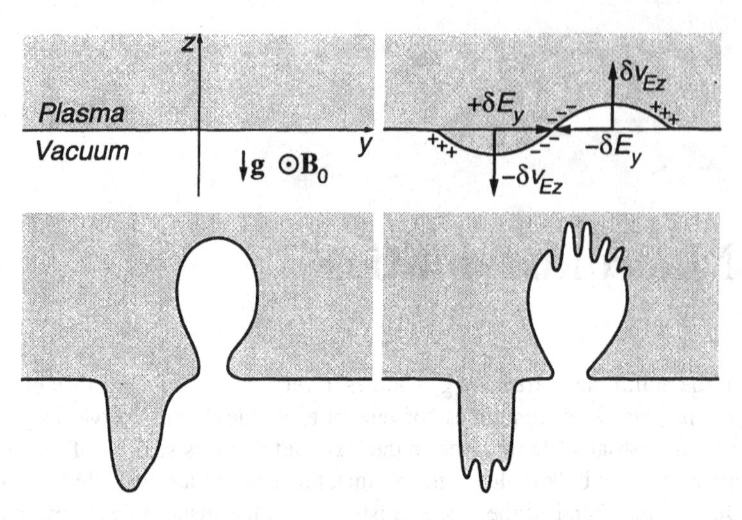

Fig. 3.1. Rayleigh-Taylor unstable plasma configuration.

$\nabla n_0 = [\partial n_0(z)/\partial z]\hat{\mathbf{e}}_z$ points upward and

$$\mathbf{g} \cdot \nabla n_0 < 0 \tag{3.1}$$

Let us for the moment neglect any thermal effects and assume that the plasma is collisionless. When the plasma boundary is distorted by a small purely electrostatic perturbation in the (x, y) plane, an instability can develop.

Consider a distortion of the boundary so that the plasma density makes a sinusoidal excursion in the z direction. The gravitational field causes an ion drift and current in the negative y direction, $v_{iy} = -m_i g/eB_0$. The electrons, because of their negligible mass, do not participate in this motion. Hence, in the region where the density disturbance causes a density enhancement, below the boundary between plasma and vacuum, the ion motion leads to a charge separation and accumulation of positive charges as shown. As a result, a charge separation electric field, δE_y, evolves. The horizontal electric disturbance field in the $+y$ direction, $+\delta E_y$, causes an upward electric field drift, $\delta v_{Ez} = +\delta E_y/B_0$, in the external magnetic field while in the region of $-\delta E_y$, the drift is downward, $\delta v_{Ez} = -\delta E_y/B_0$. These motions are in opposing directions; both of them amplify the initial distortion of the equilibrium density configuration at the plasma-vacuum magnetic field boundary.

Hence, the dilutions of the plasma caused by the initial rarefaction begin to rise up into the plasma while the initial density increases below the boundary begin to fall down. This mechanism causes light dilute plasma bubbles to rise up into the dense plasma and, under the action of gravity, it causes plasma originally supported by the magnetic field to fall down into the plasma-free magnetic field region thereby eroding the boundary and causing loss of plasma. This is shown in the sequence of Fig. 3.1. The bubbles themselves develop steep plasma boundaries which become unstable against the same Rayleigh-Taylor instability and deteriorate into smaller bubbles during the further evolution of the instability and the rise and fall of the bubbles. In the final nonlinear stage of the instability the boundary will become diffuse and the wavelength spectrum of the Rayleigh-Taylor mode becomes broadband, containing a wide range of wavelengths reaching from the long initial one to the smallest possible scales.

Dispersion Relation

In order to quantify the discussion, we linearize the cold ion equation of motion including the gravitational force and introduce the plane wave ansatz for the ion velocity and the electric field

$$\begin{aligned} \delta \mathbf{v}_i &= \delta \mathbf{v}_i(\omega, \mathbf{k}) \exp[i(\mathbf{k}\cdot\mathbf{x} - \omega t)] \\ \delta \mathbf{E} &= -i\mathbf{k}\delta\phi(\omega, \mathbf{k}) \exp[i(\mathbf{k}\cdot\mathbf{x} - \omega t)] \end{aligned} \tag{3.2}$$

to obtain

$$\left(\omega + \frac{gk_\perp}{\omega_{gi}}\right)\delta \mathbf{v}_{i\perp} = \frac{e}{m_i}\left(\mathbf{k}_\perp \delta\phi - iB_0 \hat{\mathbf{e}}_x \times \delta\mathbf{v}_{i\perp}\right) \tag{3.3}$$

Since the frequency of the disturbance will be much smaller than the ion gyrofrequency ($\omega' \ll \omega_{gi}$) the solution for the velocity disturbance is

$$\delta \mathbf{v}_{i\perp} = -\delta\phi\left[i\mathbf{k}_\perp \times \hat{\mathbf{e}}_x + \frac{\mathbf{k}_\perp}{\omega_{gi}B_0}\left(\omega + \frac{gk_\perp}{\omega_{gi}}\right)\right] \tag{3.4}$$

Using this equation to eliminate the velocity disturbance from the ion continuity equation

$$\omega\delta n_i = n_0 \mathbf{k}\cdot\delta\mathbf{v}_i - i\delta\mathbf{v}_i\cdot\nabla n_0 \tag{3.5}$$

one finds for the density disturbance

$$\delta n_i = n_0\delta\phi\left[\frac{e}{m_i}\left(\frac{k_\parallel^2}{\omega^2} - \frac{k_\perp^2}{\omega_{gi}^2}\right) + \frac{k_\perp}{B_0 L_n}\left(\omega + \frac{gk_\perp}{\omega_{gi}}\right)^{-1}\right] \tag{3.6}$$

We have introduced here the undisturbed inverse density gradient scale length

$$L_n^{-1} = \frac{d\ln n_0(z)}{dz} > 0 \tag{3.7}$$

Because the electrons are cold and do not drift in our model, the electron continuity and momentum equations yield

$$\delta n_e = -\delta\phi\frac{n_0}{B_0}\left(\frac{\omega_{ge}}{\omega}\frac{k_\parallel^2}{\omega} - \frac{k_\perp}{L_n\omega}\right) \tag{3.8}$$

For these low-frequency variations one can safely assume that the condition of quasineutrality, $\delta n_e = \delta n_i$, is satisfied. Inserting the expressions for the two oscillating densities leads to the dispersion relation

$$\frac{\omega_{gi}}{\omega}\frac{1}{k_\perp L_n}\left(1 - \frac{\omega}{\omega + gk_\perp/\omega_{gi}}\right) - \left(1 + \frac{m_i}{m_e}\right)\frac{\omega_{gi}^2 k_\parallel^2}{\omega^2 k_\perp^2} + 1 = 0 \tag{3.9}$$

This is a general dispersion relation for the Rayleigh-Taylor instability.

Growth Rate and Frequency

The highest growth rate of this instability is found for purely perpendicular propagation of the disturbance since in this case the electric field amplitude will lead to the largest vertical drift of the plasma for the longitudinal mode. Thus putting $k_\parallel = 0$ and assuming a weak gravitational effect so that $\omega \gg k_\perp g/\omega_{gi}$, one can expand the term in brackets in Eq. (3.9) to first order in this small parameter and obtains

$$\omega^2 = -g/L_n \tag{3.10}$$

which has purely imaginary solutions and thus one purely growing solution with

$$\boxed{\gamma_{0rt} = \left(\frac{g}{L_n}\right)^{1/2}} \tag{3.11}$$

It is the same expression as found for fluids under the action of gravity.

The non-oscillatory character of this instability results from the assumption of small $k_\perp g/\omega_{gi}\omega$. Expanding up to second order in Eq. (3.9) yields

$$\omega^3 + \frac{g}{L_n}\omega - \frac{g^2 k_\perp}{2\omega_{gi} L_n} = 0 \tag{3.12}$$

This equation has one real and two conjugate complex solutions. The unstable solution has frequency

$$\omega_{rt} = \gamma_{0rt}\cosh\left[\tfrac{1}{3}\cosh^{-1}\left(\frac{k_\perp L_n \gamma_{0rt}}{2\omega_{gi}}\right)\right] \tag{3.13}$$

proportional to the zero-order growth rate, γ_{0rt}. The corrected growth rate of this oscillating Rayleigh-Taylor mode is

$$\gamma_{rt} = \sqrt{3}\gamma_{0rt}\sinh\left[\tfrac{1}{3}\cosh^{-1}\left(\frac{k_\perp L_n \gamma_{0rt}}{2\omega_{gi}}\right)\right] \tag{3.14}$$

An approximate closed solution of Eq. (3.12) can be constructed for the complex frequency $\omega \to \omega_{rt} + i\gamma$. Separating the real and imaginary parts of Eq. (3.12) yields

$$\begin{aligned} \gamma^2 - 3\omega^2 &= g/L_n \\ \omega\left(\omega^2 - 3\gamma^2 + \frac{g}{L_n}\right) &= \frac{g^2 k_\perp}{2\omega_{gi} L_n} \end{aligned} \tag{3.15}$$

Together with Eq. (3.11) the first of these equations requires that $\gamma_{\rm rt} \gg \sqrt{3}\omega$, and the second equation gives for the frequency

$$\boxed{\omega_{\rm rt} = -\frac{g k_\perp}{4\omega_{gi}}} \tag{3.16}$$

Consistency with the above condition demands $k_\perp L_n \ll 4\omega_{gi}(L_n/g)^{1/2}$, or in terms of the Rayleigh-Taylor wavelength

$$\frac{\lambda_{\rm rt}}{L_n} \gg \frac{\pi}{2}\frac{\gamma_{0\rm rt}}{\omega_{gi}} \tag{3.17}$$

The right-hand side of this expression is small. So the condition of long wavelengths is easily satisfied in all cases of interest.

For oblique propagation of the disturbance, $k_\parallel \neq 0$, part of the electric field is parallel to the magnetic field and can be short-circuited by the fast electron motion along the magnetic field thus causing a field-aligned current but at the same time partially quenches the instability. Let us write the dispersion relation for this case including the second term in Eq. (3.9)

$$\omega^2 = -\frac{g}{L_n} + \frac{m_i}{m_e}\frac{k_\parallel^2}{k_\perp^2}\omega_{gi}^2 \tag{3.18}$$

When the condition

$$\frac{k_\parallel}{k_\perp} > \left(\frac{m_e}{m_i}\frac{g}{\omega_{gi}^2 L_n}\right)^{1/2} = \left(\frac{m_e}{m_i}\right)^{1/2}\frac{\gamma_{0\rm rt}}{\omega_{gi}} \tag{3.19}$$

is satisfied, the right-hand side of this expression becomes positive, and the instability ceases. Thus for a given density gradient scale, L_n, the last condition defines a marginal angle around the perpendicular direction. Inside this angle the Rayleigh-Taylor instability can evolve. This angular range is very narrow, in the ionosphere typically less than one degree and the Rayleigh-Taylor mode is nearly-perpendicular here.

Magnetospheric Growth Rates

To obtain an idea of the magnitude of the growth rate of the Rayleigh-Taylor instability in the vicinity of the Earth, we introduce the gravitational acceleration $g_0(1R_E) \approx$

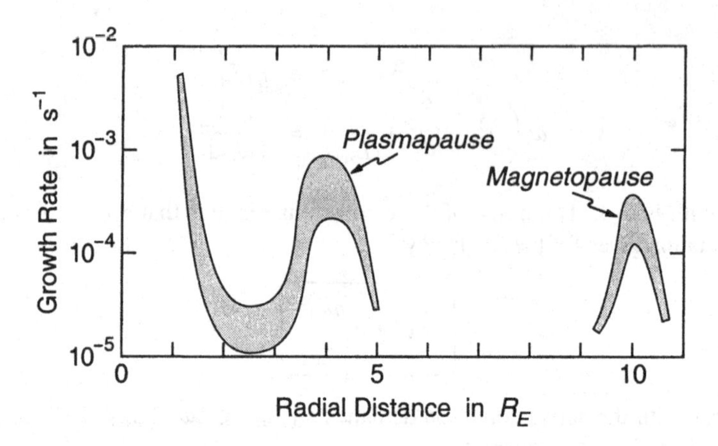

Fig. 3.2. Collisionless Rayleigh-Taylor growth rate in the magnetosphere.

10 m/s^2 at the surface and rewrite Eq. (3.11) as

$$\gamma_{\mathrm{Ort}} \approx \left(\frac{10R_E}{r^2 L_n(r)}\right)^{1/2} \tag{3.20}$$

in order to account for the radial dependence of $g(r)$ and $L_n(r)$ in the equatorial plane. The characteristic scale in the ionosphere is of the order of $L_n \approx 20$ km. But here, as shown below, a different kind of Rayleigh-Taylor instability is at work. Above 500 km altitude, L_n becomes very large. Near the plasmapause, at about 4–5 R_E, it assumes a value of about 1000 km, then becomes large until at the magnetopause it is of the order of 1000 km again. Figure 3.2 shows the schematic radial dependence of the Rayleigh-Taylor growth rate in the equatorial plane. It has two peaks at the plasmapause and at the magnetopause with the peak at the magnetopause being very narrow, restricted only to the magnetopause transition region. Characteristic growth times at the magnetopause are of the order of several hours, however. This is too long for the Rayleigh-Taylor instability to be of importance here.

Equatorial Spread-F

The Rayleigh-Taylor instability requires the presence of a non-negligible gravitational acceleration. In the vicinity of the Earth this requirement can be satisfied only at ionospheric altitudes. Following our previous discussion one must, in addition, demand that the magnetic field is perpendicular to the gravitational acceleration. Since this is vertical, the Rayleigh-Taylor instability can evolve only in the equatorial region where the magnetic field is nearly horizontal. The last requirement is that the plasma density

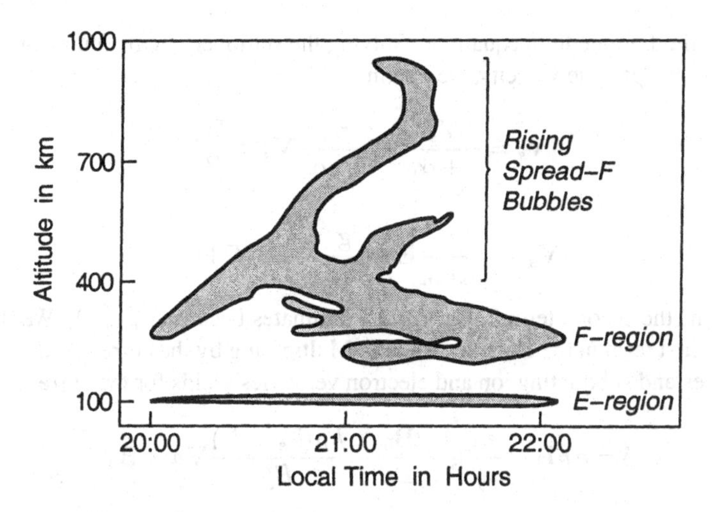

Fig. 3.3. Radar backscattered signals from rising bubbles in equatorial spread-F.

increases with altitude above the Earth's surface. This requirement is met in the equatorial electron density height profiles below the E- and F-region maxima. One may expect that these density gradients are Rayleigh-Taylor unstable and will be gradually erased.

This kind of erosion is known from radar observations of the equatorial electron density content in the F-region. Many of those observations exhibit multiple radar beam reflections from rising low density bubbles called *equatorial spread-F*. A schematic example of an observation at evening is shown in Fig. 3.3. The shadowed plumes show the backscattered radar intensity from the usual and bubble density gradients and their evolution with time. The lowest trace near 100 km altitude is the signal of the E-layer which is quite stable. The rather unstable traces at higher altitudes are the disrupting F-region structures. Bubbles rise in this case up to 1000 km height until they dissolve. Their velocity can be determined from the slope of the signal. Uprising velocities of some 100 m/s are not unusual. The very irregular shapes of the backscattered signals are caused by a mixture of the Rayleigh-Taylor instability and the various horizontal and vertical winds prevalent in the equatorial upper ionosphere.

Collisional Rayleigh-Taylor Instability

The ionospheric plasma is not perfectly collision-free. It is thus obligatory to include collisions into the Rayleigh-Taylor mode. These collisions, which are mainly collisions with neutrals, considerably modify the instability. Let us define the quantities $\alpha_s = \omega_{gs}/\nu_{sn}$, the ratio between gyro and particle-neutral collision frequency, and $D_{s\perp} = k_B T_s / m_s \nu_{sn}$, the diffusion coefficient. Moreover, we cannot neglect the pressure force term in the fluid momentum equations. Solving the stationary momentum conservation equation for the particle velocity, we obtain

$$\mathbf{v}_s = \frac{\mathbf{V}_s}{1+\alpha_s} + \frac{\alpha_s}{1+\alpha_s}\mathbf{V}_s \times \frac{\mathbf{B}_0}{B_0} \tag{3.21}$$

where

$$\mathbf{V}_s = \frac{e}{m_s \nu_{sn}}\mathbf{E} + \frac{\mathbf{g}}{\nu_{sn}} - D_{s\perp}\nabla \ln n \tag{3.22}$$

For electrons the second term in Eq. (3.21) dominates because $\alpha_e \gg 1$. We therefore neglect the first term in the electron velocity. Multiplying by the corresponding charges and densities and subtracting ion and electron velocities yields for the current

$$\mathbf{j} = \sigma_P \mathbf{E} + \frac{en}{\omega_{gi}}\mathbf{g} \times \frac{\mathbf{B}_0}{B_0} - \frac{k_B(T_e+T_i)}{B_0^2}\nabla n \times \mathbf{B}_0 \tag{3.23}$$

where σ_P is the Pedersen conductivity defined in Eq. (I.4.27) of our companion book. Due to their large mass the ions contribute most of the gravitational current permitting to approximate the plasma velocity by the ion velocity. Taking the divergence of $\mathbf{v}_i$, assuming $\mathbf{E}_0 = 0$, we find

$$\nabla \cdot \mathbf{v} = \nabla \cdot \left[\mathbf{g} \times \mathbf{B}_0 + \frac{k_B T_i}{m_i}\mathbf{B}_0 \times \nabla(\ln n)\right] = 0 \tag{3.24}$$

which can be used on the right-hand side of the continuity equation

$$\frac{\partial n}{\partial t} + \mathbf{v} \cdot \nabla n = -n\nabla \cdot \mathbf{v} = 0 \tag{3.25}$$

Hence, the flow is practically incompressible. This approximation is very good for the F-region, but breaks down in the E-region. We linearize the continuity equation and the equation of vanishing current divergence

$$\nabla \cdot \mathbf{j} = 0 \tag{3.26}$$

using Eq. (3.23) for $\mathbf{j}$. We further assume quasineutrality, $n_e = n_i = n$, for the low-frequency disturbances we expect and obtain

$$\nabla \cdot [(n_0 + \delta n)(\mathbf{g} \times \mathbf{B}_0 + \nu_{in}\delta\mathbf{E})] = 0 \tag{3.27}$$

This equation and the continuity equation together can be Fourier transformed. Introducing the gradient scale length, L_n, and the electrostatic potential, $\delta\phi$, the two transformed linear equations obtained are

$$\begin{aligned} \frac{\delta n}{n_0} + i\frac{k_\perp \nu_{in}}{g B_0}\delta\phi &= 0 \\ \left(\frac{\omega}{k_\perp} - \frac{g}{\omega_{gi}}\right)\frac{\delta n}{n_0} - \frac{1}{B_0 L_n}\delta\phi &= 0 \end{aligned} \tag{3.28}$$

Solving this homogeneous system gives the dispersion relation of collision-dominated Rayleigh-Taylor modes in the equatorial ionosphere

$$\omega = \frac{k_\perp g}{\omega_{gi}} + i\frac{g}{\nu_{in} L_n} \tag{3.29}$$

It is interesting that in this case a real part of the mode frequency appears naturally. It simply shows that the Rayleigh-Taylor mode propagates with phase velocity

$$\frac{\omega}{k_\perp} = \frac{g}{\omega_{gi}} \tag{3.30}$$

into perpendicular direction. This direction is easily found to be the eastward direction, $\mathbf{g} \times \mathbf{B}_0$, in the equatorial ionosphere since the magnetic field points toward north and the gravitational acceleration is vertically downward. The growth rate is this time given by

$$\boxed{\gamma_{rtn} = \frac{g}{\nu_{in} L_n} = \frac{\gamma_{0\mathrm{rt}}^2}{\nu_{in}}} \tag{3.31}$$

It depends on the collision frequency between ions and neutrals and is positive only when the density gradient points upward.

The Rayleigh-Taylor instability in the ionosphere is clearly collision-dominated, as is obvious from its growth rate in Eq. (3.31). For vanishing collision frequency, $\nu_{in} \to 0$, the growth rate would diverge. This non-physical behavior results from the approximations made in the present theory. Actually, in a more elaborated theory which takes into account the full particle dynamics and is not restricted to the fluid-drift approximation, one finds a smooth transition from the collisional to the collisionless Rayleigh-Taylor instability. Formally this transition can be modelled as

$$\boxed{\tilde{\gamma}_{\mathrm{rt}} = \gamma_{0\mathrm{rt}}\left[1 - \exp\left(-\frac{\gamma_{0\mathrm{rt}}}{\nu_{in}}\right)\right]} \tag{3.32}$$

a formula which describes both limits. In the limit of large collision frequency, the exponential can be expanded and yields the growth rate in Eq. (3.31), while for vanishing collisions the exponential dependence on ν_{in} disappears.

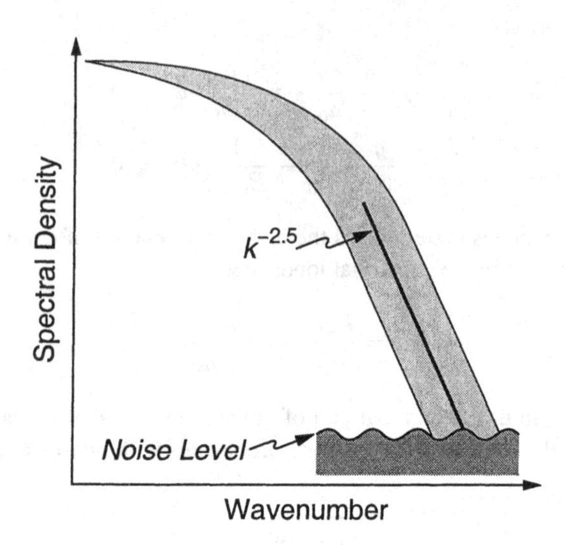

Fig. 3.4. Power spectrum of spread-F bubbles.

Rayleigh-Taylor Bubbles

The collisional Rayleigh-Taylor instability is an important instability in the ionosphere. In the E-region the instability is quenched because of the very high collision frequencies. At higher altitudes, where it becomes non-collisional, the condition that the electron density must increase with distance from the Earth is generally not satisfied. This explains why Rayleigh-Taylor bubbles develop preferentially just below the F-region maximum. Also, when the F-region rises during the night into a region where the collision frequency becomes low, the evolution of the bubbles is favored explaining the observation of strong evening and night-time high-altitude equatorial spread-F.

In the ionosphere there is a large number of possibilities to modify the ionospheric Rayleigh-Taylor mode. Convection and dynamo electric fields as well as neutral winds cause drift motions of the ionospheric plasma in both the horizontal and vertical directions. Such motions lead to deformation of bubbles and to their transport out of the unstable region. Moreover, cascading of the bubbles down to shorter wavelength and small scales produces a broad wavelength spectrum of the equatorial spread-F. Figure 3.4 shows the typical power spectrum of equatorial spread-F Rayleigh-Taylor bubbles with $|\delta n/n|^2 \propto k_\perp^{-2.5}$. At very short wavelengths comparable to the ion gyroradius the waves become damped due to diffusion. Experimental observation of spread-F bubbles by sounding rocket experiments shows that these marginal diffusive wavelengths are of the order of about 100 m.

3.2. Farley-Buneman Instability

Another instability is closely related to the ionospheric Rayleigh-Taylor instability. It is also driven by transverse currents in the collisional ionosphere, but the nature of the currents is not gravitational. Rather these currents are drift currents, and the resulting instability is a *modified two-stream instability* or *Farley-Buneman instability*. The instability arises from a difference in the drifts of electrons and ions for $\nu_{in} \gg \omega_{gi}$ (see Sec. 4.4 of our companion book).

Dispersion Relation

Let the electric field point vertically downward, $\mathbf{E}_0 = -E_0\hat{\mathbf{e}}_z$. In the crossed horizontal northward magnetic and vertical electric fields the electrons perform an eastward $E \times B$ drift motion with a velocity $v_E = -E_0/B_0$. The linearized and Fourier transformed electron continuity equation yields for the disturbance of the electron velocity

$$\delta v_{ey} = \left(\frac{\omega}{k_\perp} - v_E\right)\frac{\delta n}{n_0} \tag{3.33}$$

where quasineutrality has been assumed as usual for low-frequency waves. Neglecting electron inertia and the action of gravity on the electrons but keeping electron-neutral collisions we find after linearizing the electron equation of motion

$$\begin{aligned} \omega_{ge}\delta v_{ey} + \nu_{en}\delta v_{ez} &= 0 \\ \nu_{en}\delta v_{ey} - \omega_{ge}\delta v_{ez} &= -ik_\perp\left(\frac{e}{m_e}\delta\phi - \frac{k_BT_e}{m_e}\frac{\delta n}{n_0}\right) \end{aligned} \tag{3.34}$$

Due to the high ion-neutral collision frequency ions do not move in the vertical direction, $v_{iz} \approx 0$, and the corresponding ion equations are

$$\begin{aligned} \delta v_{iy} - \frac{\omega}{k_\perp}\frac{\delta n}{n_0} &= 0 \\ (\omega - i\nu_{in})\delta v_{iy} - k_\perp v_{\mathrm{thi}}^2\frac{\delta n}{n_0} &= \frac{e}{m_i}k_\perp\delta\phi \end{aligned} \tag{3.35}$$

This linear homogeneous system of equations is solved by setting its determinant to zero, giving the dispersion relation

$$\omega\left(1 + i\psi_0\frac{\omega - i\nu_{in}}{\nu_{in}}\right) = k_\perp v_E + i\psi_0\frac{k_\perp^2 c_{ia}^2}{\nu_{in}} \tag{3.36}$$

The combined effect of electron and ion collisions and electron and ion gyrofrequencies is contained in the quantity

$$\psi_0 = \frac{\nu_{en}\nu_{in}}{\omega_{ge}\omega_{gi}} = \frac{\nu_{en}\nu_{in}}{\omega_{lh}^2} \tag{3.37}$$

and the ion-acoustic speed is defined as $c_{ia}^2 = k_B(T_e + T_i)/m_i$. Electron and ion gyration effects lead to the appearance of the lower-hybrid resonance frequency, ω_{lh}, in the ψ_0-parameter.

Growth Rate and Frequency

As in the case of the Buneman instability the dispersion relation is a third-order equation in ω and can be solved in the same way. The weakly unstable solution yields for the frequency

$$\boxed{\omega_{\text{fb}} = \frac{k_\perp v_E}{1+\psi_0}} \tag{3.38}$$

and the growth rate

$$\boxed{\gamma_{\text{fb}} = \frac{\psi_0}{\nu_{in}} \frac{\omega_{\text{fb}}^2 - k_\perp^2 c_{ia}^2}{1+\psi_0}} \tag{3.39}$$

Buneman-Farley instability sets in whenever the wave phase velocity exceeds the ion acoustic speed, $\omega_{\text{fb}}/k_\perp > c_{ia}$ or when the vertical drift velocity exceeds the threshold

$$v_E > (1+\psi_0)c_{ia} \tag{3.40}$$

Typical Altitude Range

This instability is of importance in the equatorial electrojet region, where large vertical electric fields cause a strong horizontal drift current to flow above the equator. It lets bubbles rise into the current region and distorts the current flow in a way similar to the Buneman instability. The parameter ψ_0 depends strongly on altitude. Recalling the definition of the neutral collision frequency in Eq. (I.4.1) of our companion book, $\nu_{sn} = n_n\sigma_n\langle v_s\rangle$, one observes that ψ_0 is proportional to the square of the neutral gas density. The latter satisfies the barometric law of Eq. (I.4.29), $n_n(z) = n_0\exp(-z/H)$.

Hence, $\psi_0(z)$ decreases exponentially with z, as does the growth rate. The Farley-Buneman instability is thus restricted to lower altitudes. For a numerical value one has $\psi_0 \approx 0.22$ at $z = 105\,\text{km}$. Assuming a scale height of 10 km, it readily becomes small with altitude and can be neglected in Eqs. (3.38) and (3.39) above $z = 130 - 150\,\text{km}$. The altitude range of the Farley-Buneman instability therefore is the equatorial E-region.

3.3. Kelvin-Helmholtz Instability

Another macroinstability is generated by shear flows in magnetized plasmas. In investigating this instability, we turn away from the ionosphere to the collisionless boundary transition region between the magnetosheath and magnetosphere. The magnetosheath plasma is flowing along the magnetopause around the magnetosphere, and it is easy to imagine that any kind of contact between the flow and the magnetospheric field may cause ripples on the boundary to evolve.

This so-called *Kelvin-Helmholtz instability* is best described in magnetohydrodynamics, since it is caused by the bulk plasma flow and has wavelengths considerably longer than any of the gyroradii. Figure 3.5 shows the geometry of the problem. The left part of the figure has been drawn for the symmetric case, including a broad shear flow transition layer. Here the flow changes from positive to negative direction. The right part of the figure is an idealized model with a sharp boundary, i.e., the transition region is narrower than the wavelength, yet still wider than the ion gyroradius. Plasma density, magnetic field, and flow velocity all change abruptly across the boundary.

Dispersion Relation

Let us assume ideal conditions though usually the narrow transition layer may contain some kind of viscous interaction. Then the electric field is given by $\mathbf{E} = -\mathbf{v} \times \mathbf{B}$. From ideal one-fluid theory, assuming scalar pressure, p, eliminating the electric field, linearizing around the zero-order magnetic field and density, and introducing instead of the plasma velocity the displacement vector, $\delta\mathbf{x}$, by

$$\delta\mathbf{v} = d\delta\mathbf{x}/dt \tag{3.41}$$

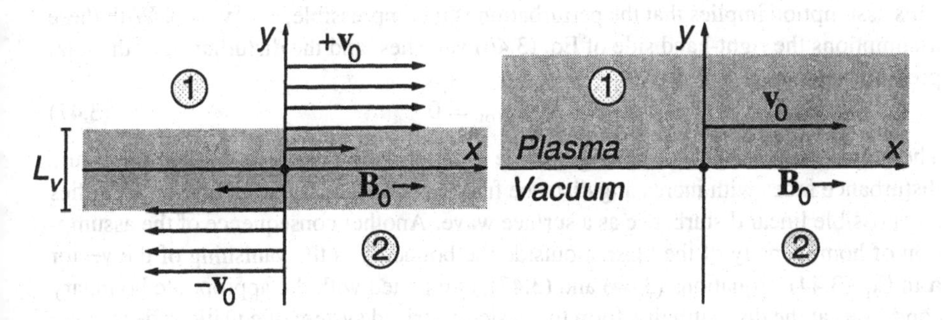

Fig. 3.5. Two configurations leading to Kelvin-Helmholtz instability.

the linearized induction and momentum conservation equations can be written as

$$\begin{aligned} \delta \mathbf{B} &= \nabla \times (\delta \mathbf{x} \times \mathbf{B}_0) = \mathbf{B}_0 \cdot \nabla \delta \mathbf{x} - \delta \mathbf{x} \cdot \nabla \mathbf{B}_0 - \mathbf{B}_0 \nabla \cdot \delta \mathbf{x} \\ \mu_0 m_i n_0 d^2 \delta \mathbf{x}/dt^2 &= -\mu_0 \nabla \delta p - \delta \mathbf{B} \times (\nabla \times \mathbf{B}_0) - \mathbf{B}_0 \times (\nabla \times \delta \mathbf{B}) \end{aligned} \tag{3.42}$$

The induction law in the above version has been integrated with respect to time t. We define the first order variation of the total pressure by

$$\mu_0 \delta p_{\mathrm{tot}} = \mu_0 \delta p + \mathbf{B}_0 \cdot \delta \mathbf{B} \tag{3.43}$$

Eliminating $\delta \mathbf{B}$ from the second Eq. (3.42), one obtains for $\delta \mathbf{x}$

$$m_i n_0 \left[(\mathbf{v}_A \cdot \nabla)^2 - \frac{\partial^2}{\partial t^2} \right] \delta \mathbf{x} = \nabla \delta p_{\mathrm{tot}} + \mathbf{a} \tag{3.44}$$

We have introduced here the Alfvén velocity, $\mathbf{v}_A = \mathbf{B}_0/\sqrt{\mu_0 m_i n_0}$, defined by the background parameters. The vector $\mathbf{a}$ on the right-hand side of this equation, which results from a combination of the vector operations in Eq. (3.42), is defined as

$$\mathbf{a} = -\delta \mathbf{B} \cdot \nabla \mathbf{B}_0 + \mathbf{B}_0 \cdot \nabla (\mathbf{B}_0 \nabla \cdot \delta \mathbf{x} + \delta \mathbf{x} \cdot \nabla \mathbf{B}_0) \tag{3.45}$$

Equation (3.44) shows the coupling of the Alfvén wave on the left-hand side to the total pressure disturbance on the right-hand side. Because the divergences of both the zero-order and the disturbed magnetic field components vanish, $\nabla \cdot \delta \mathbf{B} = \nabla \cdot \mathbf{B}_0 = 0$, Eq. (3.42) can also be manipulated into an equation for the total pressure variation

$$\nabla^2 \delta p_{\mathrm{tot}} = -m_i \nabla \cdot \left(n_0 \frac{d^2 \delta \mathbf{x}}{dt^2} \right) + \frac{1}{\mu_0} \nabla \cdot (\delta \mathbf{B} \cdot \nabla \mathbf{B}_0 + \mathbf{B}_0 \cdot \nabla \delta \mathbf{B}) \tag{3.46}$$

Now we assume that the plasma and the flow are homogeneous on both sides of the discontinuity, so that total pressure balance is satisfied to all orders outside the boundary. This assumption implies that the perturbation is incompressible, $\nabla \cdot \delta \mathbf{v} = 0$. With these assumptions the right-hand side of Eq. (3.46) vanishes, and the disturbance of the total pressure satisfies

$$\nabla^2 \delta p_{\mathrm{tot}} = 0 \tag{3.47}$$

The only change in δp_{tot} occurs right at the infinitely thin boundary, while the pressure disturbance fades with increasing distance from the boundary. This condition identifies any possible linear disturbance as a surface wave. Another consequence of the assumption of homogeneity of the plasma outside the boundary is the vanishing of the vector $\mathbf{a}$ in Eq. (3.44). Equations (3.44) and (3.47), completed with the appropriate boundary conditions at the discontinuity, form the basic linearized system of equations describing the surface waves propagating along the boundary between the two media.

The homogeneity of the problem along the interface between the two media allows one to use a plane wave ansatz in the (x, z)-plane for both variables, $\delta\mathbf{x}$ and δp_{tot}, with horizontal wavenumber, $\mathbf{k} = k_x\hat{\mathbf{e}}_x + k_z\hat{\mathbf{e}}_z$, and frequency ω. From the equation for the displacement of the boundary one has

$$\delta\mathbf{x} = \frac{\nabla\delta p_{\mathrm{tot}}}{m_i n_0\left[\omega^2 - (\mathbf{k}\cdot\mathbf{v}_A)^2\right]} \tag{3.48}$$

and the variation of the pressure as solution of Eq. (3.47) is given by

$$\delta p_{\mathrm{tot}} = p_0 \exp(-k|y|)\exp[-i(\omega t - k_x x - k_z z)] \tag{3.49}$$

with $k^2 = k_x^2 + k_z^2$. The exponential y-dependence of this solution takes into account the decay of the amplitude of the disturbance in the direction perpendicular to the interface. The physical reason behind this decay in an otherwise homogeneous medium is that the excitation of the disturbance is located at the boundary between the two media, where it must be largest. Outside the boundary the wave is evanescent, because no free energy is available there to further feed its amplitude.

The physical boundary conditions to be applied to the above solutions must also account for this fact. Hence, we choose that at a infinitely thin boundary the normal component of the displacement, $\delta\mathbf{x}_y$, must be continuous. In addition total pressure balance across the boundary is demanded, as is reasonable for a tangential discontinuity. Further, since the plasma parameters may change across the boundary we must distinguish the quantities to both sides by appropriate indices 1 and 2. Finally, in region 1 the plasma streams with constant velocity, $\mathbf{v}_0$. This implies that the wave frequency in region 1 will be Doppler-shifted to $\omega_1 = \omega_2 - \mathbf{k}\cdot\mathbf{v}_0$, where $\omega_2 = \omega$ is the non-shifted frequency in region 2. Since δp_{tot} is continuous, the condition of continuity of the vertical component of the displacement yields

$$\frac{1}{n_{02}\left[\omega^2 - (\mathbf{k}\cdot\mathbf{v}_{A2})^2\right]} + \frac{1}{n_{01}\left[(\omega - \mathbf{k}\cdot\mathbf{v}_0)^2 - (\mathbf{k}\cdot\mathbf{v}_{A1})^2\right]} = 0 \tag{3.50}$$

In deriving this equation one must take into account that the normals to both sides of the discontinuity are directed oppositely.

Equation (3.50) is the dispersion relation of the Kelvin-Helmholtz instability. Formally this relation is very similar to the dispersion relations familiar from investigation of streaming instabilities. However, in the present case the role of the plasma modes is taken over by the two Alfvén waves propagating in regions 1 and 2. Actually, the first term in Eq. (3.50) is the Alfvén wave in the non-streaming region, while the second term is the Alfvén wave in the streaming region as seen from region 2. The role of the interface is merely to couple these two Alfvén modes together. The Kelvin-Helmholtz instability may therefore be identified as a streaming magnetohydrodynamic instability which acts on Alfvén waves. Of course, we have not distinguished between the different species. Hence, it is no surprise that the dispersion relation obtained describes single fluid waves. However, the above discussion shows that the Kelvin-Helmholtz instability is one way to excite Alfvén waves in a fluid by feeding surface waves from extracting energy out of a shear flow along a boundary.

Frequency and Growth

Equation (3.50) is quadratic in frequency and yields an unstable solution

$$\boxed{\omega_{\rm kh} = \frac{n_{01}\mathbf{k}\cdot\mathbf{v}_0}{n_{01}+n_{02}}} \tag{3.51}$$

corresponding to the appearance of a conjugate complex root

$$(\mathbf{k}\cdot\mathbf{v}_0)^2 > \frac{n_{01}+n_{02}}{n_{01}n_{02}}\left[n_{01}(\mathbf{k}\cdot\mathbf{v}_{A1})^2 + n_{02}(\mathbf{k}\cdot\mathbf{v}_{A2})^2\right] \tag{3.52}$$

The Kelvin-Helmholtz instability thus occurs for a sufficiently large streaming velocity in region 1. It is the relative streaming between the two plasmas on both sides of the thin boundary which drives the mode as a surface wave. Its frequency is that of a weighted beam mode with the geometric average of the two densities as weighting coefficient. Actually, the instability condition can be easiest satisfied for waves propagating perpendicular to the unperturbed magnetic field. In the linear regime of the instability the rigidity of the magnetic field provides the dominant restoring force, which sets a threshold on the instability.

Geomagnetic Pulsations

The Kelvin-Helmholtz instability is of considerable interest for the excitation of the geomagnetic pulsations discussed in Sec. 9.7 of our companion book. It is responsible for the generation of surface wave modes at the magnetopause boundary by the fast tailward magnetosheath plasma flow along the magnetopause. The instability is restricted to the boundary, but it is not localized to a certain position at the boundary. Rather the wave is convected tailward with the flow while its amplitude grows. Hence, large amplitudes and consequently nonlinear behavior are expected only at the flanks of the magnetopause, while at the dayside magnetopause the amplitude of the surface wave is still small. The surface undulation caused by it may then trigger resonances in the magnetosphere which appear as pulsations. It is clear from the Kelvin-Helmholtz instability condition that such pulsations will arise predominantly during cases of fast solar wind or magnetosheath flows, and the faster the flow the shorter is the excited wavelength.

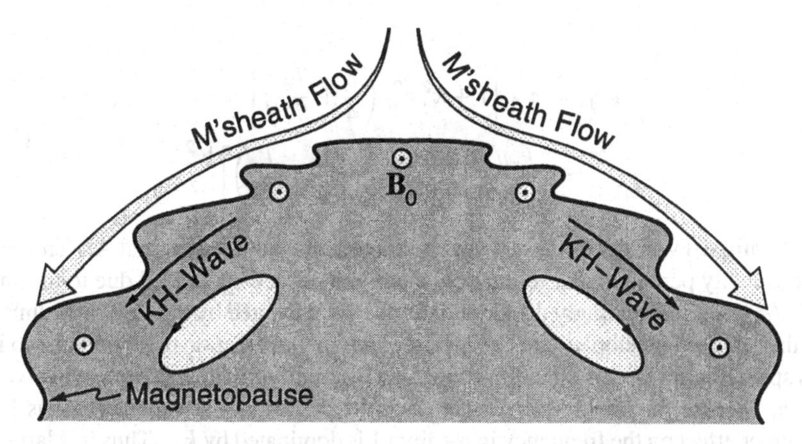

Fig. 3.6. Convective growth of magnetopause Kelvin-Helmholtz waves.

For the excitation of geomagnetic pulsations it is of interest to investigate the polarization of the Kelvin-Helmholtz vortices inside the magnetosphere. Let us assume that the wavenumber has only a very small component in the z direction. Then, from the condition of incompressibility, $\nabla \cdot \delta \mathbf{x} \approx 0$, we find that $-k\delta \mathbf{x}_y + ik_x \delta \mathbf{x}_x = 0$. Hence, we obtain for the ratio

$$\frac{\delta \mathbf{x}_y}{\delta \mathbf{x}_x} = \frac{ik_x}{(k_x^2 + k_z^2)^{1/2}} \tag{3.53}$$

which shows that the Kelvin-Helmholtz waves are elliptically polarized in just the way as used for the explanation of geomagnetic pulsations in Sec. 9.7 of our companion book. Figure 3.6 shows a schematic of a Kelvin-Helmholtz wave flowing downtail and growing convectively during its propagation.

Kinetic Alfvén Waves

The dispersion relation of the Kelvin-Helmholtz instability, Eq. (3.50), contains the dispersion relations of the Alfvén waves on both sides of the transition layer. If we assume that the transition layer is narrow, but that the variations along the boundary become comparable to the ion gyroradius, finite Larmor radius effects become important. The Alfvén wave will then be replaced by the oblique kinetic Alfvén wave (not by the shear Alfvén wave, since $\beta > m_e/m_i$ at the magnetopause; see Fig. 10.12 of our companion volume). The condition of instability for kinetic Kelvin-Helmholtz instability is still of the form of Eq. (3.52), but now with the kinetic Alfvén velocity from Eq. (I.10.179)

$$
\begin{aligned}
v_{ka\parallel} &= v_A \left[1 + k_\perp^2 r_{gi}^2 \left(\frac{3}{4} + \frac{T_e}{T_i}\right)\right]^{1/2} \\
v_{ka\perp} &= \frac{k_\parallel v_A}{k_\perp} \left[1 + k_\perp^2 r_{gi}^2 \left(\frac{3}{4} + \frac{T_e}{T_i}\right)\right]^{1/2}
\end{aligned}
\qquad (3.54)
$$

The field-aligned velocity of these waves is larger than that of the normal Alfvén waves, but the velocity perpendicular to the field along the surface is reduced due to the small ratio $k_\parallel / k_\perp$. Correspondingly the kinetic surface wave excited by the Kelvin-Helmholtz instability at the magnetopause not only propagates in the direction perpendicular to $\mathbf{B}_0$, but its speed is slow in this direction so that the instability has longer time to evolve. In addition, because for the kinetic Alfvén wave the perpendicular wavenumber is large the Doppler effect on the frequency in medium 1 is dominated by $k_\perp$. Thus it is large for perpendicular convective flow so that the instability condition can be satisfied more easily than in the ordinary Alfvén case and the growth rate of the kinetic Kelvin-Helmholtz instability will be larger than that of the ordinary instability.

Nonlinear Evolution

Application of the theory as given here to geomagnetic pulsations must be taken with care. Some of the conditions set for our derivation of the Kelvin-Helmholtz instability are barely satisfied at the magnetopause. For instance, the magnetopause plasma is clearly not incompressible. Inclusion of compressibility complicates the problem, but does not suppress the instability entirely. However, the transition from the magnetosheath to the magnetosphere has been found not to be steep. A broad boundary layer and plasma mantle is covering the magnetopause from its inner side. Here tangential plasma flows have been frequently found, and density gradients are rather gradual. Hence, one must include gradual transitions in a calculation of the Kelvin-Helmholtz instability.

Moreover, the magnetic field is not ideally tangential to the magnetopause. Normal components caused by reconnection are the rule and violate the planar geometry assumption. Finally, dissipation must be included as well as the difference between electron and ion dynamics, because the ion scales are comparable with the width of the transition layer. Both have been neglected so far.

All these effects can be considered only in a more elaborated theory, covering the nonlinear evolution of the instability. Then the Kelvin-Helmholtz instability also contributes to plasma and momentum transport across the magnetopause by mixing of the two regions. When kinetic Alfvén waves replace the magnetohydrodynamic wave mode at the magnetopause, parallel electric fields and field-aligned currents are produced by the Kelvin-Helmholtz instability at the magnetopause, too.

Auroral Kelvin-Helmholtz Instability

The Kelvin-Helmholtz instability may also occur in the upper auroral ionosphere. Here strong shear flows have been observed near auroral arcs or bands. These shear flows are accompanied by large perpendicular electric fields, which necessarily lead to $E \times B$ flows of the upper ionospheric plasma. These flows are to good approximation horizontal in the nearly perpendicular auroral zone background magnetic field. Such shear flows, when exceeding the instability condition for Kelvin-Helmholtz instability, will contribute to folding of the auroral structures. Here the incompressible approximation is quite reasonable because of the strong background magnetic field, but the existence of field-aligned currents and the narrowness of the structures suggest that kinetic Alfvén wave are involved here. In this region the waves propagate in the shear kinetic Alfvén mode (see Sec. 10.6 of our companion book).

However, let us neglect all these effects, and just assume a negatively charged layer of auroral electrons with a width $2d$ in the horizontal x direction, but of infinite extent along the horizontal y and the vertical z axis. The magnetic field is, of course, aligned with the vertical axis. Such a layer will cause strong shear flow along the y axis due to the polarization electric field directed toward the layer, parallel and antiparallel to the x axis. Because the flow is caused by the electron charges and thus by a charge separation field, this problem is fully electrostatic and can, outside of the charges, be described by Laplace's equation

$$\nabla^2 \delta\phi = 0 \tag{3.55}$$

for the electric potential field disturbance, $\delta\phi$. This disturbance is caused by the narrow charge layer, which we consider as a charged surface. A wavelike disturbance of the potential can be assumed as a surface undulation in the same way as for the conventional Kelvin-Helmholtz instability, this time for the potential field

$$\delta\phi = \delta\phi_0 \exp[\pm kx + i(ky - \omega t)] \tag{3.56}$$

The layer is uniform along the field, i.e., along the z axis, the wave amplitude decays in the $\pm x$ direction, and the wave propagates along the y axis. The solution of Laplace's equation in regions 1 and 2 and inside the layer, d, can be written as

$$\begin{bmatrix} \delta\phi_1(x,y) \\ \delta\phi_d(x,y) \\ \delta\phi_2(x,y) \end{bmatrix} = \begin{bmatrix} A\mathrm{e}^{kx} \\ B\mathrm{e}^{kx} + C\mathrm{e}^{-kx} \\ D\mathrm{e}^{-kx} \end{bmatrix} \exp[i(ky - \omega t)] \tag{3.57}$$

The constants must be determined from the boundary conditions at $x = \pm d$. These conditions are the continuity of the tangential electric field, $\delta E_y = -\partial(\delta\phi)/\partial y$, and the requirement that the jump in the normal electric field component, δE_x, is the surface charge. The latter must be determined from the continuity equation for the electrons under the condition that the flow is caused as a

$$\delta\mathbf{v} = \delta\mathbf{E} \times \mathbf{B}_0/B_0^2 \tag{3.58}$$

motion in the undisturbed magnetic field. Clearly, the flow is incompressible and the linearized continuity equation becomes

$$\frac{\partial \delta n}{\partial t} + (\mathbf{v}_0 \cdot)\nabla \delta n = -(\delta\mathbf{v}\cdot)\nabla n_0 \tag{3.59}$$

The perturbation of the electron density at the boundaries, $x = \pm d$, is calculated from the last two equations

$$\left.\frac{\delta n}{n_0}\right|_{\pm d} = \frac{\delta\phi(\pm d)}{B_0(\omega/k \mp v_0)} \left.\frac{\partial \ln n_0}{\partial x}\right|_{\pm d} \tag{3.60}$$

For sharp boundaries the derivative of $\ln n_0$ at $x = \pm d$ is discontinuous

$$\left.\frac{\partial \ln n_0}{\partial x}\right|_{\pm d} = \delta(x+d) - \delta(x-d) \tag{3.61}$$

When we use this expression and multiply by the elementary charge we obtain the surface charge, σ_s, which is the source of the electric field

$$\sigma_s(\pm d) = \pm \frac{e n_0}{B_0} \frac{\delta\phi}{\omega/k - v_0} \tag{3.62}$$

The continuity of δE_y yields

$$\begin{aligned} A &= B\exp(2kd) + C \\ D &= B + C\exp(2kd) \end{aligned} \tag{3.63}$$

The discontinuity of the normal electric field component at the two surfaces gives

$$\begin{aligned} A + B\exp(2kd) - C &= \sigma_s(+d)/\epsilon_0 \\ -D - B + C\exp(2kd) &= \sigma_s(-d)/\epsilon_0 \end{aligned} \tag{3.64}$$

Combining the algebraic equations for the four coefficients and using the expressions for the surface charge densities we arrive at the dispersion relation

$$\omega^2 = \frac{\omega_{pe}^4}{4\omega_{ge}^2}\left[\left(1 - \frac{2\omega_{ge} k v_0}{\omega_{pe}^2}\right)^2 - \exp(-4kd)\right] \tag{3.65}$$

which describes the evolution of Kelvin-Helmholtz waves in the auroral ionosphere under collisionless conditions, but in the presence of strong shear flows. The requirement for instability is that the exponential term in Eq. (3.65) is larger than the first term on the right-hand side. This yields an implicit equation for the unstable wavenumbers

$$kd < \ln\left(1 - \frac{2\omega_{ge}kv_0}{\omega_{pe}^2}\right)^{-1/2} \tag{3.66}$$

The unstable wave grows in the direction of the shear flow. It has the growth rate

$$\boxed{\gamma_{\text{akh}} = \frac{\omega_{pe}^2}{2\omega_{ge}}\left[\exp(-4kd) - \left(1 - \frac{2\omega_{ge}kv_0}{\omega_{pe}^2}\right)^2\right]^{1/2}} \tag{3.67}$$

For a sheet of auroral electrons the rotation of the vortices in the layer produced by the instability will be counterclockwise when looking against the magnetic field from the northern hemisphere upward into the aurora. The direction of the vortices agrees with the sense of gyration of the electrons.

3.4. Firehose Instability

In Chap. 9 of our companion book, *Basic Space Plasma Physics*, we found that a plasma can support low-frequency large-scale Alfvén and magnetosonic waves far below the electron- and ion-cyclotron frequencies, but did not specify which instabilities can excite these wave modes. One mechanism of generating surface Alfvén waves has been identified in the Kelvin-Helmholtz instability. We have also demonstrated that these waves may be further amplified by global resonances in closed magnetic configurations as the magnetosphere to become bulk modes (cf. Sec. 9.7 of our companion book). Another instability which excites bulk Alfvén waves is the *firehose instability*. It can arise

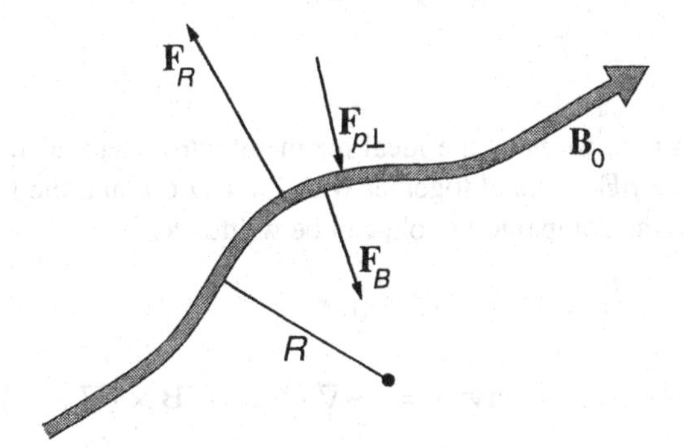

Fig. 3.7. Mechanism of the firehose instability.

in a plasma when the plasma pressure becomes anisotropic. Then Eq. (I.7.21) from the companion book

$$\mathsf{P} = p_\perp \mathsf{I} + (p_\parallel - p_\perp)\frac{\mathbf{BB}}{B^2} \tag{3.68}$$

holds. Under certain conditions a magnetic flux tube containing an anisotropic plasma can be stimulated to perform global transverse oscillations similar to the oscillations of a string brushed by a bow or of a firehose.

Mechanism

The physical mechanism of the firehose instability can be understood as follows. Imagine a magnetic flux tube and let the plasma flow along the magnetic field at average parallel velocity $\langle v_\parallel \rangle$. As shown in Fig. 3.7, whenever a flux tube is slightly bent, the plasma will exert a centrifugal force, $F_R = m_i n_0 v_{\mathrm{th}\parallel}^2/R$, on the flux tube with curvature radius, R.

The centrifugal force is directed outward and tends to increase the amplitude of the initial bending. In the absence of any restoring forces the flux tube would immediately kink. However, the thermal pressure force in the plane perpendicular to the flux tube, $F_{p\perp}$, as well as the magnetic stresses of the flux tube, F_B, resist the centrifugal force. Hence, instability sets in only for sufficiently large centrifugal forces or average parallel plasma velocities sufficiently large to overcome the restoring effects. Requiring force equilibrium in the bending zone implies

$$\frac{m_i n_0 v_{\mathrm{th}\parallel}^2}{R} = \frac{p_\perp}{R} + \frac{B_0^2}{\mu_0 R} \tag{3.69}$$

The terms on the left-hand side of this expression yield the parallel pressure force $p_\parallel/R$. Thus the condition for firehose instability is

$$p_\parallel > p_\perp + B_0^2/\mu_0 \tag{3.70}$$

Growth Rate

To derive the growth rate, we need the ideal magnetohydrodynamic equations (I.7.41) and (I.7.45), which for $\rho\mathbf{E} = 0$ and together with Eq. (I.7.60) and the frozen-in condition (I.7.54), all from the companion book, can be written as

$$\begin{aligned}
\frac{\partial n}{\partial t} + \nabla\cdot(n\mathbf{v}) &= 0 \\
\frac{\partial(nm\mathbf{v})}{\partial t} + \nabla\cdot(nm\mathbf{vv}) &= -\nabla\cdot\mathsf{P} - \frac{1}{\mu_0}\mathbf{B}\times(\nabla\times\mathbf{B}) \\
\frac{\partial\mathbf{B}}{\partial t} &= \nabla\times(\mathbf{v}\times\mathbf{B})
\end{aligned} \tag{3.71}$$

together with the anisotropic equations of state from Eqs. (I.7.27) and (I.7.30), again from our companion book

$$\begin{aligned} \frac{d}{dt}\frac{p_\perp}{nB} &= 0 \\ \frac{d}{dt}\frac{p_\| p_\perp^2}{n^5} &= 0 \end{aligned} \tag{3.72}$$

Linearizing around an equilibrium state, $n_0, p_{0\perp}, p_{0\|}, \mathbf{B}_0, \mathbf{v}_0 = 0$, and Fourier transforming, one obtains the dispersion relation

$$\omega^2 = \tfrac{1}{2}k^2 v_A^2 \left[A_1 \pm (A_2^2 + A_3^2)^{1/2}\right] \tag{3.73}$$

with

$$\begin{aligned} A_1 &= \tfrac{1}{2}\beta_{0\perp}(1+\sin^2\theta) + \beta_{0\|}\cos^2\theta + 1 \\ A_2 &= \tfrac{1}{2}\beta_{0\perp}(1+\sin^2\theta) - 2\beta_{0\|}\cos^2\theta + 1 \\ A_3 &= \beta_{0\perp}\sin\theta\cos\theta \end{aligned} \tag{3.74}$$

where θ is the angle between $\mathbf{k}$ and $\mathbf{B}_0$, and $\beta_{\perp 0} = 2\mu_0 p_{\perp 0}/B_0^2$ and $\beta_{0\|} = 2\mu_0 p_{0\|}/B_0^2$ are the parallel and perpendicular plasma beta parameters. When dividing both sides of Eq. (3.73) by k^2, the right-hand side becomes independent of k. Hence, the waves are non-dispersive. It is easy to confirm that for propagation in the perpendicular direction, $\theta = 90°$, no instability can arise. For parallel propagation, $\theta = 0°$, we have $A_3 = 0$, and there are two solutions. One is purely real

$$\omega^2 = 3k_\|^2 p_{0\|}/n_0 m_i \tag{3.75}$$

and the stable parallel-propagating ion-acoustic wave. The other has a phase velocity

$$\frac{\omega^2}{k_\|^2} = \tfrac{1}{2}v_A^2(\beta_{0\perp} + 2 - \beta_{0\|}) \tag{3.76}$$

yielding instability under the condition (3.70), or in terms of parallel and perpendicular plasma beta

$$\beta_{0\|} > \beta_{0\perp} + 2 \tag{3.77}$$

This is the condition for firehose instability which we derived above in Eq. (3.70) from simple physical considerations. Its growth rate is

$$\boxed{\gamma_{\mathrm{fh}} = \frac{k_\| v_A}{\sqrt{2}}\,(\beta_{0\|} - \beta_{0\perp} - 2)^{1/2}} \tag{3.78}$$

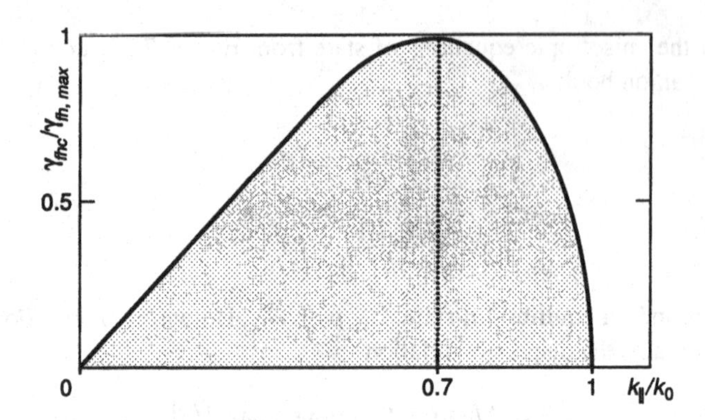

Fig. 3.8. Dependence of the firehose growth rate on wavenumber.

It is not difficult to demonstrate from the dispersion relation in Eq. (3.73) that firehose instability is also possible for oblique propagation as long as a maximum angle, $\theta_{\mathrm{fh,max}}$, is not exceeded. This angle is defined as the solution of

$$1+\sin^2\theta_{\mathrm{fh,max}} = \left[1 - 3\frac{\beta_\parallel}{\beta_\perp^2}(2+\beta_\perp-\beta_\parallel)\right]^{-1} \tag{3.79}$$

The firehose instability is a non-oscillating, purely growing mode. When the parallel thermal pressure in the magnetohydrodynamic fluid is sufficiently high, the magnetic flux tubes become unstable for transverse oscillations of the magnetic field and spontaneously excite parallel propagating Alfvén waves. In the magnetohydrodynamic approximation, where these waves are non-dispersive, the wave has no real frequency and is a very-low frequency wave in the lowest part of the Alfvén branch.

The firehose instability is a very strong instability, but it requires large parallel pressure or $\beta_\parallel > 2$, which implies that the instability is possible only in high-beta or low magnetic field plasmas as, for instance, in the solar wind. In the magnetosphere the firehose instability can arise only in the center of the tail plasma sheet, where the magnetic field is weak. This region may therefore become spontaneously excited to release fast growing Alfvén waves with amplitudes which are large compared to the neutral sheet magnetic field strength. Once excited the oscillation will propagate as an Alfvén wave along the magnetic field lines into the near-Earth magnetosphere.

Kinetic Approach

In our derivation the firehose instability turned out to be a purely growing mode with no real frequency. This is, of course, a consequence of the fluid approach, which neglects all particle effects. A more precise calculation must start from the general dispersion relation for electromagnetic waves in the very-low frequency limit, using the magnetized dielectric tensor in Eq. (I.10.147). The firehose mode is a solution of the left-hand dispersion relation given in Eq. (I.10.156), including temperature anisotropy. If this relation is expanded up to second order in $k_\| v_{ths}/\omega_{gs}$, i.e., in the ratio of gyroradius to wavelength, the wave frequency retains a non-vanishing real part of the frequency

$$\omega_{\rm fh} = \gamma_{\rm fh} k_\| / k_0 \tag{3.80}$$

where $\gamma_{\rm fh}$ is the zero-order growth rate of the purely growing firehose mode obtained above from the fluid approach, and the wavenumber k_0 describes a short wavelength cut-off of the corrected growth rate

$$\gamma_{\rm fhc} = \gamma_{\rm fh}(1 - k_\|^2/k_0^2)^{1/2} \tag{3.81}$$

which becomes effective when the wavelength of the firehose mode approaches the ion gyroradius. This cut-off wavenumber is, for large $\beta_\|$, given by

$$\frac{k_0^2}{4} = \left(\frac{\gamma_{\rm fh}}{k_\|}\right)^2 \left(1 + \sum_s \frac{\omega_{ps}^2}{\omega_{gs}^2}\right)^2 \left\{\sum_s \frac{\omega_{ps}^2}{\omega_{gs}^3}\left[\left(\frac{\gamma_{\rm fh}}{k_\|}\right)^2 + \frac{(2T_{s\perp} - 3T_{s\|})}{m_s}\right]\right\}^{-2} \tag{3.82}$$

The maximum growth rate is found at a wavenumber $k_\| = k_0/\sqrt{2}$, yielding for the maximum firehose growth rate

$$\boxed{\gamma_{\rm fh,max} = \gamma_{\rm fh}\left(k_\| = k_0/\sqrt{2}\right)/\sqrt{2}} \tag{3.83}$$

The oscillation is caused by the kinetic finite-gyroradius effect on the instability. Figure 3.8 shows the dependence of the corrected growth rate on wavenumber. The mode disappears at $k_\| > k_0$, above which it becomes heavily damped.

3.5. Mirror Instability

An instability complementary to the firehose instability is the so-called *mirror instability*. It evolves at nearly perpendicular propagation of the waves. Though this instability is a macroinstability, affecting a large plasma volume and manifesting its effects on the low-frequency scale in the macroscopic plasma parameters, it cannot be treated easily in a magnetohydrodynamic model. The reason for this difficulty is that in this instability the single particle motion along and perpendicular to the magnetic field must be taken into account, which is not contained in the fluid picture of the plasma. One therefore uses kinetic methods in the very-low frequency limit.

Dispersion Relation

The starting point is the kinetic dispersion relation of magnetized plasma waves in Eq. (I.10.147), which includes the temperature anisotropy, A_s, from Eq. (I.10.149), both given in the companion book. For extremely low frequency modes, $\omega \ll \omega_{gi}$, with very long wavelengths, $kr_{gi} \ll 1$, these equations simplify. Because there is no bulk motion of the particles, the tensor component ϵ_{s2} vanishes, and the dispersion relation with $\mathbf{k} = k_\parallel \hat{\mathbf{e}}_\parallel + k_\perp \hat{\mathbf{e}}_x$ splits into

$$\begin{aligned} N_\parallel^2 - \epsilon_{s1} &= 0 \\ (N_\parallel^2 + N_\perp^2) - \left(\epsilon_{s1} - \epsilon_{s0} + \frac{\epsilon_{s5}^2}{\epsilon_{s3}}\right) &= 0 \end{aligned} \tag{3.84}$$

The first equation describes incompressible Alfvén modes, including the anisotropy effect and is thus the kinetic equivalent to the dispersion equation of the firehose mode. We note without proof that it can be brought into the form

$$\omega^2 = k_\parallel^2 v_A^2 \left[1 - \tfrac{1}{2}\sum_s (\beta_{s\parallel} - \beta_{s\perp})\right] \tag{3.85}$$

which is the generalization of the firehose dispersion relation in Eq. (3.76) to two particle species. As before, instability sets in when the right-hand side is negative.

The second dispersion relation describes compressible slow magnetosonic modes. In the low-frequency domain for $\epsilon_{s3} \gg 1$ it can be simplified. In this case the electrons have sufficient time during one oscillation to extinguish the ion-acoustic parallel electric wave field which is still coupled into the dispersion relation by the second term in the brackets. Dropping this term we arrive at

$$(N_\parallel^2 + N_\perp^2) - (\epsilon_{s1} - \epsilon_{s0}) = (N_\parallel^2 + N_\perp^2) - \epsilon_{yy} = 0 \tag{3.86}$$

as the dispersion relation for compressible modes. In explicit form and defining $\zeta_s = \omega / k_\parallel v_{\mathrm{th}s\parallel}$ it reads

$$\epsilon_{yy} = \sum_s \left\{ \frac{\omega_{ps}^2}{\omega_{gs}^2} - \frac{N_\parallel^2}{2}\left[\beta_{s\perp} - \beta_{s\parallel} + \beta_{s\perp}\frac{k_\perp^2}{k_\parallel^2}\left(2 + \frac{\beta_{s\perp}}{\beta_{s\parallel}} Z'(\zeta_s)\right)\right]\right\} \tag{3.87}$$

Taking the very-low frequency limit and assuming that the phase velocities are small, $\omega / k_\parallel v_{\mathrm{th}i} \sim \omega / k v_A \ll 1$, we can simplify this relation even more by using the asymptotic large argument expansion for the plasma dispersion function

$$i\left(\frac{\pi}{2}\right)^{1/2} \frac{\beta_{i\perp}^2}{\beta_{i\parallel}} \frac{\omega}{k_\parallel v_{\mathrm{th}i\parallel}} = 1 + \sum_s \left(\beta_{s\perp} - \frac{\beta_{s\perp}^2}{\beta_{s\parallel}}\right) + \frac{k_\parallel^2}{k_\perp^2}\left[1 + \tfrac{1}{2}\sum_s (\beta_{s\perp} - \beta_{s\parallel})\right] \tag{3.88}$$

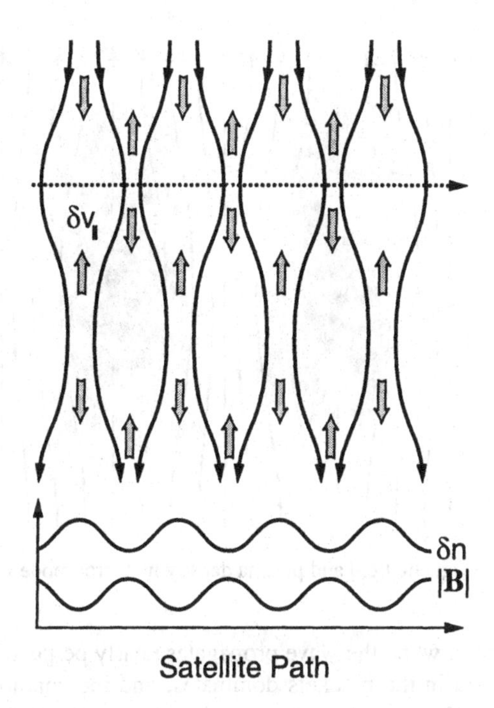

Fig. 3.9. Satellite measurements across a mirror-unstable region.

Only the ion component contributes to the imaginary part at these frequencies. The frequency of this mode is purely imaginary. It can be negative, in which case any mirror mode wave injected into the plasma will be quickly damped and will be unable to propagate. But if the bracketed term is negative, damping turns into growth.

Growth Rate

The mirror mode grows if the free energy of the pressure anisotropy is sufficiently large and must be dissipated. A change in the sign of ω can be achieved in two ways. When we assume that the waves propagate about parallel to the magnetic field, $k_\|^2 \gg k_\perp^2$, the second term in the brackets dominates, and the instability condition is

$$\sum_s \beta_{s\|} > 2 + \sum_s \beta_{s\perp} \tag{3.89}$$

which is the condition for firehose instability in Eq. (3.77). We have thus found that the firehose mode can occur for compressible nearly-parallel propagating Alfvén waves, too. This shows that any kind of Alfvén waves will under the same condition of pressure anisotropy become unstable against firehose instability.

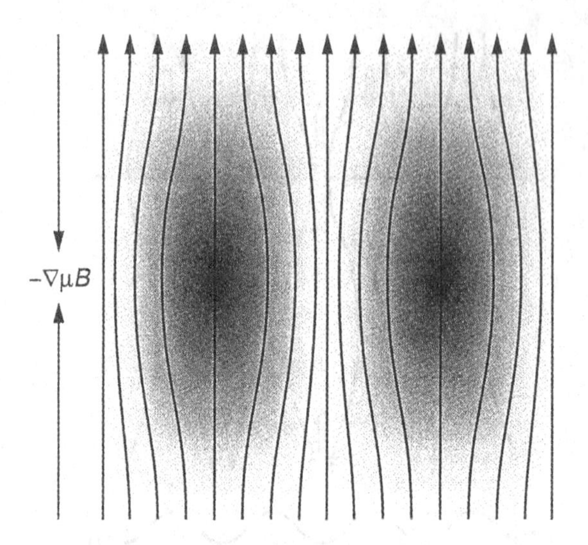

Fig. 3.10. Magnetic field and plasma density in mirror mode waves.

In the opposite case, when the wave propagates nearly perpendicular to the magnetic field, the first term in the brackets dominates, and the condition for instability excludes the firehose mode and reads

$$\sum_s \frac{\beta_{s\perp}^2}{\beta_{s\parallel}} > 1 + \sum_s \beta_{s\perp} \tag{3.90}$$

If this condition is satisfied, mirror instability sets on in an anisotropic plasma. Indeed, the condition for mirror instability cannot be satisfied at the same time as the condition for firehose instability. The two instabilities are mutually exclusive.

Both particle species contribute to the condition of instability, but the component with the strongest anisotropy in the perpendicular direction contributes most. On the other hand, the growth rate is determined by the ion anisotropy and is greater for greater parallel ion energies. This can be seen from the expression for the growth rate of the mirror instability

$$\boxed{\gamma_{\mathrm{mi}} = \sqrt{\frac{2}{\pi}\frac{\beta_{i\parallel}}{\beta_{i\perp}^2}}\left[\sum_s \beta_{s\perp}\left(\frac{\beta_{s\perp}}{\beta_{s\parallel}} - 1\right) - 1\right] k_\parallel v_{\mathrm{th}i\parallel}} \tag{3.91}$$

The mirror mode propagates perpendicular to the magnetic field. For growth it requires that the perpendicular pressure is larger than the parallel pressure. The physical mechanism behind it is that the particles become trapped in magnetic mirror configurations

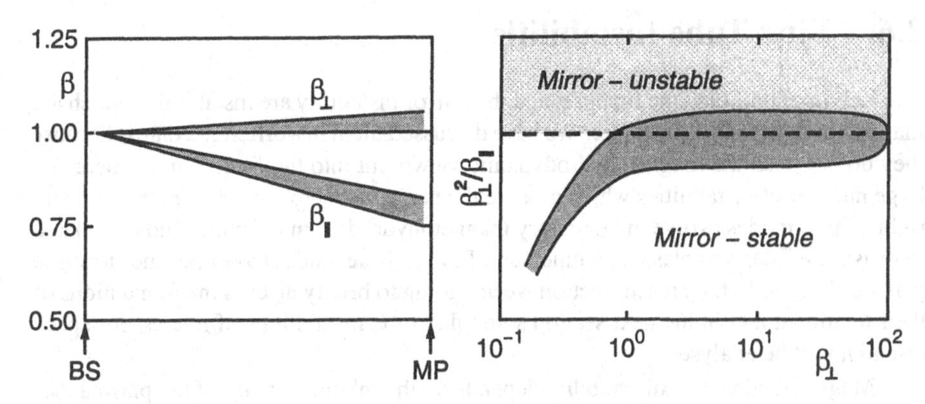

Fig. 3.11. Plasma beta values and mirror criterion in the magnetosheath.

whenever under the action of large perpendicular pressure the magnetic field locally inflates over one wavelength. In this bottle the particles perform a mirror motion between the knots of the wave. This is shown in Fig. 3.9. Particles stream into the mirror during instability. The whole region consists of magnetic mirrors, which when crossed by a spacecraft are recorded as pulsations or oscillations of the magnetic field and out-of-phase oscillations of the plasma density. Figure 3.10 shows a magnetic field and particle density plot of such a configuration.

Magnetosheath Observations

The mirror mode is a compressible slow mode. It is observed in the Earth's dayside magnetosheath, where the necessary anisotropy can develop. The shocked solar wind, on its path from the bow shock to the magnetopause, is adiabatically heated in the perpendicular direction, while at the same time field-aligned outflow toward the flanks of the magnetopause cools the plasma adiabatically in the parallel direction. These effects are shown on the left-hand side of Fig. 3.11. The main effect is parallel cooling, leading to pressure anisotropy of the plasma which increases toward the magnetopause.

In the right-hand side panel of Fig. 3.11 measurements of the left-hand side of the mirror instability criterion in Eq. (3.90) are plotted against $\beta_\perp$. The dashed horizontal line is the marginally stable case, when Eq. (3.90) holds with the equal sign. Obviously the magnetosheath is marginally mirror-unstable for high $\beta_{i\perp}$, while the mirror instability cannot develop for low $\beta_{i\perp}$. Magnetosheath mirror modes are typically excited closer to the magnetopause. The main stabilizing effect is competition of the mirror mode with other kinetic instabilities which arise in anisotropic plasmas. In particular, the electromagnetic ion-cyclotron mode, which will be discussed in Sec. 5.2, consumes the free energy available in the anisotropy before the mirror mode can grow.

3.6. Flux Tube Instabilities

The Kelvin-Helmholtz, the firehose, and the mirror instability are instabilities which are magnetohydrodynamic in nature. We have discussed them in a different context because they do not push the magnetohydrodynamic viewpoint into the foreground. There is a large number of instabilities which deserve the name magnetohydrodynamic with more right. These modes evolve in laboratory magnetohydrodynamic devices and are related to resistive effects and plasma confinement. Few of these modes have relevance to space plasma physics. In the present section we are going to briefly discuss the foundations of their treatment and in the next section some flux tube instabilities of interest for space research will be analysed.

Magnetohydrodynamic stability depends on the inhomogeneity of the plasma, because free energy in ideal magnetohydrodynamics can only be stored in inhomogeneous configurations. Good examples of such unstable situations are *solar flares* which are believed to evolve in so-called active magnetic configurations, which evolve dynamically and possibly explosively. The instability of these magnetic configurations leads to violent energy releases and causes the particle acceleration, matter ejection, optical flashes, and the various radiation processes observed in the optical, radio, x- and γ-ray energy ranges during *solar flares*. Another instability of magnetohydrodynamic nature is the *reconnection* process, first introduced in Sec. 5.1 of our companion book. To study these instabilities, it is convenient to formulate an energy principle. Such a principle is a global measure of the tendency of a magnetohydrodynamic configuration to undergo instability. In short, when the total energy variation turns out to be negative, one can expect the magnetohydrodynamic configuration to become unstable.

Energy Principle

Let us again introduce, as we have done in connection with the Kelvin-Helmholtz instability, the displacement vector, $\delta\mathbf{x}$. The general form of the non-resistive linearized magnetohydrodynamic equations, except for the momentum equation, and integrated once with respect to time is

$$\begin{aligned} \delta n &= -\nabla\cdot(n_0\delta\mathbf{x}) \\ \delta p &= -\delta\mathbf{x}\cdot\nabla p_0 - \gamma p_0\nabla\cdot\delta\mathbf{x} \\ \delta\mathbf{B} &= \nabla\times(\delta\mathbf{x}\times\mathbf{B}_0) \end{aligned} \tag{3.92}$$

Inserting these forms into the linearized momentum equation in Eq. (3.71), expressed in terms of the linear displacement gives

$$n_0 m_i d^2\delta\mathbf{x}/dt^2 = \delta\mathbf{F}(\delta\mathbf{x}) \tag{3.93}$$

where $\delta\mathbf{F}$ is defined as

$$\mu_0\delta\mathbf{F} = -\mathbf{B}_0 \times (\nabla \times \delta\mathbf{B}) - \delta\mathbf{B} \times (\nabla \times \mathbf{B}_0) + \mu_0\nabla(\delta\mathbf{x} \cdot \nabla p_0 + \gamma p_0 \nabla \cdot \delta\mathbf{x}) \quad (3.94)$$

The zero-order quantities are allowed to change in space. Since the above equations are all linear, it is convenient to eliminate the time dependence by Fourier transformation in time, yielding

$$n_0 m_i \omega^2 \delta\mathbf{x} = -\delta\mathbf{F}(\delta\mathbf{x}) \quad (3.95)$$

When we multiply this equation by the complex conjugate of the displacement vector, $\delta\mathbf{x}^*$, and integrate over all space, we obtain

$$m_i\omega^2 \int n_0 \left|\delta\mathbf{x}\right|^2 d^3x = -\int \delta\mathbf{x}^* \cdot \delta\mathbf{F}(\delta\mathbf{x}) d^3x \quad (3.96)$$

Subtract now the complex conjugate

$$m_i\omega^{*2} \int n_0 \left|\delta\mathbf{y}\right|^2 d^3y = -\int \delta\mathbf{y}^* \cdot \delta\mathbf{F}(\delta\mathbf{y}) d^3y \quad (3.97)$$

from this equation. Clearly, the integrals over all space over different dummy variables are the same. Thus one finds that the frequency is either real or purely imaginary, $\omega^2 = \omega^{*2}$, which proves our previous finding that the Kelvin-Helmholtz, the firehose and the mirror instability had only purely growing solutions. This is a general property of the magnetohydrodynamic equations.

To derive the energy principle, we multiply Eq. (3.93) by the displacement velocity, $\delta\mathbf{v} = d\delta\mathbf{x}/dt$, and integrate over space

$$\int \frac{d\delta\mathbf{x}}{dt} \cdot \left[\delta\mathbf{F}(\delta\mathbf{x}) - m_i n_0(\delta\mathbf{x})\frac{d^2\delta\mathbf{x}}{dt^2}\right] d^3x = 0 \quad (3.98)$$

Because $n_0(\delta\mathbf{x})$ and $\delta\mathbf{F}(\delta\mathbf{x})$ do not explicitly depend on time, this equation can be trivially integrated over time, t. We introduce the two quantities

$$\begin{aligned} \Delta W(\delta\mathbf{x}, \delta\mathbf{x}) &= -\tfrac{1}{2}\int \delta\mathbf{x} \cdot \delta\mathbf{F}(\delta\mathbf{x}) \\ \Delta U(\delta\mathbf{x}, \delta\mathbf{x}) &= \tfrac{1}{2}m_i \int n_0(\delta\mathbf{x}) \left|\delta\mathbf{x}\right|^2 d^3x \end{aligned} \quad (3.99)$$

and can write the integrated equation as

$$\Delta W(\delta\mathbf{x}, \delta\mathbf{x}) + \Delta U\left(\frac{d\delta\mathbf{x}}{dt}, \frac{d\delta\mathbf{x}}{dt}\right) = \text{const} \quad (3.100)$$

The constant on the right-hand side is the total displacement energy. Since this energy is unchanged and $\Delta U \geq 0$, one immediately recognizes that the disturbance energy,

ΔW, must be positive definite in order to guarantee the stability of the system, because only then the mode cannot grow over the limit $\delta U = 0$. As a consequence of this conclusion, we find in the opposite case that instability requires that

$$\Delta W(\delta \mathbf{x}, \delta \mathbf{x}) < 0 \tag{3.101}$$

This is the *energy principle* for magnetohydrodynamic instability. The proof of its necessity is a little more involved and can be found in the literature cited in the suggestions for further reading at the end of this chapter.

Energy Contributions

The energy variation, ΔW, contains three contributions, the first is the plasma contribution, ΔW_P, provided by the integral over the plasma volume. The second, ΔW_A, comes from the surface integral over the plasma surface. The third is the vacuum contribution, ΔW_V, which is the integral over all the outer space. This last contribution is

$$\Delta W_V = \frac{1}{2\mu_0} \int_{>} |\delta \mathbf{B}_V|^2 \, d^3x \tag{3.102}$$

with the integration over the outer volume of the plasma. Here one has to prescribe boundary conditions. These conditions are complicated. They are some form of the boundary conditions discussed in Chap. 8 of our companion book, *Basic Space Plasma Physics*, when one of the regions is assumed to be a vacuum. In space plasma physics when the plasma is assumed to be infinitely extended, the boundary is at infinity, and the vacuum term can be dropped. But for finite systems, as for instance solar active regions, it must be kept and introduces severe complications.

Similar arguments apply to the plasma surface contribution. It can be written as

$$\Delta W_A = \tfrac{1}{2} \oint_A \left[\nabla \left(p_0 + \frac{\mathbf{B}_0^2}{2\mu_0} \right) \right]_A \cdot d(\delta \mathbf{F}) \, |\mathbf{n} \cdot \delta \mathbf{x}|^2 \tag{3.103}$$

where $\mathbf{n}$ is the outer normal of the plasma volume and integration is over its surface. As in Chap. 8 of our companion book, the bracketed quantity here is the jump at the surface. When surface currents are absent in equilibrium pressure equilibrium requires that this jump vanishes, and the integral is zero. This is the general case of an extended plasma volume.

The most important contribution to the energy variation comes from ΔW_P

$$\begin{aligned} \Delta W_P \;=\; & \tfrac{1}{2} \int_P d^3x \, [\mu_0^{-1} |\delta \mathbf{B}|^2 + \gamma p_0 \, |\nabla \cdot \delta \mathbf{x}|^2 \\ & + \delta \mathbf{x}_\perp \cdot \nabla p_0 \nabla \cdot \delta \mathbf{x}_\perp^* + \delta \mathbf{x}^* \cdot (\delta \mathbf{B} \times \mathbf{j}_0)] \end{aligned} \tag{3.104}$$

where the perpendicular and parallel direction refer to the equilibrium magnetic field. Introducing the unit vector $\hat{\mathbf{e}}_\parallel$, the right-hand side of this formula can be written as

$$\Delta W_P = \frac{1}{2}\int_P d^3x \left\{\mu_0^{-1}|\delta\mathbf{B}_\perp|^2 + \mu_0^{-1}|\delta\mathbf{B}_\parallel - (\delta\mathbf{x}_\perp\cdot\nabla p_0)\hat{\mathbf{e}}_\parallel/B_0|^2 + \gamma p_0|\nabla\cdot\delta\mathbf{x}|^2\right.$$
$$\left. -(\mathbf{j}_{0\parallel}\cdot\hat{\mathbf{e}}_\parallel)\delta\mathbf{B}\cdot(\delta\mathbf{x}_\perp^*\times\hat{\mathbf{e}}_\parallel) - 2(\delta\mathbf{x}_\perp\cdot\nabla p_0)\left[(\hat{\mathbf{e}}_\parallel\cdot\nabla\hat{\mathbf{e}}_\parallel)\cdot\delta\mathbf{x}^*\right]\right\} \qquad (3.105)$$

For instability to set in, the positive definite terms in this expression must be sufficiently small. The last of the positive terms can be eliminated by demanding that compressibility is small compared with the other terms. Then $\nabla\cdot\delta\mathbf{x}\approx 0$ can be neglected. Moreover, $|\delta\mathbf{B}|^2\simeq|\mathbf{B}_0\cdot\nabla\delta\mathbf{x}|^2$ is small and can be dropped whenever the radial gradients of $\delta\mathbf{x}$ are much larger than the parallel variation of the displacement. This is true when field line bending can be neglected. Hence, one observes that instability occurs for large field-aligned current flow and for large perpendicular density gradients. The second case is the case of interchange modes, the first is the case of kink modes.

Pinch Instability

There are three instabilities which are purely magnetohydrodynamic instabilities. All of them are instabilities which can develop in a thin magnetic flux tube carrying a longitudinal current. These instabilities are thus caused by the $\mathbf{j}_0$-term in Eq. (3.105). These are the pinch, kink, and helical instabilities.

The physical picture of the *pinch instability* is that a field-aligned current in a flux tube of radius r produces its own azimuthal field, $B_\theta(r)\propto 1/r$, around the isolated current carrying flux tube. A slight distortion of the homogeneity of the flux tube along its direction, as for instance a minor decrease in its radius and cross-section, will immediately cause an increase in this azimuthal field due to its $1/r$-dependence on the flux tube radius. This effect is self-amplifying and thus an instability. The final state of this instability is that the current becomes pinched-off and is ultimately disrupted. Figure 3.12 shows the action of the pinch instability. The condition for instability of the pinch mode can be derived for a cylindrical straight flux tube assuming conservation of magnetic flux and current. It reads (without derivation)

$$\boxed{B_{0\theta}^2 \approx \left(\frac{\mu_0 I_0}{2\pi r}\right)^2 > 2|\mathbf{B}_{0\parallel}|^2} \qquad (3.106)$$

Here I_0 is the current in the flux tube. Only strong currents can cause the pinch to develop. In the magnetosphere the pinch instability is probably of no importance because of the strong background magnetic field. But in solar coronal active regions extremely strong field-aligned currents can evolve, which may become pinch-unstable, disrupt the current, and change the magnetic field configuration.

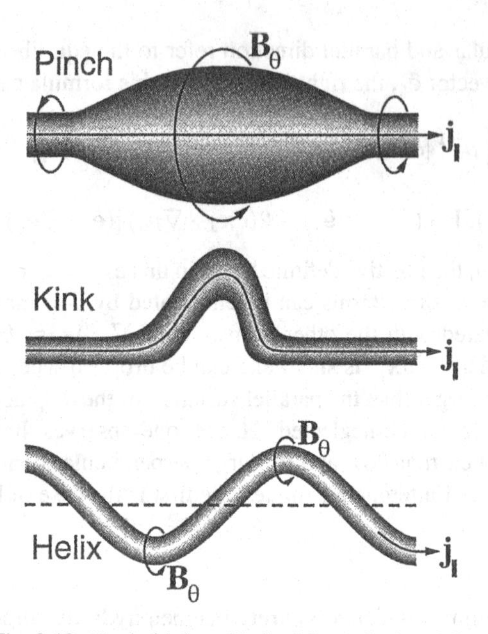

Fig. 3.12. Pinch, kink and helical flux tube instability.

Kink Instability

The second instability, the *kink instability* (see Fig. 3.12), reacts to a bending of the magnetic field lines. Its mechanism is very similar to that of the pinch mode, but in this case the curvature of the field is more important, and the field contained in the flux tube simply kinks locally under the action of the field-aligned current. This instability is also not very important in the magnetosphere, because the curvature of the field lines is weak and the field-aligned currents are faint and not able to cause a kink in the field. However during reconnection events kink instabilities may work in the vicinity of the neutral point, where the field is weak and the currents may be strong. There strong bending of the magnetic field lines in narrow flux tubes may be caused by kink instability.

The condition for instability of a narrow cylindrical flux tube with radius, a, and length, $L \gg a$, can be obtained from the energy principle (again without derivation)

$$\boxed{\ln(L/a) > |\mathbf{B}_{0\parallel}|^2/B_{0\theta}^2} \tag{3.107}$$

where $B_{0\theta}$ is the same as in Eq. (3.106). Since the ambient parallel magnetic field component is very large, this condition requires very thin tubes to yield instability.

Field-aligned currents in the inner auroral magnetosphere are unable to excite it, but again in the solar corona or in the reconnection region of the tail kinking is not excluded.

Helical Instability

The *helical instability* (see Fig. 3.12) is very common in the solar corona. Its instability condition is much lower than that of the other two so it will develop earlier if only a strong enough field-aligned current is flowing. Its effect is to transform a thin stretched magnetic flux tube into a spiral under the action of the self-magnetic effect of the field-aligned current. Flux tubes turn into helixes if it sets on. Such helical flux tubes have been observed in the x-ray light detected from the solar corona and emitted from huge solar loops which have a twisted shape. The helical mode is unstable for

$$\boxed{L/a > 2\pi|\mathbf{B}_{0\|}|/B_{0\theta}} \tag{3.108}$$

and is much easier to satisfy than the above two conditions. The latter condition favors long flux tubes to become helically unstable. Remembering the definition of $B_{0\theta}$ this condition can be written

$$\mu_0 I_0 L > 4\pi^2 a^2 |\mathbf{B}_{0\|}| = 4\pi\Phi \tag{3.109}$$

where Φ is the constant magnetic flux inside the flux tube. Since the field-aligned current strength inside the flux tube is constant, too, it is impossible to estimate at what position of the flux tube helical instability may set in.

As an example for this type of instabilities, let us outline the derivation of the dispersion equation for the helical mode. Assuming the magnetic field inside the tube to be uniform, we find from Eq. (3.46) that the Laplace equation governs the evolution of the total pressure disturbance

$$\nabla^2 \delta p_{\mathrm{tot}} = 0 \tag{3.110}$$

where p_{tot} is the total internal pressure. The displacement vector is given by the same expression as for the Kelvin-Helmholtz instability

$$\delta x = \frac{\nabla \delta p_{\mathrm{tot}}}{m_i n_0(\omega^2 - k^2 v_A^2)} \tag{3.111}$$

where k is the azimuthal wave vector. In the region outside the flux tube the magnetic disturbance is a potential field, $\delta \mathbf{B}_{\mathrm{ex}} = -\nabla\psi$, with

$$\nabla^2 \psi = 0 \tag{3.112}$$

because $\nabla \times \delta \mathbf{B}_{\text{ex}} = 0$. These equations must be completed by boundary conditions which are relatively complex. One must impose pressure balance

$$p_{\text{tot,in}} = (\mathbf{B}_{0\text{ex}} + \delta \mathbf{B}_{\text{ex}})^2 / 2\mu_0 \tag{3.113}$$

at the boundary $\mathbf{x} = \mathbf{x}_0 + \delta \mathbf{x}_n \mathbf{n}$, where $\mathbf{n}$ is the outer normal to the surface of the flux tube, and $\delta \mathbf{x}_n$ is the normal displacement. One must also impose continuity of the tangential electric field

$$\delta \mathbf{E}_t = -(\delta \mathbf{v} \times \mathbf{B}_{0\text{ex}})_t \tag{3.114}$$

Linearizing the pressure condition for small displacements, one obtains

$$2\mu_0 \delta p_{\text{tot}} = 2\mathbf{B}_{0\text{ex}} \cdot \delta \mathbf{B}_{\text{ex}} + \delta \mathbf{x}_n \left[\partial \mathbf{B}_{0\text{ex}}^2 / \partial n \right] \tag{3.115}$$

where the brackets again symbolize the jump. In cylindrical geometry this becomes

$$2\mu_0 p_{\text{tot}} = 2\mathbf{B}_{0\text{ex}} \cdot \delta \mathbf{B}_{\text{ex}} + \delta \mathbf{x}_n \left(\partial \mathbf{B}_{0\text{ex}}^2 / \partial n \right) \tag{3.116}$$

The condition on the electric field is rewritten using Faraday's law

$$\delta B_{n\text{ex}} = \mathbf{n} \cdot \nabla \times (\delta \mathbf{x} \times \mathbf{B}_{0\text{ex}}) \tag{3.117}$$

We can solve these equations in a cylindrical geometry, taking into account the boundary conditions to determine the free constants when solving for the inner and outer regions. This procedure is standard and similar to the case of the Kelvin-Helmholtz instability, but complicated by the inhomogeneity and cylindric geometry of the problem. The dispersion relation ultimately obtained is

$$\mu_0 m_i n_0 \omega^2 = k^2 B_0^2 + \left(k B_{0\|\text{ex}} + \frac{l B_{0\theta\text{ex}}}{a} \right)^2 - \frac{l B_{0\theta\text{ex}}^2}{a^2} \tag{3.118}$$

For the mode with $l = 1$ we find the unstable helical perturbation just under the condition (3.108). This is the lowest wavenumber mode. Other modes with higher wavenumbers may also exist but have higher thresholds as is immediately clear because to twist the field more requires more energy and thus stronger currents.

In the magnetosphere the helical instability does not seem to occur. Since the condition for instability is independent of the coordinate along the flux tube, we can discuss numbers for ionospheric altitudes. At auroral latitudes, where the strongest field-aligned currents are located, the magnetic field strength is $|\mathbf{B}_{0\|}| \approx 4 \cdot 10^{-5}\,\text{T}$. The length of a flux tube in a dipole field is about twice the equatorial distance from the center of the Earth, $L = 2\,L_E R_E$, and with $L_E \approx 6$ for the auroral zone, we get $L \approx 8 \cdot 10^7\,\text{m}$. Using Eq. (3.109), we estimate that the field-aligned current density, $j_0 = I_0/\pi a^2$, must be larger than $5\,\mu\text{A/m}^2$ for the onset of the helical instability. This value is about the same as the density of the strongest field-aligned current of cyclindrical geometry observed, i.e., the upward current connected to the head of the westward traveling surge (see Sec. 5.6 of our companion book). Hence, that current tube is on the edge of instability. Possibly the helical instability limits the current to this upper value. However, there is no unambiguous observational evidence for the presence of the helical instability in the auroral zone. The probable reason for this is that in the magnetosphere a flux tube is surrounded by strong background field lines which, when the tube becomes distorted, start exerting strong forces onto the flux tube which counteract instability.

Flux tubes at the magnetopause have typical sizes of $a \approx 10^3$ km, fields of 100 nT, and $L \approx 20\, R_E$. This suggests that field-aligned currents of $I_0 > 3 \cdot 10^4$ A will become unstable against the helical instability. If such currents exist, instability should develop at the outermost flux tube contacting the magnetosheath flow and should mix into the Kelvin-Helmholtz instability. Since the flow stabilizes the mode, it should probably evolve only in the vicinity of the stagnation point, where the flow velocity is low.

The helical instability also develops in the solar corona. Foot point field strengths are of the order of 0.1 T, flux tube lengths between 10^5 km and 10^7 km, and foot point diameters of the order of 10^4 km. Hence, currents of the order of 10^{11} A will force the flux tube to twist. Since highly twisted flux tubes have frequently been observed, the currents in the solar corona may be strong enough to exceed the limit of stability.

Concluding Remarks

The overview of macroinstabilities given in the present chapter is not exhaustive. It merely provides a guide to the basic ideas. Macroinstabilities, because they deal with inhomogeneous plasmas, are usually rather difficult to treat even in the linear approximation. They are important at large scales, when one needs to calculate the stability of global plasma configurations. Such scales are common in astrophysical problems and in large fusion devices, where they cause violent changes in plasma configurations during instability like plasma disruptions. As a rule, macroinstabilities do affect the global structure of a plasma, e.g., the overall configuration of an active magnetic region on the sun or that of the entire magnetosphere. Because of this reason they have attracted considerable interest in astrophysics and space physics. Macroinstabilities, because of their global properties and the inclusion of large-scale inhomogeneity into their treatment inevitably lead to the necessity of inclusion of boundary conditions. In addition to the complication introduced by the boundary conditions, this is another property which distinguishes them from the wide field of microinstabilities treated in the next chapters.

Further Reading

Macroinstabilities are covered by [2] and [4]. For a general proof of the magnetohydrodynamic energy principle for stability and instability see [1]. Here one also finds explicit expressions for the vacuum-plasma boundary conditions and the theory of the cylindrical kink and helical modes. The theory of the mirror instability is given by Hasegawa, *Phys. Fluids*, **12** (1969) 2642. For its nonlinear theory we have followed Yoon, *Phys. Fluids B*, **4** (1992) 3627. Applications of macroinstabilities to solar physics are given in [5] and [6]. Applications to the ionosphere can be found in [3], from where we have also taken the spread-F backscattered data. The measurements on which the dependences of the mirror criterion are based are found in Hill et al., *J. Geophys. Res.*, **100** (1995) 9575.

[1] D. Biskamp, *Nonlinear Magnetohydrodynamics* (Cambridge University Press, Cambridge, 1993).

[2] A. Hasegawa, *Plasma Instabilities and Nonlinear Effects* (Springer Verlag, Heidelberg, 1975).

[3] M. C. Kelley, *The Earth's Ionosphere* (Academic Press, San Diego, 1989).

[4] D. B. Melrose, *Instabilities in Space and Laboratory Plasmas*, (Cambridge University Press, Cambridge, 1991).

[5] E. Priest, *Solar Magnetohydrodynamics* (D. Reidel Publ. Co., Dordrecht, 1984).

[6] J. O. Stenflo, *Solar Magnetic Fields* (Kluwer Academic Publ., Dordrecht, 1994).

4. Electrostatic Instabilities

While macroinstabilities are caused by configuration space accumulations of free energy in spatial inhomogeneities and affect the bulk or even global plasma properties, the effects of instabilities fed by velocity space inhomogeneities, i.e., *microinstabilities*, are observed on smaller and sometimes microscopic scales. Free energy can be easily stored in deviations of the velocity distribution from its thermal equilibrium shape. As a rule the equilibrium distributions are Maxwellian distribution functions (see Sec. 6.3 of our companion book, *Basic Space Plasma Physics*). In an open system, like in space plasmas, a slight deformation of the Maxwellian distribution can be readily produced and there are many possibilities of such deviations from equilibrium. Correspondingly, a very large number of microinstabilities can arise in a plasma. The present chapter selects those which have been detected in space plasmas. We, however, restrict to electrostatic instabilities only leaving electromagnetic microinstabilities to the next chapter. Furthermore, we assume that the plasma is homogeneous which allows to neglect any effects caused by gradients and curvatures or by boundaries. The plasma is assumed to be infinitely extended, and the source of free energy is taken to be invariable. The first assumption is in many cases well satisfied because the typical scale of microinstabilities is small compared with the scales of the plasma. The second assumption, however, is valid only over the short linear time scale of the instability, unless the source of free energy is maintained constant by external means. Usually for growing waves the instability quickly reaches the nonlinear stage and the free energy is diminished. These processes are inevitable but will be reserved for discussion in later chapters.

The starting point of a theory of linear microinstabilities is the linearized Vlasov equation

$$\frac{\partial f}{\partial t} + \mathbf{v} \cdot \nabla_{\mathbf{x}} f + \frac{q}{m}(\mathbf{E} + \mathbf{v} \times \mathbf{B}) \cdot \nabla_{\mathbf{v}} f = 0 \tag{4.1}$$

or one of its derivatives introduced in Sec. 6.2 of our companion book, with the stationary non-equilibrium distribution as ingredient. The procedure to obtain the dispersion relation is the same as in Chap. 10 of the companion book. As in the case of macroinstabilities, the dispersion relation must be solved for the frequency, ω, including its imaginary part, γ. For weak instabilities the growth rate can be determined by using the methods developed in Sec. 2.1, since in these cases the growth rate yields only a small correction to the wave frequency and no new wave modes are introduced by the non-equilibrium distribution functions. Instead, the free energy is used to excite one of the known kinetic plasma modes.

We will in most cases simply use the general dispersion relations from Eqs. (1.35) and (I.10.147), which are both derived in the companion volume, rewrite them for the particular case, and model the unstable distribution as a combination of Maxwellians. For other non-thermal distributions analytical solutions do normally not exist.

4.1. Gentle Beam Instability

Electromagnetic waves may propagate from the outside into a plasma. In contrast, electrostatic waves cannot leave a plasma and cannot penetrate it. They must be excited internally. Since they are easily excited, they are the appropriate tool for the plasma to eliminate and redistribute the electric potential differences, which may arise due to local charge separations, and to rearrange particles in the distribution function to achieve thermal equilibrium. These effects arise when the plasma is unmagnetized. They arise also when particle motion is allowed only along the magnetic field. In all such cases the kinetic electrostatic waves are excited by electrostatic instabilities.

Bump-in-Tail Distribution

The simplest kinetic instability is that of an electron beam propagating on a background plasma. Figure 4.1 shows the phase space geometry of a beam plasma configuration called *gentle beam* or *bump-in-tail* configuration. The plasma is assumed thermal with fixed ions as neutralizing background and a certain electron temperature, T_e, leading to the thermal spread, $v_{\mathrm{th}e}$, of the background distribution function. Superposed on this distribution is a group of fast electrons at speed $v_b > v_{\mathrm{th}e}$ but much smaller density than the background plasma, $n_b \ll n_0$, and of narrow thermal spread, $v_{\mathrm{th}b}$.

This phase space configuration is the classical case of a non-thermal distribution subject to beam instability, but since we permit for finite beam temperatures, corresponding to deformation of the equilibrium distribution function, the instability will affect only that part of the distribution function which contains the deviation from thermal equilibrium, while the bulk distribution remains unchanged. To treat this kind of instability we must go from the fluid to a kinetic treatment.

Inverse Landau Damping

A system consisting only of electrons will be subject to Langmuir oscillations or, when including thermal effects, to propagating Langmuir waves with a dispersion given by

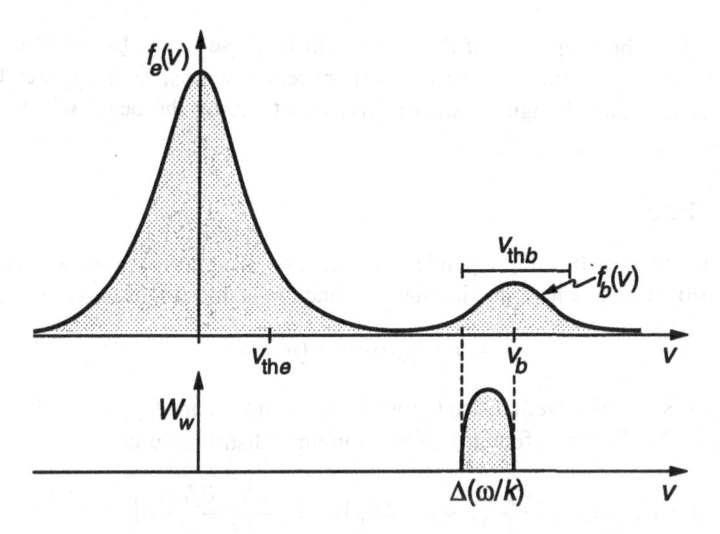

Fig. 4.1. Gentle beam plasma configuration for the bump-in-tail instability.

Eq. (I.10.32) in the companion book

$$\omega_l = \pm\omega_{pe}\left(1 + \tfrac{3}{2}k^2\lambda_D^2\right) + i\gamma_l(k) \tag{4.2}$$

These waves are Landau damped according to Eq. (I.10.48)

$$\gamma_l = \omega_l \frac{\pi\omega_{pe}^2}{2n_0k^2} \left.\frac{\partial f_{0e}(v)}{\partial v}\right|_{v=\omega/k} \tag{4.3}$$

In this equation the undisturbed electron distribution, $f_{0e}(v)$, enters as velocity derivative at the position of the wave phase velocity. If the derivative is negative here, as in thermodynamic equilibrium, the thermal particles damp the wave (see Sec. 10.2 of our companion book). When the derivative is positive, Landau damping inverts and changes sign because there will be a surplus of fast particles which can feed energy to the wave and raise its amplitude. In such a case the Langmuir wave is excited by the *gentle beam instability*.

A distribution as in Fig. 4.1 will be unstable. The undisturbed distribution has a positive derivative for velocities just below the beam speed, v_b. The background distribution falls off here with a negative derivative, but for a sufficiently fast beam far outside the bulk of the plasma this gradient is weak. Moreover, the number of particles in the bulk distribution is very small at the position of the beam. Thus damping can be neglected for waves with phase velocities $\omega/k \leq v_b$.

The waves excited by the beam are Langmuir waves of relatively high phase velocity. Since the temperature of the background electron distribution is low, $k_BT_e \ll$

$W_b = m_e v_b^2/2$, the frequency of the waves will be close to the background plasma frequency, $\omega \approx \omega_{pe}$. The condition of resonance, $\omega \approx k v_b$, then requires that the wavenumber is small. Langmuir waves driven unstable by the beam will have long wavelengths.

Growth Rate

To calculate the growth rate of gentle beam excited Langmuir waves we consider a one-dimensional case, where the distribution function is the sum of two Maxwellians

$$f_{0e}(v) = f_0(v) + f_b(v - v_b) \tag{4.4}$$

where $f_0(v)$ is the Maxwellian background distribution, and $f_b(v - v_b)$ is a shifted Maxwellian. The dielectric function of the combined distribution is

$$\begin{aligned}\epsilon(\omega, k) &= 1 - \frac{\omega_{p0}^2}{\omega^2}\left(1 + 3k^2\lambda_{D0}^2\right) - \pi i \frac{\omega_{p0}^2}{n_0 k^2} \left.\frac{\partial f_0(v)}{\partial v}\right|_{v=\omega/k} \\ &\quad - \frac{\omega_{pb}^2}{\omega^2}\left(1 + 3k^2\lambda_{Db}^2\right) - \pi i \frac{\omega_{pb}^2}{n_b k^2} \left.\frac{\partial f_b(v - v_b)}{\partial v}\right|_{v=\omega/k}\end{aligned} \tag{4.5}$$

If the beam density and temperature are low, the beam contribution to the real part of the dispersion relation can be neglected. The real part of ϵ yields the bulk plasma Langmuir wave dispersion relation, $\omega_l^2 = \omega_{p0}^2(1 + 3k^2\lambda_{D0}^2)$. The weak beam causes a weak beam instability. Thus we apply Eq. (2.9) to calculate the growth rate. The derivative of the real part of $\epsilon(\omega_l, k)$ with respect to ω_l gives

$$\frac{\partial \epsilon_r}{\partial \omega_l} = \frac{2\omega_{p0}^2}{\omega_l^3}(1 + 3k^2\lambda_{D0}^2) \tag{4.6}$$

The general expression for the growth rate is the sum of the plasma and beam terms. It can be brought into the following form

$$\begin{aligned}\gamma = \frac{\sqrt{\pi}\,\omega_l^3 \omega_{pb}^2}{(1 + 3k^2\lambda_{D0}^2)\omega_{p0}^2 k^2 v_{\text{th}b}^3} &\left\{\left(v_b - \frac{\omega_l}{k}\right)\exp\left[-\frac{(v_b - \omega_l/k)^2}{v_{\text{th}b}^2}\right]\right. \\ &\left. - \frac{n_0 T_b^{3/2}}{n_b T_0^{3/2}} \frac{\omega_l}{k} \exp\left(-\frac{\omega_l^2}{k^2 v_{\text{th}0}^2}\right)\right\}\end{aligned} \tag{4.7}$$

The first term in the brackets on the right-hand side is the beam term, while the second is the background plasma contribution. It is obvious that the plasma contributes to damping while the beam causes instability. But the growth rate is positive only for phase velocities smaller than the beam velocity, $\omega_l/k < v_b$. The phase velocity must also be close to the beam velocity, because otherwise the argument of the first exponential is large and the positive beam contribution is too small to overcome the background Landau damping. Comparing the two terms one finds that the phase velocity of the Langmuir waves must fall into the interval

$$\frac{n_b}{n_0}\left(\frac{T_0}{T_b}\right)^{3/2} < \frac{\omega_l}{kv_b} < 1 \tag{4.8}$$

For sufficiently large $\omega_l/k \approx v_b$ one can neglect the Landau damping term and write

$$\gamma_{\text{gb}} = \frac{\sqrt{\pi}\omega_l^3\omega_{pb}^2}{(1+3k^2\lambda_{D0}^2)\omega_{p0}^2k^2v_{\text{th}b}^3}\left(v_b - \frac{\omega_l}{k}\right)\exp\left[-\frac{(v_b-\omega_l/k)^2}{v_{\text{th}b}^2}\right] \tag{4.9}$$

Maximum growth is obtained if we put in the coefficient on the right-hand side

$$\omega_l/k_{\text{max}} = v_b \tag{4.10}$$

and find the maximum of the function $x\exp(-x^2)$ at $x^2 = 1/2$

$$\boxed{\gamma_{\text{gb,max}} = \left(\frac{\pi}{2\text{e}}\right)^{1/2}\frac{n_b}{n_0}\left(\frac{v_b}{v_{\text{th}b}}\right)^2\omega_l} \tag{4.11}$$

for the growth rate of the Langmuir gentle beam instability. Fast, dense and cool beams lead to large maximum growth rates.

The above calculation can be generalized to wave propagating at an oblique angle with respect to the electron beam. In this case the product kv_b is understood as a scalar product, $\mathbf{k}\cdot\mathbf{v}_b = kv_b\cos\theta$, between the wave and velocity vectors, and θ is the angle between them. We then get

$$\boxed{\gamma_{\text{gb}} = \left(\frac{\pi}{2\text{e}}\right)^{1/2}\frac{n_b}{n_0}\left(\frac{v_b\cos\theta}{v_{\text{th}b}}\right)^2\omega_l} \tag{4.12}$$

From the general expression for the growth rate in Eq. (4.9) one can demonstrate that the growth rate stays near maximum as long as the angle is sufficiently small so that $\theta \leq v_{\text{th}b}/v_b$.

Conditions for Growth

Since maximum growth of the gentle beam instability in Eq. (4.11) is obtained for $k_{\text{max}} = \omega_l/v_b$, we may insert the Langmuir frequency, $\omega_l(k)$, from Eq. (4.2) into this relation to find

$$k_{\text{max}}^2\lambda_{D0}^2 = (v_b^2/v_{\text{th}0}^2 - 3)^{-1} \tag{4.13}$$

For resonant interaction the beam velocity must exceed a threshold given by

$$v_b = \sqrt{3} v_{\text{th}0} \tag{4.14}$$

in order that the gentle beam instability can develop. If the beam is slower than this threshold value, i.e., if the beam kinetic energy is less than three times the thermal energy of the background plasma, the beam sits on the shoulder of the background distribution and will be unable to excite waves.

Let us find the width of the excited frequency spectrum. The argument of the exponential function in Eq. (4.9) can be written as $-(\omega_l - k_{\max} v_b)^2 / k_{\max}^2 v_{\text{th}b}^2$. Thus the range of frequencies over which the growth rate is close to maximum is roughly $\Delta\omega_{\text{gb}} \approx k_{\max} v_{\text{th}b}$. We can therefore define the bandwidth of the gentle beam instability

$$\Delta\omega_{\text{gb}} \approx \omega_l v_{\text{th}b} / v_b \tag{4.15}$$

This expression shows that the growth rate of the gentle beam instability is less than the bandwidth of the excited waves as long as

$$n_b / n_0 \leq (v_{\text{th}b} / v_b)^3 \tag{4.16}$$

Comparison with the corresponding condition for the weak beam instability in Eq. (2.38) shows that these two conditions are complementary. Hence, for large beam densities violating the last condition, the gentle beam instability will go over into the weak beam mode. But when the beam becomes too cold, the theory breaks down, the bandwidth of the instability vanishes, and the beam-to-plasma density ratio exceeds the range of values allowed in Eq. (4.16). In this case the weak beam instability is excited, too.

Upstream Solar Wind

Since weak electron beams will readily be generated in both unmagnetized and magnetized plasmas (in the latter case they propagate along the magnetic field) by slight low-frequency potential differences, which may arise in travelling waves like ion-acoustic waves or kinetic Alfvén waves, the gentle beam instability can evolve naturally in any space plasma. By extracting energy from the beam it will ultimately slow the beam down and deplete the source of free energy (see Sec. 8.3).

Figure 4.2 shows an example of a measurement of beam-excited Langmuir waves in the solar wind in front of the Earth's bow shock. The electrons, which excite these waves, stream back from the Earth's bow shock along the magnetic field into the solar wind at about twice the initial velocity in the solar wind frame, roughly about 1200 km/s. This velocity is about marginal for Langmuir instability. The plasma frequency is close to 30 kHz. The measured bandwidth is found to be $\Delta\omega \approx 6$ kHz. From

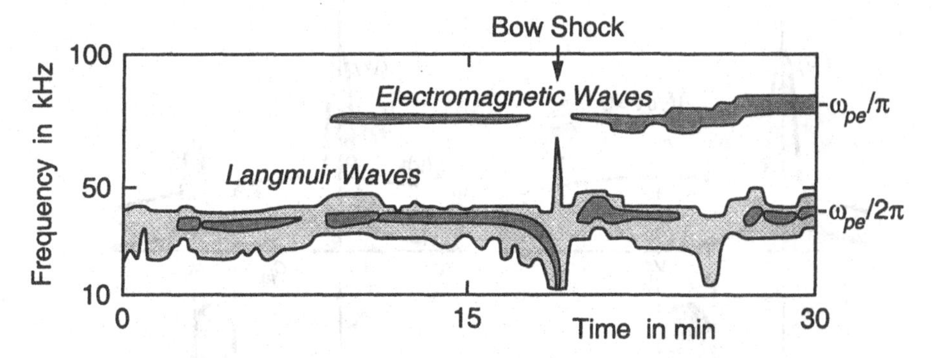

Fig. 4.2. Signature of gentle beam-excited Langmuir waves in the solar wind.

this value we deduce a beam velocity roughly five times the beam thermal velocity spread or $v_{\mathrm{th}b} \approx 200$ km/s. Since the gentle beam instability requires low beam densities, we can use the above restriction to estimate that the beam density was smaller than $n_b < 8{\cdot}10^{-3} n_0$. With this relatively high upper limit we find that the instability growth rate given in Eq. (4.11) is $\gamma < 0.15\omega_{p0}$, corresponding to about 0.2 ms growth time.

4.2. Ion-Acoustic Instabilities

Ion-acoustic waves, the other unmagnetized plasma eigenmode, can be excited in the first place by electron currents or by ion beams flowing across a plasma. This instability is also a kinetic instability. Since it requires much lower velocities than the Langmuir gentle beam instability it can be driven unstable by weak currents. In addition, it is not restricted to low background electron temperatures.

Current-Driven Instability

The idea behind the electron current-driven *ion-acoustic instability* is that the combined equilibrium distribution function, consisting of the drifting hot electron background and the cold ions, exhibits a positive slope where resonance between waves and particles can occur.

These conditions are sketched in Fig. 4.3. The left-hand part of the figure refers to bulk current flow with drift velocity, v_d. The drifting electrons have a hot shifted Maxwellian velocity distribution while the ions are cold and immobile. Instability can occur in the region of positive slope of the electron distribution function. The right-hand part of the figure shows the case of a fast ion beam crossing the plasma. The hot electron distribution is slightly shifted to the left of the origin to compensate for

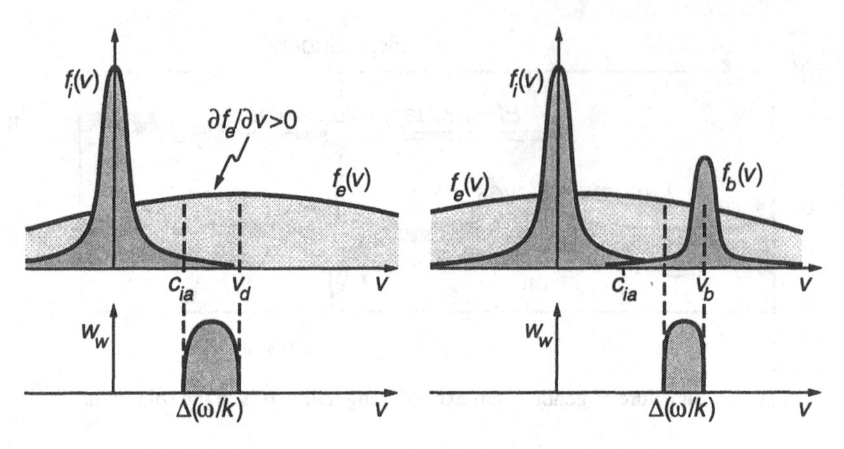

Fig. 4.3. Ion-acoustic-unstable velocity space distributions.

the weak current, $j_b = en_b v_b$, caused by the ion beam. Because of the small density, $n_b \ll n_0$, of the beam the condition $j = e(n_b v_b - n_0 v_{de}) = 0$ yields for this shift $v_{de} = n_b v_b / n_0$ which is negligible as long as the ions have moderate beam velocities. Again, instability can arise in the positive slope of the beam ion distribution function.

Growth Rate

We are looking for an electrostatic current-driven instability, neglecting the electromagnetic effect of the current. The assumed distribution functions are a shifted one-dimensional Maxwellian

$$f_{e0}(v - v_d) = \frac{n_0}{\pi^{1/2} v_{\text{the}}} \exp\left[-\frac{(v - v_d)^2}{v_{\text{the}}^2}\right] \tag{4.17}$$

for the electrons and a Maxwellian distribution at rest for the ions

$$f_{i0}(v) = \frac{n_0}{\pi^{1/2} v_{\text{thi}}} \exp\left(-\frac{v^2}{v_{\text{thi}}^2}\right) \tag{4.18}$$

with unlike thermal velocities. We also assume that the thermal velocity of the electrons is much larger than their drift velocity, $v_d \ll v_{\text{the}}$. In the opposite case the plasma is unstable against the Buneman instability as demonstrated in Eq. (2.47). The dielectric response function (I.9.62) for this one-dimensional case is obtained from Eq. (I.10.83) as

$$\epsilon(\omega, k) = 1 + \frac{2}{k^2 \lambda_D^2}\left[1 + \zeta_e Z(\zeta_e)\right] + \frac{2}{k^2 \lambda_{Di}^2}\left[1 + \zeta_i Z(\zeta_i)\right] \tag{4.19}$$

where the arguments of the plasma dispersion function (see App. A.7 of our companion book) are

$$\begin{aligned} \zeta_e &= (\omega - kv_d)/kv_{\text{the}} \\ \zeta_i &= \omega/kv_{\text{thi}} \end{aligned} \tag{4.20}$$

From inspection of the left part of Fig. 4.3 it is clear that instability can arise only in a restricted range of wave phase velocities, ω/k. The phase velocity must be larger than the background ion thermal velocity

$$|\omega/k| \gg v_{\text{thi}} \tag{4.21}$$

in order to escape ion Landau damping, but it must fall within the positive slope of the electron distribution which implies that

$$|\omega/k - v_d| \ll v_{\text{the}} \tag{4.22}$$

The ion Z function can thus be expanded in the large argument limit, while for the electron Z function the small argument limit is appropriate. Assuming weak instability, $\gamma/\omega \ll 1$, the real part of the dielectric function reproduces the ion-acoustic relations Eqs. (I.10.54) and (I.10.55), both given in the companion volume. The above frequency range thus supports propagation of ion-acoustic waves, which travel on the background of the ions at rest, but with a phase velocity close to the electron drift speed. For weak growth we find using Eq. (2.9)

$$\begin{aligned} \gamma = \left(\frac{\pi}{8}\right)^{1/2} \frac{1}{k^2\lambda_D^2} \frac{\omega_{ia}^4}{kc_{ia}\omega_{pi}^2} & \left\{ \left(\frac{m_e}{m_i}\right)^{1/2} \left(\frac{kv_d}{\omega_{ia}} - 1\right) \exp\left[-\frac{(\omega - kv_d)^2}{k^2 v_{\text{the}}^2}\right] \right. \\ & \left. - \left(\frac{T_e}{T_i}\right)^{3/2} \exp\left(-\frac{\omega_{ia}^2}{k^2 v_{\text{thi}}^2}\right) \right\} \end{aligned} \tag{4.23}$$

Instability arises when the electron term dominates the ion term and when the electron current drift velocity exceeds the wave phase velocity, $\omega/k < v_d$, as we concluded qualitatively from inspection of Fig. 4.3. We can use the dispersion relation of ion-acoustic waves given in Eq. (I.10.55) and expand the electron exponential to simplify the growth rate

$$\begin{aligned} \gamma_{ia} = \left(\frac{\pi}{8}\right)^{1/2} \frac{\omega_{ia}}{(1 + k^2\lambda_D^2)^{3/2}} & \left[\left(\frac{m_e}{m_i}\right)^{1/2} \left(\frac{kv_d}{\omega_{ia}} - 1\right) \right. \\ & \left. - \left(\frac{T_e}{T_i}\right)^{3/2} \exp\left(-\frac{T_e}{T_i(1 + k^2\lambda_D^2)}\right) \right] \end{aligned} \tag{4.24}$$

The ion contribution to the growth rate corresponds to Landau damping of the waves and thus counteracts instability. It is large whenever the ion temperature is large, because in this case the wave phase velocity resonance falls into the decaying shoulder of the ions. In particular, for $T_i \approx T_e$, the damping rate is comparable to the wave frequency, $\gamma_{ia} \approx \omega$, and there is no instability. But for large electron temperatures, $T_e \gg T_i$ ion damping can entirely be neglected and the second term in the growth rate disappears. The basic requirement for ion-acoustic current-driven instability is thus a cold-ion hot-electron plasma with drift velocity exceeding the ion sound velocity

$$v_d > c_{ia} \quad \text{or} \quad k^2\lambda_D^2 \ll 1 \tag{4.25}$$

In the opposite case when $k^2\lambda_D^2 \gg 1$, the frequency of the wave is close to the ion plasma frequency, $\omega \approx \omega_{pi}$. Here Landau damping is strong and the instability can hardly occur. The condition of instability for this case is

$$v_d > v_{\mathrm{th}i}(1-\omega^2/\omega_{pi}^2) \quad \text{or} \quad 1 < k^2\lambda_D^2 \ll T_e/T_i \tag{4.26}$$

The upper limit on the wavenumber results from the requirement that ion Landau damping be as weak as possible. The condition on the drift velocity is anyway satisfied by the initial requirement in Eq. (4.21). But then Eq. (4.22) requires $v_d \ll v_{\mathrm{the}}$ or very hot electrons for instability.

The dominant range of ion-acoustic instability is therefore realized when the electron temperature exceeds the ion temperature, $T_e \gg T_i$, the wavelength is much longer than the Debye length, $k^2\lambda_D^2 \ll 1$, and the current drift velocities, $v_d > \omega/k \approx c_{ia}$, exceed the ion sound speed. Under these conditions Landau damping is unimportant, and the growth rate assumes the simple form

$$\boxed{\gamma_{ia} = \left(\frac{\pi m_e}{8m_i}\right)^{1/2} \frac{\omega_{ia}}{(1+k^2\lambda_D^2)^{3/2}} \left(\frac{kv_d}{\omega_{ia}} - 1\right)} \tag{4.27}$$

where $\omega/k \approx c_{ia}$, and the drift velocity in the ion frame of reference can be expressed through the current density, $\mathbf{v}_d = -\mathbf{j}/en_0$. In general this is a vector quantity. Again, as for the gentle beam instability, the only direction given in space is the direction of the current. Introducing the angle θ between $\mathbf{k}$ and $\mathbf{v}_d$, the ion-acoustic instability is restricted to a cone around the direction of the current, $0 < \theta < \theta_{\max}$, where

$$\cos^2\theta > \cos^2\theta_{\max} = c_{ia}^2/v_d^2 \tag{4.28}$$

The ion-acoustic waves excited by the electron current are beamed in the direction of the current but propagate at a speed which is slightly less than that of the current itself.

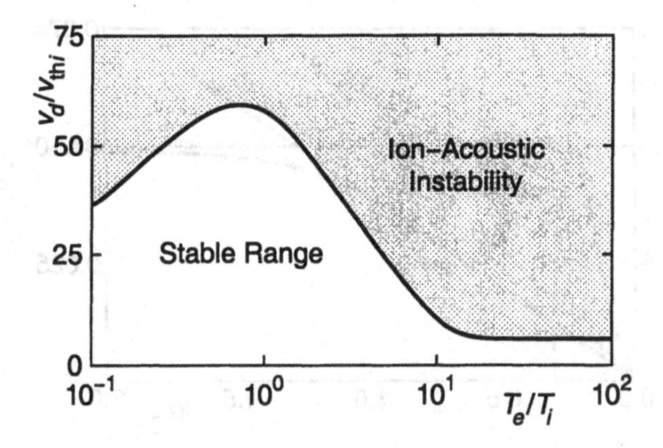

Fig. 4.4. Threshold velocity curve for the ion-acoustic instability.

Parametric Investigation

The growth rate of the ion-acoustic instability in Eq. (4.24) shows a strong dependence on the temperature ratio, T_e/T_i. Figure 4.4 shows the numerical result of a calculation of the velocity threshold in dependence on this ratio. For very large electron temperatures the threshold is just the ion sound speed. For lower ratios the threshold gets fairly high, but there is still weak instability for low electron temperatures. However, in this regime other instabilities like ion-cyclotron waves are more important.

The maximum growth rate can be obtained by maximizing $\gamma_{ia}(k)$ for fixed sufficiently large drift velocities, $v_d \gg c_{ia}$

$$\gamma_{ia,\max} \approx 0.32\omega_{pi} \left(\frac{\pi m_e}{8m_i}\right)^{1/2} \left(\frac{v_d}{c_{ia}} - 1\right) \tag{4.29}$$

This growth rate is attained at the small wavenumber, $k_{\max}^2 \lambda_D^2 \approx 0.1(1 - c_{ia}/v_d)$. For instance, assuming $v_d/c_{ia} = 10$ and a hydrogen plasma, this growth rate is of the order of $\gamma_{ia,\max} \approx 0.04\omega_{pi}$. Figure 4.5 shows the schematic relation between the ion-acoustic dispersion relation and the growth rate of the instability.

Auroral Ionosphere

The ion-acoustic instability occurs in the topside auroral ionosphere. Since the currents propagate along the magnetic field, we can use the unmagnetized case. For a plasma density of $10^2\,\mathrm{cm}^{-3}$ and oxygen as the dominant ion, the ion plasma frequency is about $\omega_{pi} \approx 3.3\,\mathrm{kHz}$, yielding a maximum growth rate of $\gamma_{ia,\max} \approx 35\,\mathrm{s}^{-1}$ and maximum

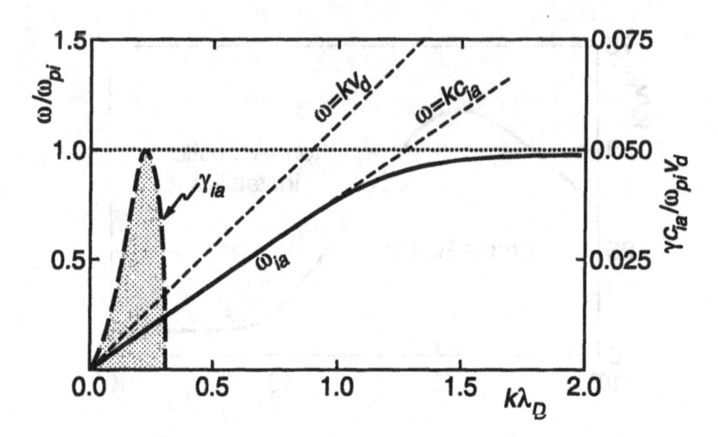

Fig. 4.5. Frequency and growth rate of the ion-acoustic instability.

growing wavelengths $\lambda_{ia} \approx 20\lambda_D$ or about 30 m. Thus one expects that auroral field-aligned currents of sufficient strength will excite ion-acoustic waves of several tenths of meter wavelength along the magnetic field. These waves propagate into the direction of the current carrier's velocity, i.e., downward for currents carried by precipitating auroral electrons and upward for bulk auroral return currents.

Since ion-acoustic waves are electrostatic waves, they are accompanied by fluctuations of the plasma density. Thus the appearance of density irregularities along the auroral field lines in the upper ionosphere during field-aligned current flow is the consequence of the excitation of such long wavelength auroral ion-acoustic waves. However, the threshold for excitation of ion-acoustic waves is relatively high and requires high velocities of the current carriers to exceed the ion-acoustic velocity and thus strong field-aligned currents. Other waves might be driven unstable before the ion-acoustic instability sets on. Such waves are the ion-cyclotron waves which we will discuss below in context with magnetized electrostatic instabilities.

Beam-Driven Instability

A similar setup for ion-acoustic instabilities is given on the right-hand side of Fig. 4.3, where an ion beam crosses an unmagnetized plasma or flows along the magnetic field. Beams of this kind are found in the auroral ionosphere during active auroras and substorm conditions. They flow out of the ionosphere upward into the magnetosphere and have kinetic energies of a few keV. For oxygen ions this energy corresponds to beam velocities of 100–200 km/s. Since the theory of the ion-acoustic instability is Galilei-invariant and because for dilute beams we can neglect the electron velocity shift caused by them, the theory of ion-acoustic waves excited by the ion beams remains unchanged. The beam will excite ion-acoustic waves propagating close to the beam velocity if far enough away from the thermal electron distribution for not being Landau damped. The condition for instability is

$$\omega/k \gg v_{\rm the} \tag{4.30}$$

and the growth rate can be written as

$$\begin{aligned}\gamma_{iab} &= \left(\frac{\pi}{8}\right)^{1/2}\frac{\omega^4}{\omega_{pi}^2|k|^3c_{ia}\lambda_D^2}\left\{\frac{n_{ib}}{n_0}\left(\frac{T_e}{T_{ib}}\right)^{1/2}\right.\\ &\left.\left(\frac{kv_b}{\omega}-1\right)\exp\left[-\frac{(\omega-kv_b)^2}{k^2v_{\rm thib}^2}\right]-\left(\frac{m_e}{m_i}\right)^{1/2}\exp\left(-\frac{\omega^2}{k^2v_{\rm the}^2}\right)\right\}\end{aligned} \tag{4.31}$$

where we neglected the contribution of the background ions, and the total density is $n_0 = n_{e0} = n_{i0} + n_{ib}$. The condition for beam ion-acoustic instability is the same as for current instability with replacement of the current drift speed by the ion beam velocity, $v_{ib} > c_{ia}$. Compared to the electron current-driven instability this growth rate is smaller by the ratio of beam to plasma densities. Thus the ion beam driven mode is a weaker instability than the electron current ion-acoustic instability.

Heat Flux-Driven Instability

Ion-acoustic waves can also be destabilized by heat currents. These currents play a role similar to real electric currents in a plasma. But they depend on the presence of heat flows and thus on the presence of temperature gradients in a plasma. Such temperature gradients are familiar from observations in the solar wind, at the magnetopause and in the various boundary layers of the magnetosphere. One may therefore expect that heat flux-driven ion-acoustic modes will exist in these regions.

In most cases the heat flux is carried by electrons, since they move faster along the magnetic field. The mechanism of the instability is exactly the same as that for the electron current-driven ion-acoustic instability. The only difference is that the electron drift speed is now calculated from the parallel heat flux, $q_\|$, under the condition that the electric current vanishes, $\mathbf{j}_\| = 0$. The two equations for the parallel heat flux and current in a zero-order parallel electric field, $E_{0\|}$, and parallel temperature gradient $\nabla_\| T_e$, are

$$\begin{aligned} q_{e\|} &= \eta_{e\|}E_{0\|} - \kappa_{e\|}\nabla_\| T_e \\ j_{e\|} &= \sigma_{e\|}E_{0\|} + \zeta_{e\|}\nabla_\| T_e = 0 \end{aligned} \tag{4.32}$$

Heat fluxes can evolve only in the presence of non-vanishing transport coefficients (see App. B.2 in our companion book). We therefore allow for non-ideal conditions and finite electric and thermal conductivities, $\sigma_{e\|}, \kappa_{e\|}$, and thermoelectric conductivities, $\eta_{e\|}, \zeta_{e\|}$. From the definition of the transport coefficients one knows that $\zeta_{e\|} \approx 0.7\sigma_{e\|}/e$. The vanishing current equation yields for the electric field

$$E_{0\|} \approx -0.7\nabla_{\|}T_e/e \tag{4.33}$$

This electric field enters the heat flux equation so that

$$q_{e\|} \approx -(\kappa_{e\|} - 0.7\eta_{e\|}/e)\nabla_{\|}T_e \tag{4.34}$$

from where it is seen that it is the temperature gradient which gives rise to the heat flux or, vice versa, it is the heat flux which causes the parallel electric field. This electric field accelerates the electrons until a stationary state is reached between the acceleration and collisional retardation due to the non-zero electron-ion collision frequency, $\nu_{ei} \neq 0$. Equating the two forces

$$eE_{0\|} = m_e\nu_{ei}v_d \tag{4.35}$$

the drift velocity enforced by the heat flux and being used in the growth rate of the ion-acoustic instability in Eq. (4.27) is given by

$$v_d \approx 0.7\nabla_{\|}T_e/m_e\nu_{ei} > c_{ia} \tag{4.36}$$

This is the condition for ion acoustic instability.

Solar Wind

In a nearly collisionless plasma like the solar wind the collision frequency is small and the instability threshold is easily exceeded, even for weak temperature gradients. There is no unambiguous experimental confirmation of a heat flux-driven instability in the solar wind from observations of ion-acoustic waves, but heat fluxes are connected with most deviations of the parallel electron equilibrium distribution from Maxwellian symmetry. Such non-symmetric shapes are a general feature of solar wind electron distributions and can be the cause of heat flux-excited ion-acoustic waves.

4.3. Electron-Acoustic Instability

To excite electron-acoustic waves a drifting hot electron component is needed in addition to the cold background electrons. The ions are assumed to neutralize the charge. Moreover, zero current flow is assumed for the electrons, $n_c\mathbf{v}_{dc} + n_h\mathbf{v}_{dh} = 0$. For vanishing drifts and low cold electron temperatures three weakly damped modes can exist. These are Langmuir waves, ion-acoustic waves and electron-acoustic waves. The frequency of the latter is close to the cold electron plasma frequency, $\omega_{ea} \approx \omega_{pc}$. When the two electron components exhibit a relative drift but no current flows according to the above zero-current condition, the ion-acoustic and the electron-acoustic wave can both become unstable. The former propagates into the direction of the cold electron relative drift, as was discussed above, while the electron-acoustic wave propagates parallel to the hot electron drift velocity.

Frequency and Growth Rate

If we assume that all three components are Maxwellians, the dielectric function of this combination of particle distributions is

$$\epsilon(\omega, k) = 1 + \frac{2}{k^2\lambda_{Dc}^2}\left[1 + \zeta_{ec}Z(\zeta_{ec})\right] + \frac{2}{k^2\lambda_{Dh}^2}\left[1 + \zeta_{eh}Z(\zeta_{eh})\right] + \frac{2}{k^2\lambda_{Di}^2}\left[1 + \zeta_i Z(\zeta_i)\right] \tag{4.37}$$

where the arguments are defined similarly to the case of the ion-acoustic instability

$$\begin{aligned} \zeta_{ec} &= (\omega - \mathbf{k}\cdot\mathbf{v}_{dc})/kv_{\mathrm{thc}} \\ \zeta_{eh} &= (\omega - \mathbf{k}\cdot\mathbf{v}_{dh})/kv_{\mathrm{th}h} \\ \zeta_i &= \omega/kv_{\mathrm{th}i} \end{aligned} \tag{4.38}$$

If one assumes that the ions and cold electrons are much colder than the hot electrons, $T_i \approx T_c \ll T_h$, one can expand the hot component plasma dispersion function in the small amplitude limit, $\zeta_{eh} \ll 1$, and finds the frequency of the electron-acoustic mode

$$\omega_{ea} = \mathbf{k}\cdot\mathbf{v}_{dc} + \omega_{pc}/(1 + 1/k^2\lambda_{Dh}^2)^{1/2} \tag{4.39}$$

In the weakly unstable limit the weak instability growth rate becomes

$$\boxed{\gamma_{ea} \approx \left(\frac{\pi}{8}\right)^{1/2}\frac{\omega_{pc}}{k^2\lambda_{Dh}^2}\left(\frac{\mathbf{k}\cdot\mathbf{v}_{dh}}{kv_{\mathrm{th}h}} - \frac{\omega_{ea}}{kv_{\mathrm{th}h}}\right)\exp\left[-\frac{(\mathbf{k}\cdot\mathbf{v}_{dh} - \omega_{ea})^2}{k^2v_{\mathrm{th}h}^2}\right]} \tag{4.40}$$

This instability requires that $\mathbf{k}\cdot\mathbf{v}_{dh} > \omega_{ea}$, which also requires that the drift speed of the cold component is small against the speed of the hot component. Moreover, numerical calculations show that the threshold condition is $k\lambda_{Dc} \approx 5$. The above threshold of the electron-acoustic instability is maintained as long as the relative electron drift speed, $|v_{dh} - v_{dc}|$, is smaller than the thermal speed of the hot electrons. For higher velocities one recovers the electron-electron beam-fluid two-stream instability, where the negative part of the frequency is no more of kinetic origin.

Parametric Investigation

Figure 4.6 shows the domain of the kinetic electron-acoustic instability in the parameter space of relative electron drift velocity and density ratio. This figure shows that low cold

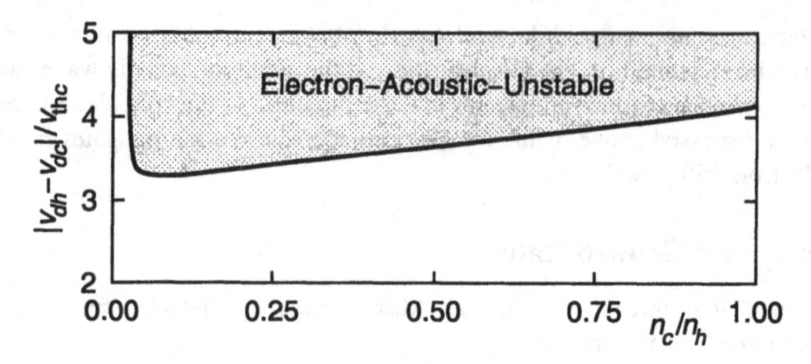

Fig. 4.6. Parameter space of the electron-acoustic instability.

electron densities and relative drifts a few times the cold electron thermal velocity favor the excitation of the instability. This threshold is much higher than the ion-acoustic instability threshold for low cold electron densities, but lower than the latter for high cold electron densities. Therefore, for small cold electron densities the ion-acoustic wave excited by the drift of the electrons will be the dominant instability, while for higher cold electron densities, yet still below the hot electron density, the electron-acoustic instability is of greater importance.

Foreshock

Electron-acoustic instability is expected to occur in the Earth's foreshock region, where diffuse hot electrons are superimposed on the solar wind plasma and have a relatively large drift against the solar wind ions and electrons. These waves will occur near the cold electron plasma frequency, but will be beam waves according to the dispersion relation in Eq. (4.39). They can exist also in the shock ramp, where hot electrons from the shock mix into the cold solar wind electron beam, and may contribute to shock dissipation.

4.4. Current-Driven Cyclotron Modes

Even in the electrostatic limit, currents do not only drive Langmuir and acoustic modes unstable. They can also excite electrostatic ion-cyclotron and lower-hybrid modes.

Ion-Cyclotron Instability

The field-aligned current instability competing most with the ion-acoustic instability is the electrostatic ion-cyclotron instability. For propagation purely perpendicular to the magnetic field its dispersion relation is that of Eq. (I.10.133) and the wave is an ion Bernstein mode. To find an instability we must allow parallel wave propagation. In this case the dispersion relation is a version of Eq. (I.10.104), modified due to the presence of the field-aligned current. Assuming that this current is carried exclusively by the electrons, its dielectric function becomes

$$\epsilon(\omega, \mathbf{k}) = 1 - \sum_s \sum_{l=-\infty}^{\infty} \frac{\omega_{ps}^2 \Lambda_l(\eta_s)}{k^2 v_{\mathrm{th}s}^2} \left[Z'(\zeta_s) - \frac{2l\omega_{gs}}{k_\parallel v_{\mathrm{th}s}} Z(\zeta_s) \right] = 0 \tag{4.41}$$

We assume isotropic conditions for both kinds of particles, electrons and ions. The arguments of the plasma dispersion functions are

$$\begin{aligned} \zeta_i &= (\omega - l\omega_{gi})/k_\parallel v_{\mathrm{th}i} \\ \zeta_e &= (\omega + l\omega_{ge} - k_\parallel v_d)/k_\parallel v_{\mathrm{th}e} \end{aligned} \tag{4.42}$$

The above dispersion relation allows instability at every ion-cyclotron harmonic frequency, $\omega = l\omega_{gi}$. For purely parallel propagation and $T_e \gg T_i$ it also contains ion-acoustic waves, since this case corresponds to the field-free case. Hence, to cover ion-cyclotron waves we consider the wavenumber range $k_\parallel \ll k_\perp$, and temperature ratios $T_e \approx T_i$. We may expect that the strongest instability will occur for the lowest harmonic, $l = 1$, because it requires the smallest amount of energy to increase its amplitude. The ion-cyclotron wave solution

$$\omega \approx \omega_{gi} \left[1 + \frac{\Lambda_1(\eta_i)}{1 + T_i/T_i - G} \right] \tag{4.43}$$

where $G = \Lambda_1 + (1 - \Lambda_0)/\eta_i$, is obtained under the conditions that $\eta_e \ll 1$ and, as for the ion-acoustic instability, $\zeta_e \ll 1$, while the ion plasma dispersion function is expanded in the large argument limit.

Since $G < 1, \Lambda_1 < 1$, the correction term on the cyclotron frequency is usually smaller than 0.5, so that the frequency of the ion-cyclotron wave is close to ω_{gi}, but shifts away from ω_{gi} for decreasing T_e/T_i. On the other hand, when $T_e \gg T_i$ and $\eta_i \ll 1$, the term $G \to 1$, and the correction increases. In the limit $\eta_i \to 0$ the frequency becomes

$$\omega \approx \omega_{gi}(1 + k_\perp^2 c_{ia}^2 / 2\omega_{gi}^2) = \omega_{gi}(1 + \Delta) \tag{4.44}$$

valid in the long wavelength limit and applicable to perpendicular propagation.

For oblique propagation the expression for the growth rate is complicated. It is more interesting to determine the marginally unstable current drift velocity for the ion cyclotron wave above which the instability sets in. The procedure is to put the growth rate $\gamma = 0$ in the dielectric response, setting its real part $\epsilon_r(\omega, k_\parallel, k_\perp) = 0$, and solving

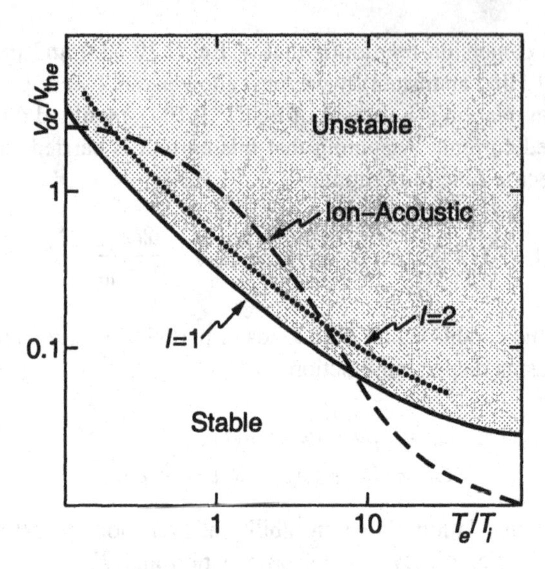

Fig. 4.7. Current drift velocity threshold for the ion-cyclotron wave.

for the drift velocity, v_d, as function of the parameters, e.g., the temperature ratio T_e/T_i. The general expression for the drift velocity obtained from $\gamma = 0$ is

$$\frac{k_\| v_d}{\omega} = \left[1 + \left(\frac{m_i T_e}{m_e T_i}\right)^{1/2} \sum_s \frac{\sum_l \Lambda_l(\eta_s) \mathrm{Im} Z(\zeta_s)}{\sum_l \Lambda_l(\eta_e) \mathrm{Im} Z(\zeta_e)}\right] \tag{4.45}$$

This equation is an implicit equation for the drift velocity, because v_d is still contained in the electron plasma dispersion function on the right-hand side. For the ion-cyclotron mode this expression reduces, with the help of the above dispersion relation, to

$$\frac{v_d}{v_{\text{thi}}} = \frac{\omega}{k_\| v_{\text{thi}}} \left\{1 + \left(\frac{m_i T_e}{m_e T_i}\right)^{1/2} \Lambda_1(\eta_i) \exp\left[-\frac{(\omega - \omega_{gi})^2}{k_\|^2 v_{\text{thi}}^2}\right]\right\} \tag{4.46}$$

Using the l=1 frequency in Eq. (4.44), this becomes

$$\frac{v_d}{v_{\text{thi}}} = \left(1 + \frac{1}{\Delta}\right) \frac{(\omega - \omega_{gi})}{k_\| v_{\text{thi}}} \left\{1 + \left(\frac{m_i T_e}{m_e T_i}\right)^{1/2} \Lambda_1(\eta_i) \exp\left[-\frac{(\omega - \omega_{gi})^2}{k_\|^2 v_{\text{thi}}^2}\right]\right\} \tag{4.47}$$

One can minimize this equation with respect to one of the components of the wave vector to obtain the critical current drift velocity, v_{dc}, above which the wave undergoes instability. This procedure yields an expression

$$\frac{v_{dc}}{v_{\mathrm{th}i}} = \left(1 + \frac{1}{\Delta(\eta_i^*)}\right) \left\{ \ln \left[2 \left(\frac{m_i T_e}{m_e T_i} \right)^{1/2} \Lambda_1(\eta_i^*) \right] \right\}^{1/2} \tag{4.48}$$

where η_i^* is the value of $k_\perp$ where the drift velocity minimizes. Figure 4.7 shows this marginally unstable current drift velocity as function of the temperature ratio obtained from numerical solution of the above equation. Here the drift speed has been normalized to the electron thermal velocity. The important result is that the critical drift velocity is relatively low for low temperature ratios in a range where the ion-acoustic instability has higher threshold. But for increasing harmonic number, $l > 1$, the threshold increases, too. Higher harmonic ion-cyclotron waves are less unstable than the l=1 mode, as had been expected from the very beginning. Moreover, for high temperature ratios the ion-acoustic instability takes over, and the ion-cyclotron instability ceases.

Modified Two-Stream Instability

The instabilities considered so far are driven by either field-aligned currents or field-aligned beams. Magnetic fields have little effect on these electrostatic instabilities. The only exception is the electrostatic current-driven ion-cyclotron instability, which has a frequency close to the ion-cyclotron frequency and where the wave propagates nearly perpendicular to the magnetic field.

A similar current-driven instability exists also for currents which flow perpendicular to the magnetic field. Independent of their origin, such currents serve as free energy sources. There are several possibilities for generating perpendicular currents. The most important are pressure gradient drifts or polarization field drifts. An instability connected with these currents is the *lower-hybrid drift instability*, which we will derive in detail when discussing the effects of inhomogeneities on kinetic instabilities.

The *modified two-stream instability* takes advantage of the very large ion gyroradius compared to the electron gyroradius. Thus for wavelengths, $\lambda < v_{\mathrm{th}i}/\omega_{gi} = r_{gi}$, the ions behave unmagnetized and can be considered to propagate along straight lines. On the other hand, the electrons are strongly magnetized in this case. The instability arising from the interaction of the two particle populations, when one of them is drifting across the magnetic field, turns out to have a frequency close to the lower-hybrid frequency. This instability is the modified two-stream instability. It is very important in all regions where transverse currents can exist in a plasma, as, for instance, the various boundaries across which the magnetic field changes its value or direction.

If we assume that the ions are drifting into the x direction with velocity v_d and have Maxwellian distribution function while the electrons are cold and strongly magnetized, the wave obeys the dispersion relation

$$\epsilon(\omega, k_\parallel, k_\perp) = 1 + \frac{\omega_{pe}^2}{\omega_{ge}^2} - \frac{k_\parallel^2}{k^2}\frac{\omega_{pe}^2}{\omega^2} + \frac{2\omega_{pi}^2}{k_\perp^2 v_{\mathrm{th}i}^2}\left[1 + \zeta_i Z(\zeta_i)\right] \tag{4.49}$$

where the ion argument is given by $\zeta_i = (\omega - k_\perp v_d)/k_\perp v_{\mathrm{th}i}$. This argument is assumed to be large enough, $\zeta_i \gg 1$, for the large argument expansion of the plasma dispersion to be applicable so that the dispersion relation becomes

$$1 - \frac{\omega_{lh}^2}{(\omega - k_\perp v_d)^2} - \frac{m_i k_\parallel^2}{m_e k^2}\frac{\omega_{lh}^2}{\omega^2} = 0 \tag{4.50}$$

an expression very similar to the Buneman dispersion relation. The frequency, $\omega_{lh} = \omega_{pi}/(1 + \omega_{pe}^2/\omega_{ge}^2)^{1/2}$, appearing in the numerators of the above equation, is the lower-hybrid frequency. The similarity of Eq. (4.50) to the Buneman dispersion relation

$$1 - \frac{\omega_{pi}^2}{\omega^2} - \frac{\omega_{pe}^2}{(\omega - kv_0)^2} = 0 \tag{4.51}$$

suggests the following replacements in the equations for the latter

$$\begin{aligned} \omega_{pe} &\rightarrow \omega_{lh} \\ \omega_{pi}^2 &\rightarrow (m_i k_\parallel^2 / m_e k^2)\omega_{lh}^2 \end{aligned} \tag{4.52}$$

By the same reasoning used to derive the threshold condition in Eq. (2.50) for the Buneman instability we find that the frequency of the modified two-stream instability is

$$\omega_{\mathrm{mts}} = k_\perp v_b \left[1 + \left(\frac{m_e k^2}{m_i k_\parallel^2}\right)^{1/3}\right]^{-1} \tag{4.53}$$

and as condition on the drift velocity for modified two-stream instability

$$\frac{k_\perp^2 v_b^2}{\omega_{lh}^2} < \left(\frac{m_e k^2}{m_i k_\parallel^2}\right)^{1/3}\left[1 + \left(\frac{m_i k_\parallel^2}{m_e k^2}\right)^{1/3}\right]^3 \tag{4.54}$$

As in the case of the Buneman instability, this condition sets a limit on the combination of wavenumber and drift velocity. Since the right-hand side of this condition depends on the lower-hybrid frequency, which is lower than the electron plasma frequency, the modified two-stream instability is driven unstable by much lower velocities than the Buneman instability.

Since in the discussion of the Buneman instability the densities of the particles were equal, the maximum growing wavenumber now is found at

$$\max\left(\frac{k_\parallel^2}{k_\perp^2}\right) = \frac{m_e}{m_i} \ll 1 \tag{4.55}$$

Hence, the parallel wavelength of the modified two-stream wave is much longer than the perpendicular wavelength and the wave propagates nearly perpendicular to the magnetic field. Its maximum growth rate

$$\boxed{\gamma_{\mathrm{mts,max}} = \omega_{lh}/2} \tag{4.56}$$

is obtained at

$$k_{\perp\mathrm{max}} v_d \approx \omega_{lh}/2 \tag{4.57}$$

if the growth rate is maximized with respect to $k_\perp$. This growth rate is nearly the lower-hybrid frequency and thus nearly the wave frequency itself.

The modified two-stream instability is a fairly strong instability which will easily be driven unstable by ion streams propagating perpendicular to the magnetic field. Knowing the drift velocity of these ions or the corresponding current, the wave frequency at maximum growth assumes the value

$$\boxed{\omega_{\mathrm{mts,max}} = k_{\perp\mathrm{max}} v_d/2} \tag{4.58}$$

Magnetopause

The modified two-stream instability can be excited at the magnetopause, during times when it is a tangential discontinuity. Assuming a non-zero magnetosheath field of about 30–40 nT, the jump in the magnetic field across the magnetopause is of the order of $\Delta B \approx 60$ nT. The width of the magnetopause is of the order of 1000 km or less. These values suggest a current of $4.8 \cdot 10^{-8}$ A/m^2. For a plasma density of $n \approx 30\,\mathrm{cm}^{-3}$ the drift velocity is $v_d \approx 10$ km/s. The lower-hybrid frequency is of the order of 65 Hz. This corresponds to a maximum perpendicular wavenumber of $k_\perp \approx 0.082$ and a wave growth rate of $200\,\mathrm{s}^{-1}$.

Hence, fast growth at the lower-hybrid frequency is expected at the magnetopause with wavelengths of the order of 70 m, shorter than the ion gyroradius. These waves account for lower-hybrid wave spectra observed during magnetopause crossing and will be able to dissipate part of the kinetic energy of the magnetopause current. This current is continuously restored by the interaction of solar wind and geomagnetic field. The energy fed into the modified two-stream lower-hybrid waves originates from this interaction and thus from the solar wind kinetic energy.

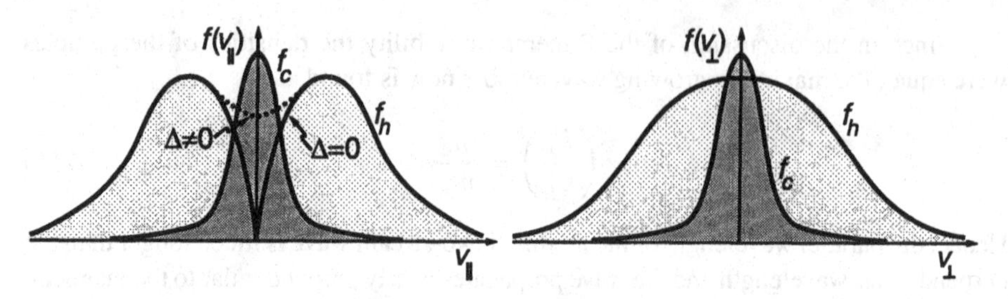

Fig. 4.8. Loss cone-unstable parallel and perpendicular distributions.

4.5. Loss Cone Instabilities

Aside from the lower- and upper-hybrid waves, the electrostatic electron- and ion-cyclotron waves are the most important electrostatic waves in homogeneous magnetized plasmas. They cover a wide range of cyclotron harmonics and, because of being principle plasma resonances, they contribute to energy exchange between waves and particles. In the following we discuss a few ways to drive these waves unstable. Since these waves, except for the lowest harmonics, are purely of kinetic origin, the instabilities connected with them are also kinetic requiring particular shapes of the undisturbed particle distribution functions. The two types of distributions leading to the excitation of electrostatic cyclotron waves are temperature anisotropies and loss cone distributions.

Modeling temperature anisotropies in the equilibrium distribution function is comparably simple. One uses an anisotropic Maxwellian distribution as the one given in Eq. (I.6.37) of our companion book, *Basic Space Plasma Physics*. For loss cone distributions one takes either the Dory-Guest-Harris form of Eq. (I.6.40) or the partially-filled loss cone distribution of Eq. (I.6.41), again given in our companion book. The former is, however, much more difficult to treat. We will therefore exclusively use the latter

$$f(v_\parallel, v_\perp; \Delta, \beta) = \frac{n}{(\pi^3 \langle v_\parallel \rangle^2 \langle v_\perp \rangle^4)^{1/2}} \exp\left(-\frac{v_\parallel^2}{\langle v_\parallel \rangle^2}\right) G(v_\perp, \Delta, \beta) \tag{4.59}$$

The first part of this function is a parallel Maxwellian. The information about the loss cone is contained in the function $G(v_\perp, \Delta, \beta)$, given in Sec. 6.3, and the graphical representation of the distribution (4.59) is found in Fig. 6.7, both in our companion book. Loss cones imply that particles with small perpendicular velocities are lost and are therefore missing in the distribution function. Hence, loss cones and temperature anisotropies are distribution functions which are not thermodynamically stable, but carry the free energy for exciting instabilities.

Because this excess free energy is stored in the perpendicular particle motion, i.e., in the gyration of the particles comprising the distribution function, distributions of this kind are particularly suited for exciting waves related to the cyclotron motion of the

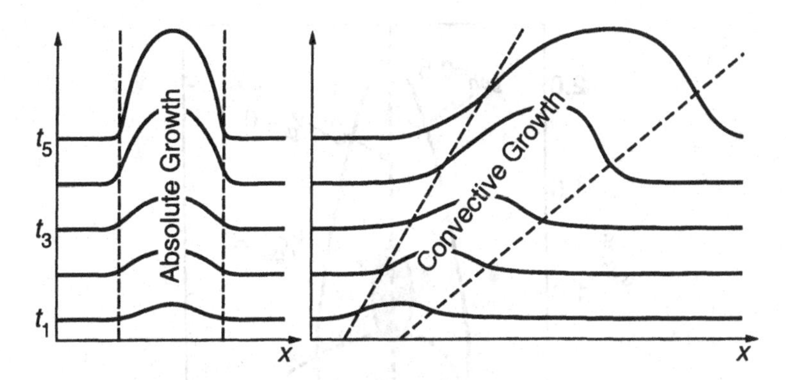

Fig. 4.9. Absolute and convective growth of wave amplitudes.

particles. These waves want to reduce the energy stored in the gyration, which can most easily been done by resonance with the gyrating particles at their gyrofrequency, ω_{gs}. This is the physical reason for excitation of electrostatic and electromagnetic cyclotron waves by loss cone and anisotropic Maxwellian distributions with

$$A_s = T_{s\perp}/T_{s\|} - 1 > 0 \tag{4.60}$$

Electron-Cyclotron Instability

We first discuss the mechanism by which electron-cyclotron harmonics can be excited by loss cone distributions. The resulting instability is the *electrostatic electron-cyclotron loss cone instability*. The theory is mathematically complicated, since there are infinitely many electron-cyclotron harmonics, $\omega \approx l\omega_{ge}$, which can be driven unstable. One must ultimately turn to numerical tools.

Let us assume that the plasma in the magnetosphere is composed of a neutralizing ion background, which at the high electron-cyclotron frequencies is immobile and needs not be taken into account, a cold electron background, which can be assumed isotropic and Maxwellian, and the hot and diluted anisotropic loss cone component superimposed on the background plasma. Calculating the parallel and perpendicular temperatures by taking the second-order moment of the partially filled loss cone distribution function given in Eq. (4.59), yields the anisotropy $\mathcal{A}_{eh}$

$$\mathcal{A}_{eh} = \frac{\int (v_\perp^2 - v_\|^2) f_{eh}(v_\|, v_\perp; \Delta, \beta) v_\perp^2 \, dv_\perp dv_\|}{\int v_\|^2 f_{eh}(v_\|, v_\perp; \Delta, \beta) v_\perp^2 \, dv_\perp dv_\|} \tag{4.61}$$

where a description of the loss cone parameters, Δ and β, has been given in Sec. 6.3 of our companion book. The ratio of the two integrals warrants to forget the normalization factors in front of the hot electron distribution, $f_{eh}(v_\|, v_\perp; \Delta, \beta)$. It is easy to solve

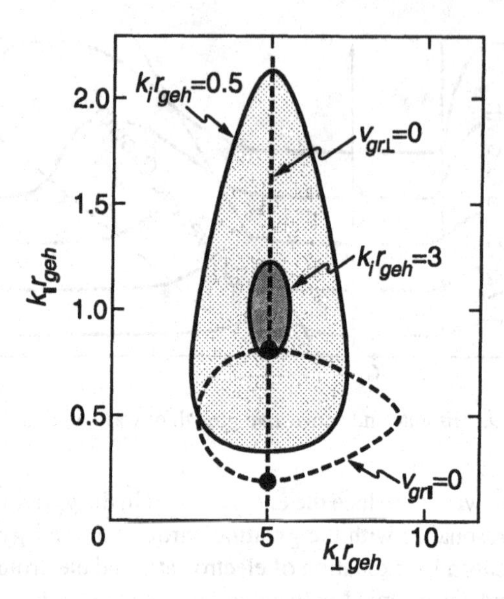

Fig. 4.10. Amplification length of the loss cone electron-cyclotron instability.

the integrals because this distribution has been constructed such that it consists only of Maxwellians, yielding

$$\mathcal{A}_{eh} = A_e\Delta + (1-\Delta)\left[A_e + (A_e+1)\beta\right] \tag{4.62}$$

where $A_e = T_{e\perp}/T_{e\parallel} - 1$ is the ordinary anisotropy. Hence, for $\Delta = 1$ one has $\mathcal{A}_{eh} = A_e$, while for $A_e = 0$ there is still some anisotropy $\mathcal{A}_{eh} = (1-\Delta)\beta$ retained, which is a pure loss cone effect. The combined hot and cold electron distribution functions are shown in Fig. 4.8.

The magnetized electrostatic response function of the hot electron contribution is to be calculated from Eq. (I.10.96) using $f_{eh}(v_\parallel, v_\perp; \Delta, \beta)$. To this response function one must add the cold electron isotropic response function. The cold electrons will cyclotron-damp the electrostatic waves because they are a sink of energy. The combined effect depends heavily on the ratio of cold to hot electron densities, n_c/n_h, with largest growth rates when no cold electrons are present.

Parametric Investigation

For parametric investigations one has to distinguish between the two different ways of wave growth sketched in Fig. 4.9. A wave can be excited in a local position so that the energy does not flow out from the region where the wave is excited but stays there and accumulates. This kind of instability is an *absolute* or *non-convective instability*. Under most conditions, however, the wave will transport energy out of the region of excitation. Hence, if the region of excitation is narrow, the wave will not be strongly amplified. It can be amplified only over that distance where the growth rate is positive. Such instabilities are *convective instabilities*.

This distance is given by the amplification length, which can be calculated from the growth rate and the group velocity of the wave as the imaginary part of the wavenumber $\mathbf{k} = \mathbf{k}_r + i\mathbf{k}_i$. For a plane wave proportional to $\exp[-i(\omega t - \mathbf{k}\cdot\mathbf{x})]$ amplification will proceed as long as $k_i < 0$ and

$$|k_i| = \gamma\left[\left(\frac{\partial\omega}{\partial k_\perp}\right)^2 + \left(\frac{\partial\omega}{\partial k_\parallel}\right)^2\right]^{-1/2} \tag{4.63}$$

The instability becomes absolute at that position in parameter space, where the parallel and the perpendicular components of the group velocity vanish simultaneously. Hence, one calculates numerically the growth rate, γ, as function of the density and temperature ratios and possible other parameters such as Δ and β. Then one determines the parallel and perpendicular group velocities and finds the locations where both vanish simultaneously. Here the amplification length turns zero and k_i diverges. An example of this procedure applied to electrostatic electron-cyclotron wave excitation in the magnetosphere is shown in Fig. 4.10. The parameters for this particular case are $\Delta = 0$, $T_{eh} = 100T_{ec}$, $n_c = 0.2n_h$, and $\omega_{uhc}/\omega_{ge} = 1.8$.

The region of large imaginary k_i coincides with one of the two crossing points of the two curves, where the two components of the group velocity vanish, $\partial\omega/\partial k_\parallel = \partial\omega/\partial k_\perp = 0$. In these two points, i.e., for the corresponding wavenumbers parallel and perpendicular to the magnetic field, the wave instability becomes absolute, and the energy accumulates in the region of instability. This calculation is performed for the fundamental harmonic, $l = 1$, of the electron-cyclotron frequency. At higher harmonics similar instability is found. The parametric dependence of the regions of absolute instability for the first and fourth harmonics is shown in Fig. 4.11.

Electrostatic electron-cyclotron harmonics can become non-convectively unstable in hot loss cone plasmas when a small amount of cold electrons is admixed. For the lowest cyclotron harmonic the unstable region in parameter space is largest. With increasing harmonic number it shifts to increasing cold upper-hybrid frequencies, ω_{uhc}, but at the same time there is a narrow region of high cold-to-hot density ratios where electron-cyclotron harmonic waves can be destabilized by hot electron loss cones. The most probable region of non-convective instability centers around ratios of $n_c/n_h \approx 0.4$. Here a large number of harmonics may become destabilized. An example of many electron cyclotron harmonics excited in the nighttime equatorial magnetosphere just outside the plasmasphere is given in Fig. 4.12. Harmonics up to the sixth number are visible in the electric wave field spectral density for several minutes.

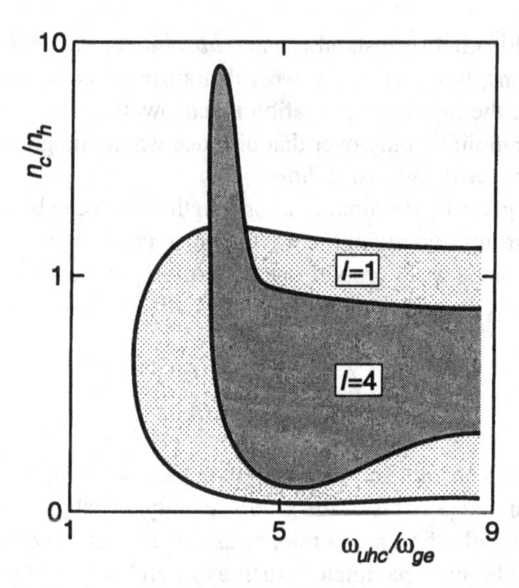

Fig. 4.11. Regions of instability for electron-cyclotron harmonics.

Ion-Cyclotron Instability

The cyclotron resonance symmetry between electrons and ions suggests that an *electrostatic ion-cyclotron loss cone instability* should complement the electrostatic electron-cyclotron instability discussed. This is indeed the case. The technique to determine the non-convective instability in parameter space is the same as that for the electrostatic electron-cyclotron waves. The only difference is found in exchanging the hot electron loss cone distribution with the hot ion loss cone distribution in Eq. (I.10.96), but in addition one must retain the electron contribution to the dielectric response function

$$\begin{aligned}
\epsilon(\omega, \mathbf{k}) &= 1 + \frac{\omega_{pe}^2}{\omega_{ge}^2}\frac{k_\perp^2}{k^2} - \frac{1}{k^2\lambda_D^2}Z'(\zeta_e) \\
&\quad + \frac{\omega_{pi}^2}{k^2}\sum_l \int_{-\infty}^{\infty} dv_\parallel \frac{G_{li}(v_\parallel) f_{0i}(v_\parallel, v_\perp; \Delta, \beta)}{\omega - k_\parallel v_\parallel - l\omega_{gi}} = 0
\end{aligned} \tag{4.64}$$

where the dimensionless operator function, G_{li}, is defined as

$$G_{li} = 2\pi \int_0^\infty v_\perp dv_\perp J_l^2(k_\perp r_{gi}) \left[k_\parallel \frac{\partial}{\partial v_\parallel} + \frac{l\omega_{gi}}{v_\perp}\frac{\partial}{\partial v_\perp}\right] \tag{4.65}$$

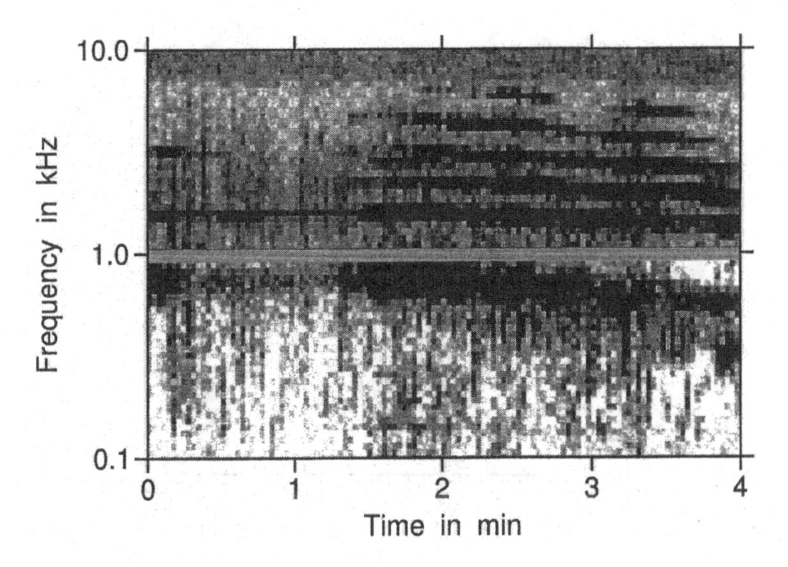

Fig. 4.12. Electron-cyclotron harmonics in the magnetosphere.

and the undisturbed non-equilibrium ion distribution function is the sum of the hot loss cone and cold isotropic ion distributions

$$f_{0i} = f_{ic}(v_\parallel, v_\perp) + f_{ih}(v_\parallel, v_\perp; \Delta, \beta) \tag{4.66}$$

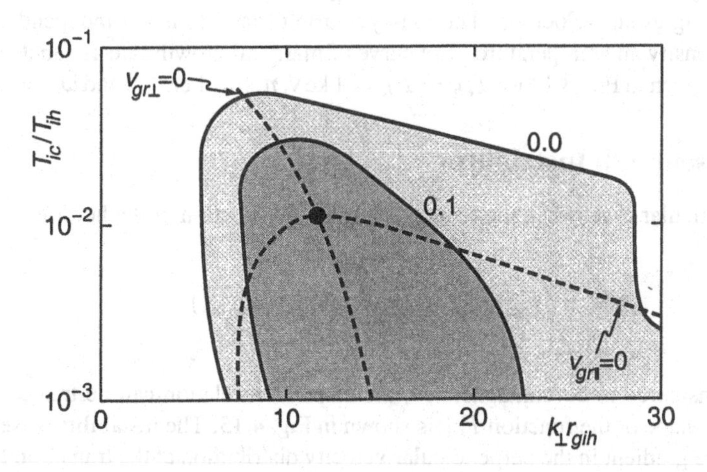

Fig. 4.13. Regions of convective and absolute ion-cyclotron instability.

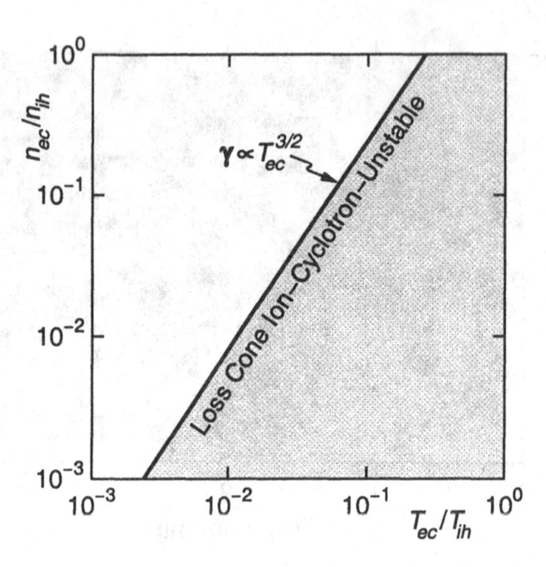

Fig. 4.14. Region of ion cyclotron instability in parameter space.

Figure 4.13 shows the result of a parametric investigation of the region of non-convective growth of the electrostatic ion-cyclotron loss cone instability. The conditions for absolute instability are satisfied at the crossing point of the parallel and perpendicular group velocities. The particular parameters used in this calculation are $n_{ic}/n_{ih} = 0.1$, $T_{ih} = T_{eh} = 1$ keV, and $k_{\|}r_{gih} = 0.6$. The two boundaries for the inverse growth length, $k_{i\perp}r_{gih}$, are 0 and 0.1, with $k_{i\perp}$ increasing toward the crossing point of the vanishing group velocities. The ion-cyclotron instability is not independent of the electron density and temperature. The curve of marginal growth rate as function of n_{ec} and T_{ec} is given in Fig. 4.14 for $T_{eh} = T_{ih} = 1\,\text{keV}$, $n_{ih} = 1\,\text{cm}^{-3}$, and $\omega_{ge} = 5.3\,\text{kHz}$.

Post-Rosenbluth Instability

In one particular case, when the reduced distribution function of the hot ions

$$F_{i\perp}(v_\perp) = 2\pi \int_{-\infty}^{\infty} dv_{\|} f_{ih}(v_\perp, v_{\|}) \tag{4.67}$$

can be considered to be unmagnetized, the dispersion relation can be treated analytically. The shape of the function $F_{i\perp}$ is shown in Fig. 4.15. The instability arises due to the positive gradient in the perpendicular velocity distribution at the transition from the loss cone to trapped particles. The dispersion relation including electrons then simpli-

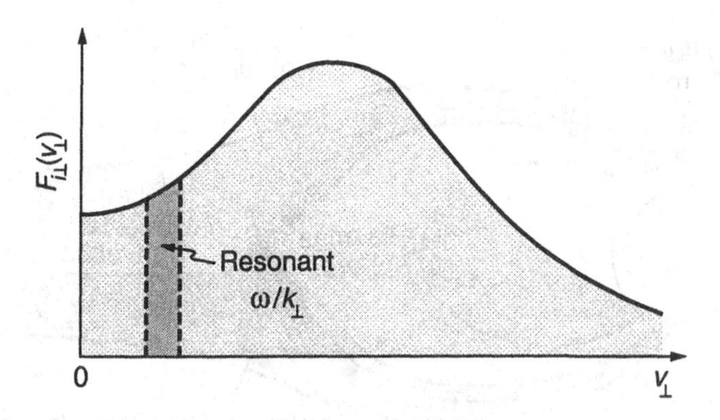

Fig. 4.15. Unmagnetized ion loss cone distribution function.

fies to

$$\epsilon(\omega, \mathbf{k}) = 1 + \frac{\omega_{pe}^2}{\omega_{ge}^2} - \frac{k_\parallel^2}{k^2}\frac{\omega_{pe}^2}{\omega^2} + \frac{1}{k^2\lambda_{Di}^2}\left[F_{i\perp}(0) + G\left(\frac{\omega}{k_\perp}\right)\right] = 0 \qquad (4.68)$$

where $F_{i\perp}(0)$ is the value of the distribution function inside the loss cone, and

$$G\left(\frac{\omega}{k_\perp}\right) = \int_0^{\omega/k_\perp} \frac{\partial F_{i\perp}(v_\perp)/\partial v_\perp}{(1 - k_\perp^2 v_\perp^2/\omega^2)^{1/2}} dv_\perp \qquad (4.69)$$

The resonant region of the perpendicular velocity space is defined by $\omega = k_\perp v_\perp \cos\phi$, which covers the entire region $v_\perp^2 > \omega^2/k_\perp^2$. The non-resonant region, $\omega \neq k_\perp v_\perp \cos\phi$, is restricted to $v_\perp^2 < \omega^2/k_\perp^2$. Separating the real and imaginary parts of ϵ and assuming weak instability, the unstable wave frequency of the ion loss cone instability is

$$\omega_{\mathrm{ilc}}^2 = \frac{k_\parallel^2\lambda_{Di}^2}{[k^2\lambda_{Di}^2(1 + \omega_{pe}^2/\omega_{ge}^2) + F_{i\perp}(0) + G]}\omega_{pe}^2 \qquad (4.70)$$

The growth rate can be expressed through an integral over the entire resonant region as

$$\gamma_{\mathrm{ilc}} = \frac{\omega_{\mathrm{ilc}}}{2[k^2\lambda_{Di}^2(1 + \omega_{pe}^2/\omega_{ge}^2) + F_{i\perp}(0) + G]} \int_{\omega/k_\perp}^{\infty} \frac{\partial F_{i\perp}(v_\perp)/\partial v_\perp}{(k_\perp^2 v_\perp^2/\omega^2 - 1)^{1/2}} dv_\perp \qquad (4.71)$$

This unmagnetized ion loss cone instability is called *Post-Rosenbluth instability*. It is a lower-hybrid wave instability. In the modified two-stream instability we already met one member of this family. There the instability was driven by streaming ions or electrons. In the present case the free energy is provided by the ion loss cone only, but

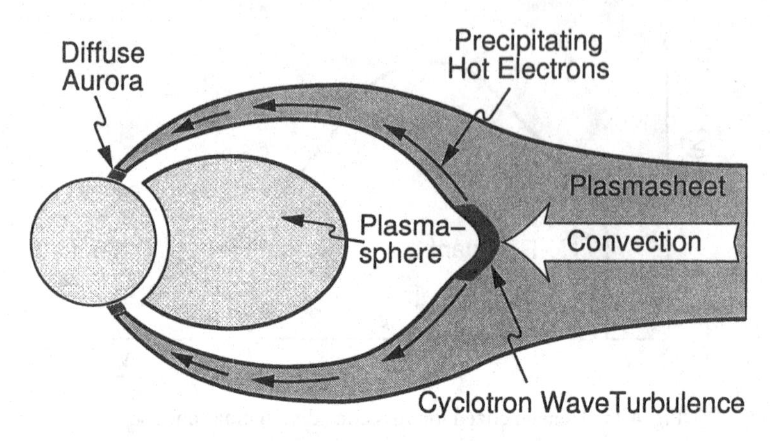

Fig. 4.16. Ring current particle precipitation due to electrostatic cyclotron waves.

the resulting wave is again a wave close to the lower-hybrid frequency. Compared to the low harmonics of the ion-cyclotron frequency, this wave is a high-frequency instability and it grows close to the ion-electron resonance at ω_{lh}. This can be understood from the fact that due to the high frequency the ions are unmagnetized and cannot have ion-cyclotron resonances, but the effect of the magnetic field is introduced via the magnetization of the electrons.

4.6. Electrostatic Cyclotron Waves

Loss cones are a general feature of trapped particle distributions in mirror-type magnetic configurations. These configurations are encountered in the global magnetospheric fields of planets, the global loop magnetic fields of the solar corona and, on smaller scales, in local field depletions in the polar cusps and magnetotail. In all these places electrons and ions are trapped locally and bounce back and forth between the magnetic mirror points and electrostatic cyclotron waves are excited if the density of a cold background distribution is not too high.

Ring Current

In the ring current, outside the plasmasphere, the conditions for excitation of electrostatic cyclotron waves by loss cone distributions are ideal, especially when plasma injections caused by enhanced convection from the tail into the inner magnetosphere enhance the ring current. Figure 4.12 shows electron-cyclotron waves excited during enhanced convection at a distance of 6–8 R_E in the post-midnight magnetosphere. Ion-cyclotron waves are excited in the ring current region, too. But their frequencies are

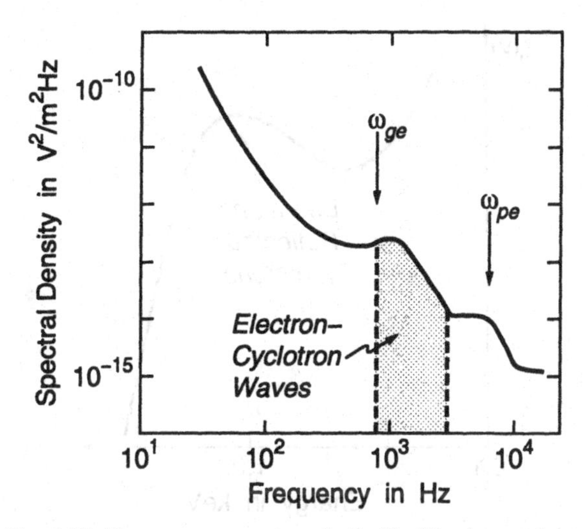

Fig. 4.17. Plasma wave spectrum in the Earth's plasma sheet.

much lower than those of the electrons and are barely resolved by current instrumentation. One can only observe their secondary effects on the ion distribution.

Cyclotron waves play an important role in the dynamics of the magnetospheric plasma. They extract the free energy from the loss cone distribution, thereby filling the loss cone. The interaction between the electrostatic cyclotron waves and the hot component causes a continuous flux of particles, electrons and ions, to be scattered into the loss cone, where they will precipitate into the ionosphere and are lost by collisions.

The region in the equatorial plane where this happens lies outside the plasmapause and overlaps with the auroral zone. This suggests that part of the precipitating auroral electron flux is caused by the excitation of electrostatic electron-cyclotron waves in the nightside equatorial ring current region, and that the excitation mechanism of the diffuse aurora is electrostatic electron-cyclotron wave excitation by ring current electron loss cone distributions. Figure 4.16 sketches the geometry of this interaction.

Plasma Sheet

Wave measurements in the undisturbed plasma sheet exhibit the presence of a broad band of electron-cyclotron waves reaching from the electron gyrofrequency to the electron plasma frequency (see Fig. 4.17). Due to instrumental resolution the harmonics are not resolved, but the bump on the spectrum is a clear indication of their presence. That electron-cyclotron harmonics are excited in the plasma sheet indicates that local magnetic mirror configurations with electron loss cone distributions must have evolved there. Moreover, the plasma must consist of two populations, hot plasma sheet electrons

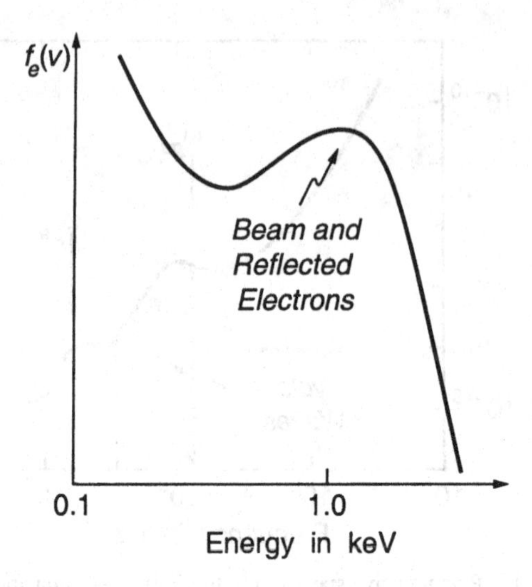

Fig. 4.18. Auroral electron energy spectrum.

and a dilute cold electron component. Since the electrostatic electron-cyclotron waves have frequencies up to the cold upper-hybrid frequency, one can use the high-frequency cut-off of the bump on the spectrum to estimate the ratio between cold and total electron density. In the present case we have $n_{ec}/n_e \approx 0.15$.

Electrostatic Hiss

Electrostatic cyclotron waves produce a continuous flux of charged particles into the loss cone and thus a continuous flux of particles precipitating along the magnetic field into the auroral ionosphere. Outside the equatorial wave excitation region, these particle fluxes are excess fluxes parallel to the magnetic field and will give rise to the excitation of beam-driven high-frequency electrostatic waves and also of low-frequency broad-band emissions called *electrostatic hiss*.

Figure 4.18 shows a precipitating auroral electron beam distribution function. The bump in the distribution is caused by the superposition of the primary electron beam and the secondary electrons backscattered from the ionosphere. Clearly, a positive slope exists on the distribution function in this three-component system consisting of dense ionospheric cold electrons, the beam and the backscattered electrons. This positive slope will spontaneously emit waves by the Cherenkov mechanism in every frequency range where the Cherenkov condition in Eq. (2.12) is satisfied. This condition requires large indices of refraction. For particles of auroral energies $N_\parallel = \omega/k_\parallel c \approx 10^{-2}$. In a dense plasma the emission will be predominantly in the R- and X-modes below the

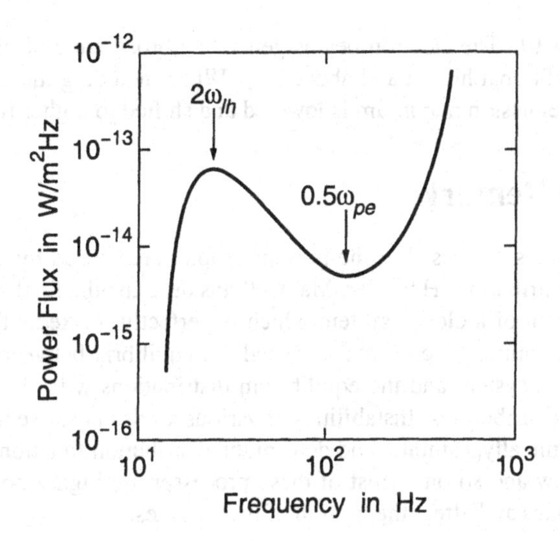

Fig. 4.19. Power spectrum of whistler-band auroral hiss.

upper-hybrid frequency and in the whistler mode near the resonance cone angle

$$\cos^2\theta_{\mathrm{res}} = \left(1 + \frac{\omega_{ge}^2}{\omega_{pe}^2}\right)\frac{\omega^2}{\omega_{ge}^2} - \frac{m_e}{m_i}\left(1 - \frac{\omega^2}{\omega_{ge}^2}\right) - \frac{\omega^4}{\omega_{ge}^2\omega_{pe}^2} \tag{4.72}$$

These waves propagate between the lower-hybrid frequency, ω_{lh}, and ω_{ge} or ω_{pe}. For dense plasmas the upper limit of the emission is ω_{ge}, for dilute plasmas ω_{pe}.

The growth rate of the low-frequency Cherenkov emission can be calculated from the dielectric response function which for the three components in the inhomogeneous ionospheric plasma is a complicated expression. The waves propagate on the dense background, which entirely determines their propagation properties, while the beam is responsible for their emission. Hence, one can assume weak instability and obtain

$$\gamma = \frac{\omega^3}{2k^2\cos^2\theta}\sum_s \frac{n_s}{n_0}\frac{\mathrm{Im}Z'(\zeta_s)}{v_{\mathrm{th}s}^2} \tag{4.73}$$

where s counts the beam electrons, the backscattered electrons, and the ionospheric ions. This expression breaks down near the two resonances at the electron-cyclotron and the lower-hybrid frequency.

The field-aligned propagation of the waves in the inhomogeneous ionosphere can be taken into account by assuming that the growth rate and frequency change in the same way as the density and the magnetic field. Then one can calculate the emitted power from Eq. (2.15) for given velocity distributions. A numerical calculation is shown in Fig. 4.19. The electron beams generate a broad band of electrostatic hiss above ω_{lh} and additional hiss at and above ω_{pe}. When including more secondaries in the electrons, the emission maximum is lowered and shifted to higher frequencies.

Concluding Remarks

The microinstabilities discussed in the present chapter are caused by deviations from the Maxwellian distribution. However, Maxwellians describe the final state of thermodynamic equilibrium of a closed system which is perfectly relaxed. This situation is hardly realized in nature. The normal case will be equilibria between energy inflow and dissipation in a system and the equilibrium distributions will often not resemble ideal Maxwellian distributions. Instabilities of various kinds may arise in these equilibria, compete dynamically, saturate, and deform the distribution function, causing other instabilities to grow and so on. Most of these processes are highly nonlinear, couple different wave modes and affect the motion of the particles.

Further Reading

A number of microinstabilities are reviewed in [1], [4], and [5], where the latter is a standard reference for most of the linear plasma instabilities. Ionospheric instabilities are discussed in connection with electrojet theory in [2]. The current-drift velocity thresholds for ion-cyclotron modes were first calculated by Kindel and Kennel, *J. Geophys. Res.* **76** (1971) 3055. The linear and nonlinear theory of electrostatic loss cone-driven cyclotron modes is found in [3]. Cyclotron waves and instabilities are found in [6]. Power spectra of whistler-band auroral hiss were calculated by Maggs, *J. Geophys. Res.* **34** (1976) 1707, and the plasma sheet wave spectra are from Baumjohann et al., *J. Geophys. Res.* **95** (1990) 3811.

[1] A. Hasegawa, *Plasma Instabilities and Nonlinear Effects* (Springer Verlag, Heidelberg, 1975).

[2] M. C. Kelley, *The Earth's Ionosphere* (Academic Press, San Diego, 1989).

[3] C. F. Kennel and M. Ashour-Abdalla, in *Magnetospheric Plasma Physics*, ed. A. Nishida (D. Reidel Publ. Co., Dordrecht, 1982), p. 245.

[4] D. B. Melrose, *Instabilities in Space and Laboratory Plasmas* (Cambridge University Press, Cambridge, 1991).

[5] A. B. Mikhailovskii, *Theory of Plasma Instabilities I* (Consultants Bureau, New York, 1974).

[6] T. H. Stix, *Waves in Plasma* (American Institute of Physics, New York, 1992).

5. Electromagnetic Instabilities

We now turn to the wide field of electromagnetic instabilities caused by velocity space inhomogeneities or a deformation of the phase space distribution function. We assume homogeneous plasma conditions and straight field lines. The examples chosen below are selected from the space plasma physics viewpoint.

5.1. Weibel Instability

In addition to electrostatic instabilities, electromagnetic instabilities may also develop in an unmagnetized plasma if specific conditions are satisfied. The only electromagnetic mode is the ordinary mode. The instability causing its growth is known as the *Weibel instability*. It is driven by a particular electron velocity distribution in the presence of immobile neutralizing ions. In the nonrelativistic case, one can model it as

$$f_e(v_\perp, v_\parallel) = \frac{n_0\delta(v_\parallel)}{2\pi v_{\mathrm{the}\perp}^2}\exp\left(-\frac{v_\perp^2}{2v_{\mathrm{the}\perp}^2}\right) \tag{5.1}$$

where the directions are given with respect to the direction of the wave vector, since this is the only preferred direction in an unmagnetized non-streaming plasma.

Dispersion Relation

Using the above distribution and the dielectric response function from Eq. (I.9.59) in our companion book, *Basic Space Plasma Physics*

$$\begin{aligned} \epsilon_{\mathrm{L}}(\omega, \mathbf{k}) &= [\mathbf{k}\cdot\boldsymbol{\epsilon}(\omega,\mathbf{k})\cdot\mathbf{k}]/k^2 \\ \epsilon_{\mathrm{T}}(\omega, \mathbf{k}) &= [\mathrm{tr}\boldsymbol{\epsilon}(\omega,\mathbf{k}) - \epsilon_{\mathrm{L}}(\omega,\mathbf{k})]/2 \end{aligned} \tag{5.2}$$

it is easy to calculate the transverse response function

$$\epsilon_{\mathrm{T}}(\omega, \mathbf{k}) = 1 - \frac{\omega_{pe}^2}{\omega^2}\left[1 + \frac{k^2 v_{\mathrm{the}\perp}^2}{\omega^2}\left(1 - N^{-2}\right)\right] \tag{5.3}$$

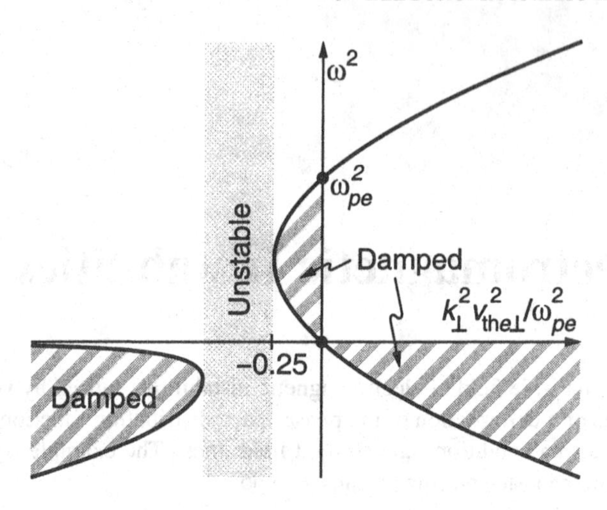

Fig. 5.1. Unstable dispersion branch of the Weibel instability.

which inserted into the dispersion relation, $N^2 = \epsilon_T(\omega, \mathbf{k})$, yields

$$\frac{\omega^4}{\omega_{pe}^4} - \frac{\omega^2}{\omega_{pe}^2}\left(1 + \frac{k^2c^2}{\omega_{pe}^2}\right) - \frac{k^2 v_{\text{the}\perp}^2}{\omega_{pe}^2} = 0 \tag{5.4}$$

For $k = 0$ the solution of this equation is the Langmuir oscillation, $\omega = \pm\omega_{pe}$. For $k \neq 0$ the unstable branch of the dispersion relation is schematically shown in Fig. 5.1. For negative square of the wavenumber instability may arise. This is a convective instability, because the wave moves at light velocity and cannot be excited locally.

Growth Rate

Solving for the unstable mode at frequency near $\omega = \omega_{pe}/2$, one finds under the condition that $|kv_\perp/\omega_{pe}| > 0.5$

$$\boxed{\gamma_{\text{wei}} \approx \sqrt{2}kv_{\text{the}\perp}\left(1 + k^2c^2/\omega_{pe}^2\right)^{-1/2}} \tag{5.5}$$

This growth rate is small, since it depends only on the perpendicular thermal velocity of the plasma. Moreover, the electron inertial length enters the growth rate, suggesting that the instability is related to the penetration depth of the wave into the plasma. Finally, a parallel thermal spread of the electron velocity distribution will tend to suppress the instability, because the parallel electric wave field will cause Landau damping. For non-zero parallel electron temperature, $T_{e\|}$, we can write the dispersion relation

$$k^2c^2 = \omega_{pe}^2 A_e - (A_e + 1)\omega_{pe}^2 \zeta_e Z(\zeta_e) \tag{5.6}$$

where $\zeta_e = i\gamma/kv_{\mathrm{the}\parallel}$. In the small argument limit of the plasma dispersion function, $Z \approx -2\zeta_e$, suggested by weak instability condition, this dispersion relation can be solved for the growth rate

$$\gamma^2 = \frac{k^2 v_{\mathrm{the}\parallel}^2}{2} \frac{k^2}{k_0^2} \left(1 - \frac{k_0^2}{k^2}\right) \frac{A_e}{A_e + 1} \tag{5.7}$$

Here $k_0 = A_e^{1/2}\omega_{pe}/c$ is the short wavelength cut-off of the Weibel instability. Its growth rate increases for long wavelengths, but has a maximum at $k = k_0/\sqrt{2}$

$$\boxed{\gamma_{\mathrm{wei,max}} = \frac{k_0 v_{\mathrm{the}\parallel}}{2\sqrt{2}} \left(\frac{A_e}{A_e + 1}\right)^{1/2}} \tag{5.8}$$

The short wavelength cut-off is given by the anisotropy and the electron inertial length

$$k_0^2 = A_e \omega_{pe}^2 / c^2 \tag{5.9}$$

There is an interesting similarity between the Weibel instability growth rate and the behavior of the firehose instability growth rate at finite frequency. Both instabilities have short wavelength cut-offs and are driven by temperature anisotropies and, although their nature is very different, Fig. 3.8 applies also to the Weibel instability.

5.2. Anisotropy-Driven Instabilities

Electromagnetic waves at frequencies below the electron-cyclotron frequency are cyclotron waves. These waves can be excited by several means. The most interesting are excitation by temperature anisotropies and loss cones. Below we will treat only the former one. As we have shown loss cones and temperature anisotropies are closely related.

The most famous electromagnetic velocity space instability is the instability of parallel propagating low-frequency electromagnetic waves of right- and left-circular polarization, the R- and L-modes, introduced in Eqs. (I.9.122) and (I.9.127) of the companion volume

$$N_{\mathrm{R,L}}^2 = 1 - \omega_{pe}^2 / [\omega(\omega \mp \omega_{ge})] \tag{5.10}$$

where the negative and positive sign stand for the right-hand and left-hand polarized mode, respectively. The electric wave vector of these waves is perpendicular to the magnetic field and rotates during propagation of the wave along the magnetic field in the same sense as the corresponding particle component. In particular, the R-mode electric field vector rotates in the same sense as electrons gyrate around the magnetic field, and the L-mode has the same sense of rotation as the gyration of ions.

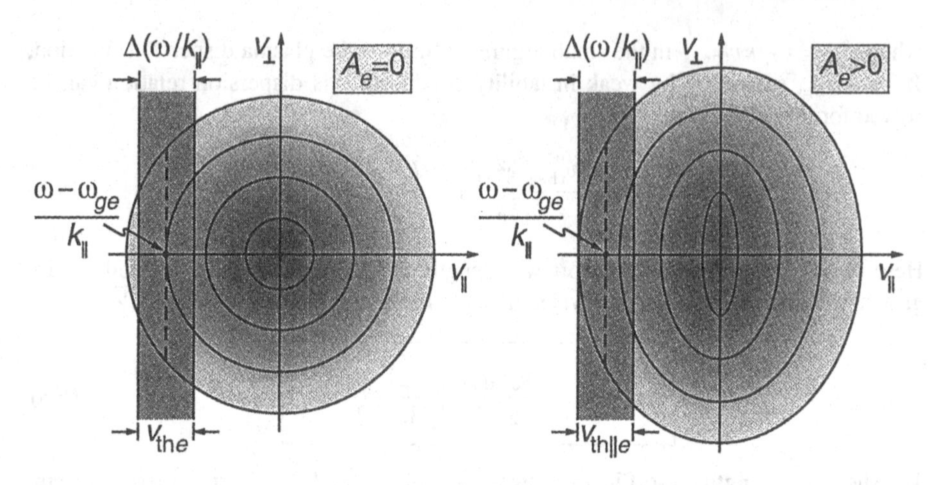

Fig. 5.2. Resonant region of whistlers in velocity space.

Cyclotron Resonance

Hence, it is easy to imagine that particles of the right species with a particular parallel velocity will see a constant perpendicular electric wave field in their frame of reference and will undergo strong interaction with the wave. This is the nature of *cyclotron resonance*. In order to provide unstable conditions one must chose a distribution function such that it has an excess of particles with a higher momentum than the wave. These particles will be retarded by the wave and decelerated in the wave electric field, thereby feeding the instability. The resonance condition is simply that

$$k_{\|}v_{\|} = \omega - l\omega_{gs} \tag{5.11}$$

the parallel velocity of the particle in resonance equals the parallel phase velocity of the wave, Doppler shifted by the corresponding lth harmonic of the cyclotron frequency. In other words, in the frame of the particle moving with resonant velocity the frequency of the wave is equal to the lth harmonic of the cyclotron frequency, and the electric field vector rotates at l times the rate of the particle rotation around the magnetic field. Note that for l=0 the resonance condition becomes the usual Landau resonance.

We have already used this resonance condition when introducing electrostatic cyclotron waves. Clearly, for e.g., $l = 1$ and in perfect resonance, $\omega = \omega_{gs}$, the particle will be at rest in the wave frame, all the time seeing the wave at the same phase. But if there is a finite difference in the frequencies the resonance condition requires that the difference picks out such a phase of the wave rotation that the particle is maintained just in phase with the electric field, thus experiencing maximum electric effect. Implicit in this assumption is that the perpendicular wavelength of the wave is larger than

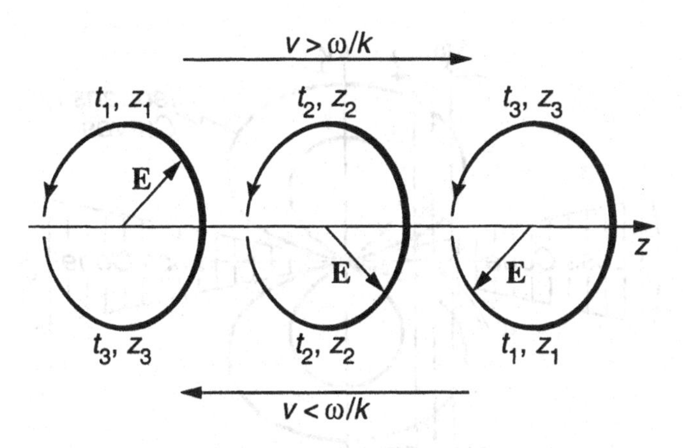

Fig. 5.3. Cyclotron resonance mechanism.

the gyroradius of the particles, warranting that the particles are magnetized. But this condition is irrelevant for strictly parallel propagation, because then the wave field is homogeneous in the perpendicular direction with infinite perpendicular wavelength.

The resonance condition is not as easy to understand intuitively as Landau resonance. This has to do with the coupling of parallel and perpendicular particle motion in the resonance. Firstly, it cannot be satisfied for arbitrary velocities. For instance, for electron whistlers with $l = 1$ and no harmonics at larger l, the whistler frequency given in Eq. (I.9.125), $\omega_{\mathrm{w}} < \omega_{ge}$, is smaller than the electron-cyclotron frequency, and $\omega_{\mathrm{w}} - \omega_{ge}$ is negative, implying that electrons in resonance with whistlers propagate opposite to the whistler along the magnetic field.

In a velocity space representation like that of Fig. 5.2, the resonant region is located in the negative $v_{\|}$ plane. For an isotropic Maxwellian plasma with $A_e = 0$ and $T_{e\perp} = T_{e\|}$, the isodensity contours of the velocity are circles, and the width of the resonant region can be estimated to be of the order of the isotropic electron thermal velocity, v_{the}. For an anisotropic distribution the width is the parallel thermal speed, $v_{\mathrm{the}\|}$. In the isotropic case there is no free energy available and no instability can arise.

The physical mechanism of cyclotron damping and instability is sketched in Fig. 5.3. A right-hand circularly polarized wave is moving along the magnetic field, in the z direction. The electric field vector rotates as shown when the wave moves from left to right. In the electron frame, where the electron is stationary, the wave is passing to the right if the electron is slow with respect to the wave and to the left if the electron is faster than the wave. A slow electron will see wave vectors in the succession shown by the arrow in the lower part of the figure. Hence, the electric field experienced by the electron rotates in the same direction as the electron. In this case the electron in resonance will be accelerated by the wave field in the perpendicular direction, leading

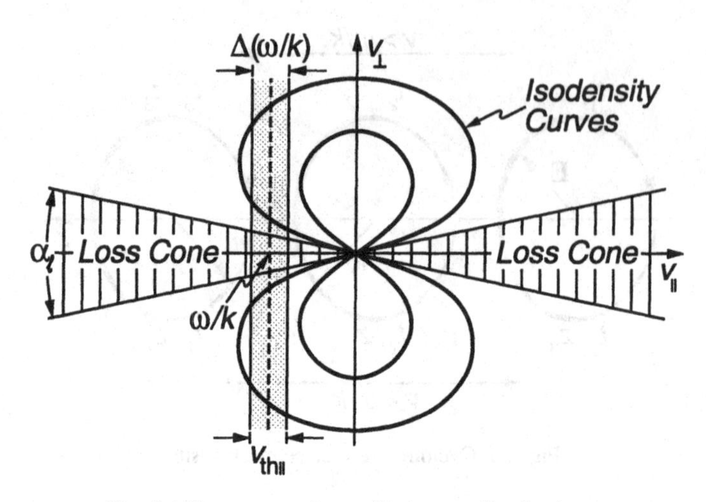

Fig. 5.4. Resonant region and loss cone distribution.

to wave damping. A fast electron will see the wave rotating as shown in the upper part of the figure, which is opposite to the sense of rotation of the electron. Hence though the electron is in resonance with the wave, the rotational sense is wrong and the electron does not see a stationary electric field and does not interact.

Resonant cyclotron interaction is thus favored if there are many particles at low velocities. Figures 5.2 and 5.4 show two examples of phase space isodensity plots of electrons in interaction with whistlers. Figure 5.2 shows the cases of an isotropic and of an anisotropic distribution. It will be shown below that a temperature anisotropy with excess in the perpendicular velocity, like in the right-hand side distribution, leads to cyclotron instability, with the free energy stored in the temperature anisotropy as the source. Figure 5.4, a loss cone distribution, is another example.

Whistler Instability

Before turning to the explicit calculation of the whistler growth rate, we may ask ourselves what the cyclotron resonance condition Eq. (5.11) implies for the resonant energies of the electrons. Squaring and multiplying it by $m_e/2$ and using the whistler dispersion relation in Eq. (5.10) in order to eliminate the parallel wavenumber, we can construct the parallel resonant electron energy

$$W_{e\|\mathrm{res}} = W_B \frac{\omega_{ge}}{\omega}\left(1 - \frac{\omega}{\omega_{ge}}\right)^3 \qquad (5.12)$$

where we introduced the magnetic energy per particle, $W_B = B^2/2\mu_0 n$. Thus the particles in resonance with the whistler have a particular parallel energy.

Let us assume that the plasma contains a hot electron component with positive temperature anisotropy, $A_e > 0$. This component should be of sufficiently low density, $n_h \ll n_0$, in order to not disturb the propagation of the whistler mode. In such a case the dispersion of the whistler wave is still described by the dispersion relation (I.10.164). In other words, the whistlers propagate on the background of the cold and isotropic plasma component. Since we require this background electron plasma to be sufficiently cold, the cyclotron damping term in Eq. (I.10.165) shows that the waves are practically undamped for $v_{\mathrm{thec}} \to 0$. Whistler instability may arise due to the free energy stored in the temperature anisotropy of the hot electron component.

Dispersion Relation

Retaining the anisotropy forces allows to use the right-hand dispersion relation of a anisotropic Maxwellian plasma derived in Eq. (I.10.159) of our companion volume

$$\frac{k_\parallel^2 c^2}{\omega^2} = 1 + \sum_{s=c,h} \frac{\omega}{k_\parallel v_{\mathrm{th}\parallel s}} \frac{\omega_{ps}^2}{\omega^2} \left[Z(\zeta_{s,1}) - \frac{k_\parallel v_{\mathrm{th}\parallel s}}{\omega} A_{es} Z'(\zeta_{s,1}) \right] \tag{5.13}$$

We need to consider only the imaginary part of this dispersion relation when looking for weak growth entirely provided by the hot component. Using the large argument expansion of the hot electron plasma dispersion function, the imaginary part of the dispersion relation can be written as

$$D_i(\omega, k_\parallel) = \frac{\sqrt{\pi}\omega}{|k_\parallel| v_{\mathrm{th}h\parallel}} \frac{\omega_{ph}^2}{\omega^2} \left[1 - A_e \left(1 - \frac{\omega_{ge}}{\omega} \right) \right] \exp\left[-\frac{(\omega - \omega_{ge})^2}{k_\parallel^2 v_{\mathrm{th}h\parallel}^2} \right] \tag{5.14}$$

The real part of the dispersion relation is taken in its simplified version, since we assumed a cold dense background plasma

$$D_r = \frac{k_\parallel^2 c^2}{\omega^2} - 1 + \frac{\omega_{pc}^2}{\omega(\omega - \omega_{ge})} \tag{5.15}$$

Growth Rate

Weak instability is determined by the relation $\gamma = -D_i/(\partial D_r/\partial\omega)|_{\gamma=0}$. Performing the simple algebraic calculations yields the whistler growth rate

$$\gamma_{\mathrm{w}} = \frac{\sqrt{\pi}\omega_{ph}^2}{|k_\parallel| v_{\mathrm{th}h\parallel}} \left[1 - A_e \left(1 - \frac{\omega_{ge}}{\omega} \right) \right] \exp\left[-\frac{(\omega - \omega_{ge})^2}{k_\parallel^2 v_{\mathrm{th}h\parallel}^2} \right] \left[2 + \frac{\omega_{pc}^2 \omega_{ge}}{\omega(\omega - \omega_{ge})^2} \right]^{-1} \tag{5.16}$$

Two interesting conclusions can be drawn from this expression. First, since the whistler frequency depends on the cold plasma frequency, the growth rate is proportional to the ratio of hot-to-cold electron densities, $\gamma \propto n_h/n_c$. The second and more important conclusion is that the growth rate is positive only under the condition that

$$\boxed{A_e > A_{ec} = \omega/(\omega_{ge} - \omega)} \tag{5.17}$$

where A_{ec} is the critical anisotropy for instability. This condition can also be written as a condition on the unstable frequency range for a given anisotropy

$$\boxed{\omega < \omega_c = \omega_{ge} A_e/(A_e + 1)} \tag{5.18}$$

where ω_c is the critically unstable frequency. Using this expression in the low-frequency whistler dispersion relation gives a lower limit on the whistler wavelength

$$\boxed{k_\parallel^2 < k_{\parallel c}^2 = \omega_{pc}^2 A_e/c^2} \tag{5.19}$$

For a given temperature anisotropy of the hot electron distribution function there will always be an unstable range of frequencies and wavenumbers, which will be excited and will deplete the temperature anisotropy. Clearly, this frequency range is below the electron-cyclotron frequency and may be found at very low frequencies. The lowest reasonable frequency, up to which the ion effects can be neglected, is the lower-hybrid frequency, ω_{lh}. Hence, this theory is valid up to anisotropies as small as $A_e \approx 0.02$.

Threshold

We are now in the position to estimate the lowest threshold for the resonant energy of the electrons for instability. Inserting the critical frequency into Eq. (5.12) produces

$$\boxed{W_{e\parallel \mathrm{res}} > \frac{W_B}{A_e(A_e + 1)^2}} \tag{5.20}$$

For resonance and instability the electron energy must exceed the limit set by this condition. Otherwise instability will be inhibited. Low anisotropies require very high electron energies to drive whistlers unstable. In the magnetosphere the anisotropies are relatively low, of the order of $A_e \approx 0.1 - 0.5$ in the electron radiation belts and even lower in the near-Earth plasma sheet.

Parallel energies must thus exceed the Alfvén energy several times in order to generate whistlers. But this condition is easily satisfied by the trapped radiation belt particle component and, under certain favorable conditions, also in the near-Earth plasma sheet. In these regions broadband low-frequency whistler noise is generated and frequently observed as one of the fundamental electromagnetic low-frequency emissions. These waves have a strong effect on the trapped and quasi-trapped particle distributions of the radiation belt and the plasma sheet electrons, leading to enhanced precipitation of hot electrons during excitation of whistler mode noise due to enhanced anisotropies.

Since enhanced anisotropies are caused by convection of plasma from the tail into the inner magnetosphere, as we have shown in Eq. (I.2.63) of our companion book, increase of the cross-tail electric field during substorms will often be associated with whistler mode noise excitation in the near-Earth plasma sheet and enhanced hot electron precipitation into the diffuse aurora. Similarly, trapping of large amounts of very energetic electrons in the Earth's radiation belts will also lead to enhanced whistler noise generation at lower latitudes and cause radiation belt electrons to precipitate. Hence, whistler mode noise is a very important resonator for anisotropic energetic electrons in the magnetosphere.

Ion-Cyclotron Waves

An entirely equivalent theory can be developed for the left-hand polarized parallel propagating mode at low frequencies, the electromagnetic ion-cyclotron wave. It is easy to follow the same reasoning as presented for whistlers, using only the different definition of the ion-cyclotron dispersion relation (I.9.140) given in the companion volume, to find that ion-cyclotron waves of frequency $\omega < \omega_{gi}$ also experience instability in anisotropic hot ion plasmas superimposed on a cold isotropic ion background. The dispersion relation in this case includes the effect of the electrons and becomes

$$\begin{aligned} D(\omega, k_\parallel) &= 1 - \frac{k_\parallel^2 c^2}{\omega^2} + \frac{\omega_{pe}^2}{\omega_{ge}^2} - \frac{\omega_{pe}^2}{\omega_{ge}\omega} + \\ &+ \frac{\omega_{pi}^2}{\omega^2}\left\{\frac{\omega}{k_\parallel v_{\mathrm{th}i}} Z(\zeta_i) + A_i\left[1 + \zeta_i Z(\zeta_i)\right]\right\} \end{aligned} \tag{5.21}$$

where $\zeta_i = (\omega - \omega_i)/k_\parallel v_{\mathrm{th}i}$ and charge neutrality requires that

$$\frac{\omega_{pe}^2}{\omega_{ge}} = -\frac{\omega_{pi}^2}{\omega_{gi}} \tag{5.22}$$

When neglecting the effect of the electrons, the solution to this dispersion relation simply parallels that of the electron whistler case. Ion-cyclotron instability driven by temperature anisotropies of hot ions is found in the frequency range

$$\boxed{\omega < \omega_c = \omega_{gi} A_i/(A_i + 1)} \tag{5.23}$$

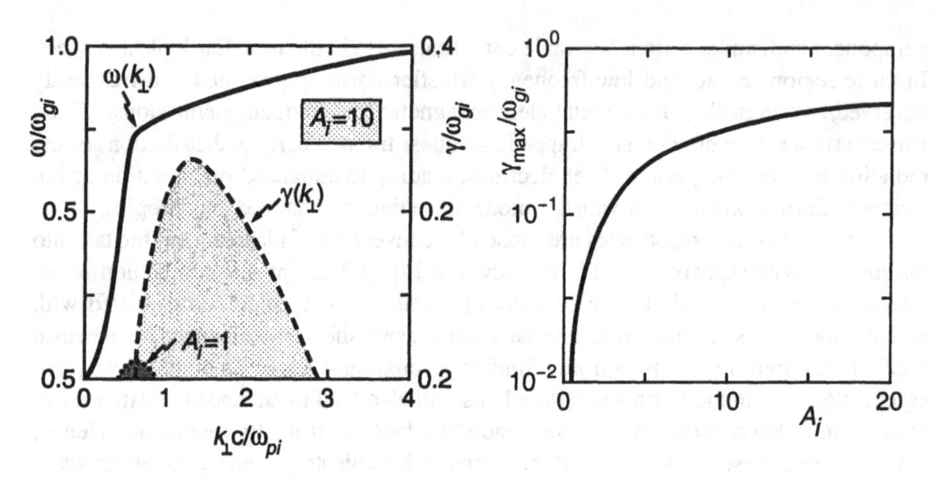

Fig. 5.5. Frequency and growth rate of the anisotropic ion-cyclotron instability.

The particle energy condition is slightly modified, because of the differences in the electron and ion dispersion relations

$$W_{i\parallel\mathrm{res}} > \frac{W_B}{A_i^2(A_i+1)} \tag{5.24}$$

For very large anisotropy one can drop the parallel temperature and find the approximate dispersion relation

$$k_\parallel^2 c^2 = \frac{\omega_{pi}^2}{\omega_{gi}}\frac{\omega^2}{(\omega-\omega_{gi})} - \frac{k_\parallel^2 c^2 \beta_{i\perp}}{2}\frac{\omega_{gi}^2}{(\omega-\omega_{gi})^2} \tag{5.25}$$

where $\beta_{i\perp} = 2\mu_0 n k_B T_{i\perp}/B_0^2$ is the perpendicular ion plasma beta. Solution of this cubic equation yields instability at maximum growth for $k_\parallel^2 \gg \omega_{pi}^2/c^2$, or wavelengths much shorter than the ion inertial length. This very short wavelength waves have growth rates

$$\gamma_{\mathrm{aic}} \approx \omega_{gi}\sqrt{\beta_{i\perp}/2} \tag{5.26}$$

Growth rates calculated for the more realistic case $T_{i\parallel} \neq 0$ are schematically shown for the *anisotropic ion-cyclotron instability* in Fig. 5.5 for $\beta_{i\perp} = 1$. Maximum growth is obtained at wavelength close to the ion inertial length, with rapidly decreasing growth rate for lower anisotropies at longer wavelengths. These lower anisotropies are more realistic in space plasmas. The dependence of the maximum growth rate on the anisotropy

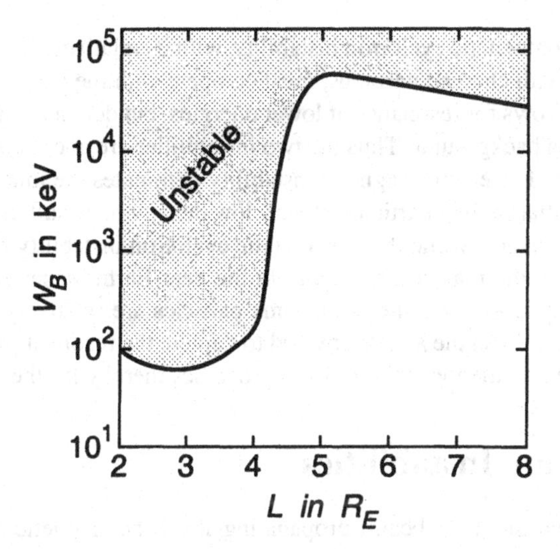

Fig. 5.6. Threshold resonant magnetic energy in the equatorial magnetosphere.

under the same conditions demonstrates the steep cut-off of the instability at low ion anisotropies. At such small anisotropies the wave becomes a very-low frequency wave.

The ion energies required for resonance turn out to be considerably higher than for electrons because of the A_i^{-2} dependence in Eq. (5.24). Only the very energetic radiation belt ions generate ion-cyclotron waves due to trapped ion anisotropies and provide a mechanism of controlling energetic ion precipitation into the mid-latitude ionosphere.

Electromagnetic vs. Electrostatic

Figure 5.6 shows the approximate variation of the magnetic energy per particle in the equatorial plane of the Earth's magnetosphere (in this figure we used the average densities shown in Fig. 5.7 of our companion book). In the plasmasphere the magnetic energy is comparably low. In this region resonance of electrons and ions with whistlers and ion-cyclotron waves is easier than outside the plasmasphere. Particles having resonant energies below 100 keV can become unstable near $L = 3$. During injection events and compression of the plasmasphere the threshold energy decreases, and instability may occur for even lower energies. The sensitivity of the threshold resonance energy on the background density is rather impressive. Even modest cold plasma injection into a region outside the plasmasphere will cause a drastic decrease in the resonant particle energy and will cause both electrons and protons entering cyclotron resonance with the electromagnetic whistler and ion-cyclotron waves.

But the electromagnetic cyclotron instability, with its relatively high energy threshold for resonance with charged particles, must compete with the electrostatic cyclotron instability which allows for resonance at lower energies, but depends on the presence of a denser hot plasma background. Thus the two instabilities are about complementary in the magnetosphere. The electromagnetic instability dominates the inner plasmasphere, where energetic radiation belt particles of very low density have sufficiently high energies to fall into resonance, while the electrostatic cyclotron instability dominates in the ring current, outside the plasmasphere, and in the near-Earth plasmasheet, where the cold plasma density is low and the hot plasma particles are relatively frequent. Both instabilities try to diminish the anisotropy and to deplete the resonant particles, causing them to precipitate into the ionosphere. Both processes thereby fill the loss cone.

5.3. Ion Beam Instabilities

In this section we consider ion beams propagating along the magnetic field as a source of low-frequency electromagnetic waves. In the linear regime these waves are beam-excited ion-cyclotron modes, which at low frequencies make the transition to Alfvén waves and the two other magnetohydrodynamic modes, the fast and slow modes. These waves and their ionic excitation mechanisms are important in all places where shock waves appear. It is widely believed that they are responsible for shock formation and regeneration. We briefly discuss how instability can arise in these modes. There are three types of instabilities, the R-resonant, L-resonant, and the non-resonant beam instabilities. Figure 5.7 shows schematically the ion velocities for the two resonances. The resonances are found at

$$v_{\mathrm{R,L},res} = (\omega \pm \omega_{gi})/k_{\|} \tag{5.27}$$

Ion-Ion R-mode Instability

Let us assume that the plasma consists of three components, a hot Maxwellian electron distribution and two drifting Maxwellian ion distributions. These must not necessarily be of same temperature or density. The core distribution may be denser, the beam distribution more dilute. This implies $n_{ic} \gg n_{ib}$. The core distribution is also assumed to be slow, $v_{ic} \ll v_{ib}$. We are interested only in the parallel wave propagation. In this subsection we consider merely the right-hand mode which can be excited by the ion interaction. As before, for parallel propagation of waves the only possible resonance is the l=±1 cyclotron resonance, which implies that resonant interaction occurs for $\zeta_{s,1} \leq 1$, with $s = e, c, b$.

When the ion beam is cool, $v_b \gg v_{\mathrm{th}b}$, only the beam is resonant, and both the electrons and core ions satisfy $\zeta_{e,1} \gg \zeta_{c,1} \gg 1$. The resonance condition for the beam

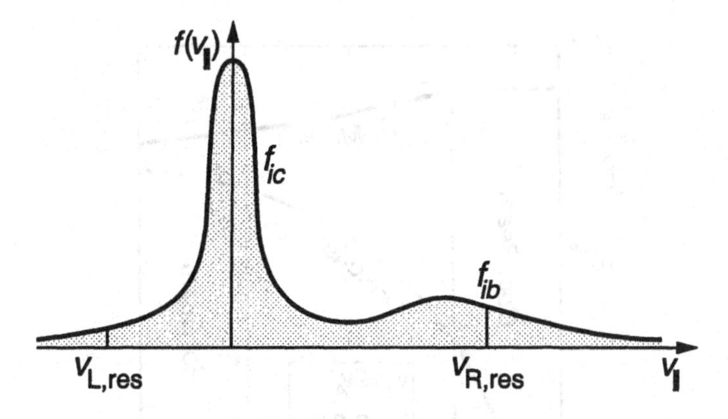

Fig. 5.7. Ion beam resonances with R- and L-modes.

with the R-mode is

$$\omega = k_\parallel v_b - \omega_{gi} \tag{5.28}$$

and the mode propagates along the beam with $k_\parallel > 0$. The growth rate for this instability is found numerically to be of the order of the frequency, $\gamma \approx \omega$. The real part of the dispersion relation under these conditions is

$$\frac{k^2c^2}{\omega^2} = 1 - \sum_s \frac{\omega_{ps}^2(\omega - k_\parallel v_{sb})}{\omega^2(\omega \pm \omega_{gs} - k_\parallel v_{sb})} \tag{5.29}$$

and the weakly unstable growth rate for drifting Maxwellians is obtained by replacing $\omega \rightarrow \omega - k_\parallel v_{sb}$ in the imaginary part of the dispersion relation. This yields

$$\gamma = \omega\frac{\sqrt{\pi}}{2}\sum_s \frac{\omega_{ps}^2}{\omega^2}\left(\frac{v_{sb}}{v_{\mathrm{th}s}} - \frac{\omega}{k_\parallel v_{\mathrm{th}s}}\right)\exp\left[-\frac{(\omega \pm \omega_{gs} - k_\parallel v_{sb})^2}{k_\parallel^2 v_{\mathrm{th}s}^2}\right] \tag{5.30}$$

Under the above assumptions and for $v_{ib} \gg v_A$, the maximum growth rate is

$$\boxed{\gamma_{\mathrm{iib,max}} = \omega_{gi}\left(\frac{n_b}{2n_0}\right)^{1/3}} \tag{5.31}$$

This electromagnetic *ion-ion beam R-mode instability* resembles the electrostatic beam instability in Eq. (2.37). Wave growth is obtained above a threshold beam speed of

$$v_{ib} \geq v_A \tag{5.32}$$

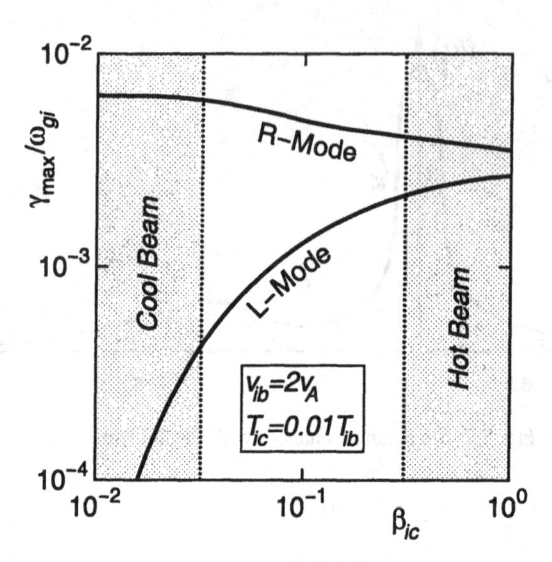

Fig. 5.8. Maximum growth rates for ion beam-resonant modes.

The excited wave is essentially a right-hand circularly polarized Alfvén whistler with long wavelength dispersion

$$\omega \approx k_{\parallel} v_A \tag{5.33}$$

and positive helicity. As discussed earlier, for larger angles of propagation this mode goes smoothly over into the magnetosonic mode. For larger wavenumbers it becomes the usual whistler.

Ion-Ion L-mode Instability

The other important ion beam instability is the resonance with the left-hand polarized mode, which at long wavelength and small wavenumbers is the ion whistler or ion-cyclotron mode. It has negative helicity and propagates parallel to the beam. Furthermore, the electron and the core distributions are non-resonant, such that the plasma dispersion function is of the large-argument type. The numerical solution of the dispersion relation for this case shows that the threshold speed for instability is, as for the right-hand mode, also the Alfvén velocity.

However, it is easier to excite the right-hand mode under cool beam conditions because at the low thermal velocities of the beam there are only few ions which can resonate with the left-hand mode. Therefore cold beams will predominantly generate right-hand waves. An increase of the beam temperature raises the number of particles which can resonate with the L-mode, and its growth rate becomes comparable to that of

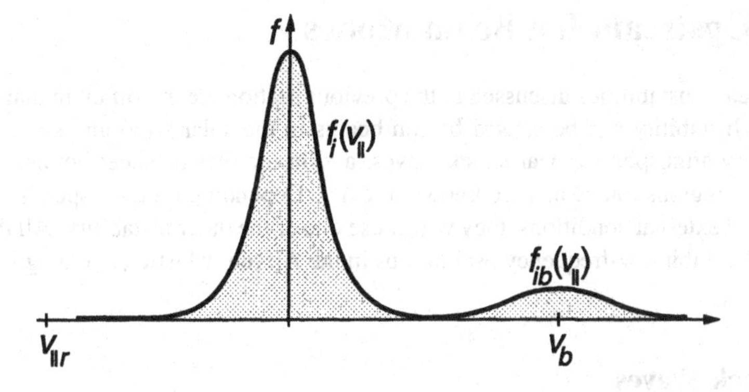

Fig. 5.9. Configuration of the non-resonant ion beam instability.

the R-mode. Figure 5.8 shows the maximum growth rates for both modes under ion-ion beam excitation as function of the cold background ion plasma beta, β_{ic}.

Non-resonant Ion Beam Instability

The last example of an ion beam instability possibly important for space plasmas is a non-resonant instability as shown in Fig. 5.9. The resonant velocity, $v_{\parallel r}$, for a resonant instability would lie far outside of any of the distributions, either background or beam. Because of the condition of non-resonance, the plasma dispersion functions for all three components are to be expanded in the large argument limit, $\zeta \gg 1$.

This non-resonant mode will propagate in the direction opposite to the ion beam. It has negative helicity and small phase velocity. The instability is basically a firehose instability, caused by the inertia of the fast ion beam which exerts a centrifugal force on the bent magnetic field. Its maximum growth rate is

$$\boxed{\gamma_{\mathrm{inr,max}} \approx \frac{n_{ib}}{2n_0}\frac{v_{ib}}{v_A}\omega_{gi}} \tag{5.34}$$

For fast beams this growth rate is quite substantial, larger than the resonant ion beam growth rate, and if the ions in the beam have a larger mass than the background, the instability grows even faster due to its firehose-like mechanism. On the other hand, the instability has a larger threshold, since it has to overcome the restoring forces of perpendicular pressure and magnetic tension. But in regions of weak magnetic fields, as the foreshock region of the Earth's bow shock, the non-resonant instability is important.

5.4. Upstream Ion Beam Modes

The ion beam instabilities discussed in the previous section are important in many space plasmas. Instability can be caused by ion beams in the solar wind and solar corona, in cometary atmospheres, near shock waves, and in the plasma sheet boundary layer. In all these regions ion beams are known to exist. Depending on their speeds, thermal spreads and external conditions, they will cause one or the other instability. All these regions will exhibit low-frequency oscillations in the Alfvén, whistler, and magnetosonic modes.

Foreshock Waves

In the Earth's foreshock region ions are reflected from the shock front (see Sec. 8.5 of our companion book) and propagate upstream at moderate velocity on the background of the cold solar wind beam. This is the classical case of an electromagnetic counter-streaming beam situation. The beam is less dense than the solar wind, $n_{ib} \approx 0.01 n_0$, but becomes warm due to scattering at solar wind fluctuations and at the self-generated waves. Although the solar wind ions are cool, the plasma beta is relatively high. Hence, the R-mode instability is the fastest growing mode. It causes large fluctuations in the

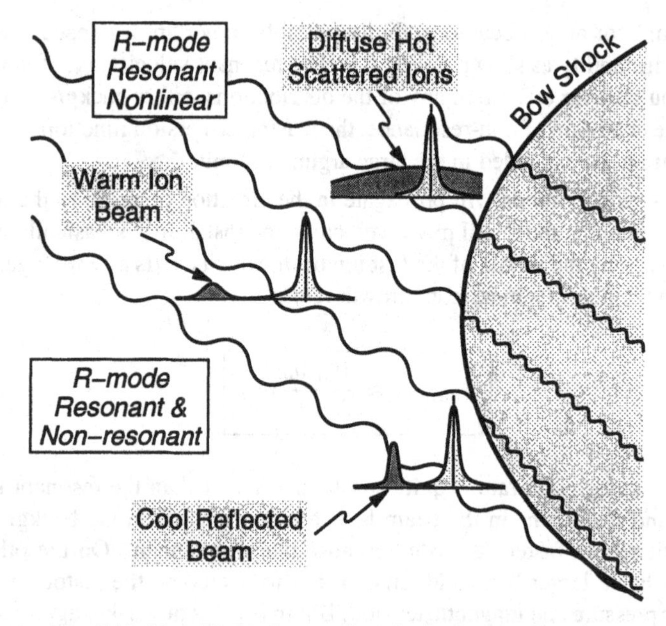

Fig. 5.10. Unstable Ion beam effects in the foreshock region.

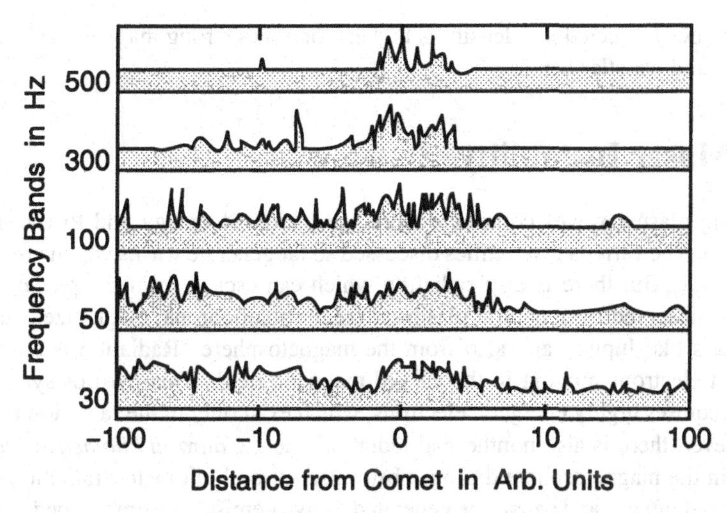

Fig. 5.11. Magnetic power density of cometary electromagnetic ion beam waves.

foreshock solar wind magnetic field, which have important effects on the reflected and backstreaming beam ions and on the shock formation process itself. Figure 5.10 shows schematically what may happen in the foreshock region of the Earth's bow shock.

Scattering of the ion beams by broad-spectrum electromagnetic waves will heat the ion beams diffusely, while in monochromatic waves the beams will become partially trapped and thus phase-bunched. Both effects have been observed. In addition, the waves may reach such large amplitudes that nonlinear effects appear. As the large amplitude waves are convected downstream toward the shock, they steepen, accumulate at the shock front and modify it.

Cometary Waves

Electromagnetic ion beam instabilities are also caused by injection of cometary ions into the solar wind. Comets evaporate neutral gases which are ionized by solar UV radiation and by collisions in the cometary atmosphere. These fresh ions move relative to the solar wind and form beams. The beams are cool and dense, consist of heavy ions, and are nearly at rest in the frame of the comet. Their relative speed in the solar wind frame is initially of the order of the solar wind velocity, much larger than the Alfvén speed.

Such beams can excite the non-resonant ion beam instability at low frequencies. Near comets one observes these Alfvénic and magnetosonic oscillations in the R-mode, next to other electrostatic beam-excited waves like the lower-hybrid mode. Figure 5.11 shows the magnetic spectra measured in the vicinity of a comet. At closest approach, where the newly injected ion density is highest, one sees strong magnetic emissions in the Alfvén and whistler bands.

5.5. Maser Instability

Propagating plasma waves contribute to redistribution of energy and to transport of information. The various instabilities discussed so far generate waves which are trapped in the plasma. But there is also radiation which can escape from the plasma. Such radiation is observed from natural plasmas, like the solar corona, magnetized stars, the large planets like Jupiter, and also from the magnetosphere. Radiation is emitted by accelerated electrons moving in the curved magnetic field. This gyro or synchrotron radiation requires highly energetic electrons, which exist only in the radiation belts.

However, there is also nonthermal radiation like the *auroral kilometric radiation* observed in the magnetosphere during substorms. Auroral kilometric radiation is very impulsive and intense and cannot be generated as gyro-emission from trapped particles. Thus one needs a linear excitation mechanism, an instability in either the O-mode or the X-mode. Mechanisms of this kind can exist only under extreme plasma conditions, because the escaping branches of both modes propagate at very high speeds and therefore need very long amplification lengths to reach reasonable amplitudes and require relativistic electrons to interact with.

Relativistic electrons modify the resonant orbits in the $(v_\parallel, v_\perp)$ plane and favor resonance with one of the escaping modes. Numerically, a strong instability can be obtained if the weakly relativistic electron distribution has a loss cone. Such distributions are common in magnetospheres, in the solar atmosphere, and in mirror-type magnetic field configuration. That the electrons have to be relativistic is only a weak restriction. Even 10 keV electrons are sufficiently relativistic to yield an instability. Also weakly relativistic parallel electron beams, as observed in the aurora, are capable of direct amplification of the escaping modes.

In effect, the physical mechanism of the direct electromagnetic mode instability is an inversion of the absorption coefficient of the plasma by the combination of the relativistic effect and the presence of the loss cone. The excess energy stored in the particles outside the loss cone and the positive gradient in the perpendicular velocity distribution of the particles are responsible for this turn-around. Because of this reason the plasma starts behaving not as an absorber but as a nearly coherent emitter. This is the reason why one is speaking of a *cyclotron maser instability*.

Cyclotron Emissions

All mechanisms of direct cyclotron radiation are based on linear instability of the plasma in one of the free space modes. Such instabilities depend on the resonance condition

$$k_\parallel v_\parallel - \omega + l\omega_{ge} = 0 \tag{5.35}$$

where only the electron-cyclotron frequency enters for the high-frequency modes. The ions serve as a neutralizing background, as was the case for the solutions of the dispersion relation (I.9.115) given in the companion volume. The above resonance condition contains resonances at all harmonics. The l=0 resonance is the Cherenkov resonance, which requires $N^2\cos^2\theta = k_\parallel^2 c^2/\omega^2 > 1$ for emission and thus very fast particles. Electrostatic Cherenkov emission (see Secs. 2.1 and 4.6) is of no interest here, but the l=0 resonance for parallel or oblique propagation will always cause Landau damping.

Cyclotron emission is obtained for $\omega \approx l\omega_{ge} \gg |k_\parallel v_\parallel|$. Actually, this condition is appropriate for escaping waves, because for these waves $N^2 < 1$ and thus for nonrelativistic particles $k_\parallel v_\parallel = N(v_\parallel/c)\omega\cos\theta \ll 1$. For nonrelativistic resonant particles leading to emission of waves one speaks of *cyclotron emission*, mildly relativistic particles produce *gyro-emission*, and ultra-relativistic particles lead to *synchrotron emission*. In both relativistic cases the relativistic dependence of the cyclotron frequency on the velocity must be taken into account. This requires redefinition of ω_{ge}

$$\omega_{ge} \to \omega_{ge}/\gamma_R \tag{5.36}$$

The right-hand side contains the relativistic gamma factor

$$\gamma_R = \left(1 - v^2/c^2\right)^{-1/2} \tag{5.37}$$

which shows that the resonance condition (5.35) becomes a complicated function of both components of the particle velocity, which implies that it now describes a full *resonance curve* in the velocity plane. The straight resonance strips of the low-frequency R- and L-modes thereby deform into resonance regions for the high-frequency modes.

For mildly relativistic electrons γ_R can be expanded, and Eq. (5.35) states that

$$k_\parallel v_\parallel - \omega + l\omega_{ge}\left[1 - (v_\parallel^2 + v_\perp^2)/2c^2\right] = 0 \tag{5.38}$$

which is a quadratic equation for the resonant frequency. One obtains two resonant values for $v_\parallel$ depending on $v_\perp$, instead of the single nonrelativistic resonance, $v_\parallel = (\omega - l\omega_{ge})/k_\parallel$. Since $v/c < 1$, the physical solutions of the resonance condition are restricted to the region inside a circle of radius v/c in the $(v_\parallel/c, v_\perp/c)$ plane. To identify the curve described by the above resonance condition, we rewrite it as

$$\frac{v_\perp^2}{c^2} + \frac{(v_\parallel - k_\parallel c^2/l\omega_{ge})^2}{c^2} = \frac{v_r^2}{c^2} \tag{5.39}$$

In three-dimensional velocity space this equation describes a sphere centered at $v_{c\perp} = 0$ and $v_{c\parallel}/c = k_\parallel c/l\omega_{ge}$ with radius

$$\frac{v_r}{c} = \frac{v_{c\parallel}}{c}\left[1 - \frac{2c^2(\omega - l\omega_{ge})}{l\omega_{ge}v_{c\parallel}^2}\right]^{1/2} \tag{5.40}$$

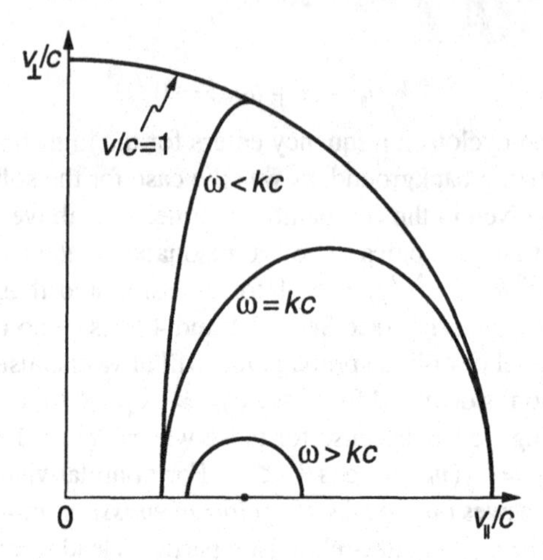

Fig. 5.12. Maser resonance ellipses.

The general resonance condition (5.35), including the full relativistic replacement of the gyrofrequency, is more complicated. It describes a resonant ellipsoid which cuts the $(v_\parallel, v_\perp)$ plane in a resonance ellipse. The resonant ellipse parameters, eccentricity, ε, center position, $v_{c\parallel}/c$, on the $v_{c\perp}$ axis, and major semi-axis, v_r/c, are given by

$$\begin{aligned} \varepsilon &= [k_\parallel^2 c^2/(k_\parallel^2 c^2 + l^2 \omega_{ge}^2)]^{1/2} \\ v_{c\parallel}/c &= \omega k_\parallel c/(k_\parallel^2 c^2 + l^2 \omega_{ge}^2) \\ v_r/c &= [(k_\parallel^2 c^2 - \omega^2 + l^2 \omega_{ge}^2)/(k_\parallel^2 c^2 + l^2 \omega_{ge}^2)]^{1/2} \end{aligned} \tag{5.41}$$

For $\omega^2 > k_\parallel^2 c^2$ this ellipse lies entirely inside the unit sphere of radius $v/c = 1$, for $k_\parallel^2 c^2 = \omega^2$ it touches it on the $v_\parallel$ axis, and for $\omega^2 < k_\parallel^2 c^2$ it crosses the unit sphere somewhere outside of the axis. Figure 5.12 illustrates this situation.

Growth Rate

To calculate the maser growth rate for a given velocity space distribution of the particles, one must integrate over these resonant parts of the resonance ellipses inside the unit circle of radius v/c. Any external part of an ellipse does not contribute, because it would be unphysical to take it into account. But before performing this calculation let us add some remarks on the nonrelativistic case. In the nonrelativistic case the resonant curve is a straight line parallel to the $v_\perp$ axis at

$$v_\parallel/c = v_{nr}/c = (\omega - l\omega_{ge})/k_\parallel c \tag{5.42}$$

This straight line is a tangent to one of the ellipses in the resonant plane at small $v_\|$ and, of course, also for small $v_\perp$. But since v_{nr} must be nonrelativistic, one requires in addition that the relativistic shift of the center of the resonant curve must be large so that $|v_{c\|}| \approx v_r \gg |v_{nr}| \approx |v_r - v_{c\|}|$. That this tangent is not a good approximation in many cases is immediately obvious from the shapes of the resonance curves. Moreover, the integration in the nonrelativistic case is performed along this straight line up to infinity, where the resonance curve has long deviated from it.

In the mildly relativistic case, when one integrates along the shifted-circle resonance curve, integration is performed only over a limited range of perpendicular velocities, $v_\perp < v_r$. Let us write the variables of integration along the resonant circle as

$$\begin{aligned} v_\| &= v_{c\|} - w\cos\psi \\ v_\perp &= w\sin\psi \end{aligned} \tag{5.43}$$

This transformation allows to rewrite the resonant delta function as

$$\delta(k_\| v_\| - \omega + l\omega_{ge}/\gamma_R) \approx (c^2/v_r l\omega_{ge})\delta(w - v_r) \tag{5.44}$$

In calculating the resonant growth rate, we make the usual assumption of weak instability, linearize the Vlasov equation, and use the cold plasma dispersion relation for the free space electromagnetic O- and X-modes. But one cannot use the conventional approach of the plasma dispersion function. Instead, one must explicitly integrate along the resonant circle over the given equilibrium distribution function, $f_{e0}(v_\|, v_\perp)$. This can be simplified by replacing the variables in the velocity integrals with the help of the above definitions as follows

$$2\pi\int v_\perp dv_\perp dv_\| = 2\pi\int w^2 dw \int_0^\pi d\psi \tag{5.45}$$

The limits on the w integration are given by the resonant delta function. It simply requires that the integrand has to be taken at $w = v_r$, the radius of the resonant circle for the mildly relativistic electrons.

In symbolic form the growth rate of the cyclotron maser instability can be expressed as the integral over the angular variable ψ over the resonant half-circle

$$\gamma_{\rm cm} = -\sum_{l=-\infty}^{\infty} \frac{4\pi^2 v_r \omega_{pe}^2 R(\mathbf{k})}{l\omega_{ge}\omega n_0} \int_0^\pi |A_l|^2(\mathbf{k},\mathbf{v})\, \mathcal{O}_l f_{e0}(v_\|, v_\perp)\Big|_{w=v_r} d\psi \tag{5.46}$$

The operator $\mathcal{O}_l$ and the factor $R(\mathbf{k})$ are defined as

$$
\begin{aligned}
\mathcal{O}_l &= k_\parallel \frac{\partial}{\partial v_\parallel} - \frac{l\omega_{ge}}{v_\perp}\frac{\partial}{\partial v_\perp} \\
R(\mathbf{k}) &= \frac{1}{2}\left(1+\frac{K_\sigma^2}{1+T_\sigma^2}\right)\left[N\frac{\partial(\omega N)}{\partial \omega}\right]^{-1}
\end{aligned}
\tag{5.47}
$$

with $N = kc/\omega$ the refraction index of the emitted wave mode, which can be taken from the cold plasma wave dispersion relation in Eq. (I.9.115). The factor A_l, which contains the various contributions of the harmonics and the polarization of the emitted wave mode, is given by

$$
A_l(\mathbf{k},\mathbf{v}) = (1+K_\sigma^2+T_\sigma^2)^{-1/2}\begin{bmatrix} K_\sigma \sin\theta + T_\sigma\cos\theta \\ -i \\ K_\sigma\cos\theta - T_\sigma \sin\theta \end{bmatrix} \cdot \begin{bmatrix} l v_\perp^2 J_l(\eta)/\eta \\ i v_\perp J_l'(\eta) \\ v_\parallel J_l(\eta) \end{bmatrix}
\tag{5.48}
$$

Here $\eta = k_\perp r_{ge}$, the index $\sigma = \pm 1$ identifies the wave mode, with $\sigma = 1$ for the O-mode and $\sigma = -1$ for the X-mode. Explicit expressions for N, K_σ, T_σ in the cold plasma approximation are in conventional notation as used in magneto-ionic theory with the abbreviations $X = \omega_{pe}^2/\omega^2, Y = \omega_{ge}/\omega$

$$
\begin{aligned}
N^2 &= 1 - XT/(T - Y\cos\theta) \\
T_\sigma &= -(Y^2\sin^2\theta - 2\sigma\Delta)/[2Y(1-X)\cos\theta] \\
K_\sigma &= XYT\sin\theta/[(1-X)(T-Y\cos\theta)] \\
\Delta^2 &= Y^4\sin^4\theta/4 + (1-X)^2Y^2\cos^2\theta
\end{aligned}
\tag{5.49}
$$

and T is the solution of the quadratic equation

$$
T^2 + \frac{Y\sin^2\theta}{(1-X)\cos\theta}T - 1 = 0
\tag{5.50}
$$

Furthermore, for $X < 1$ and oblique but nearly parallel propagation of the emitted wave

$$
\begin{aligned}
N_\sigma^2 &= 1 - X/(1+\sigma Y|cos\theta|) \\
K_\sigma &= \omega_{pe}^2\omega_{ge}\sin\theta/[\omega(\omega^2-\omega_{pe}^2)(1+\sigma|\cos\theta|\omega_{ge}/\omega)] \\
T_\sigma &= -\sigma \mathrm{sgn}(\cos\theta)\mathrm{sgn}(1-\omega_{pe}^2/\omega^2)
\end{aligned}
\tag{5.51}
$$

For nearly perpendicular propagation one has $T_{+1} \to \infty, T_{-1} \to 0$ and

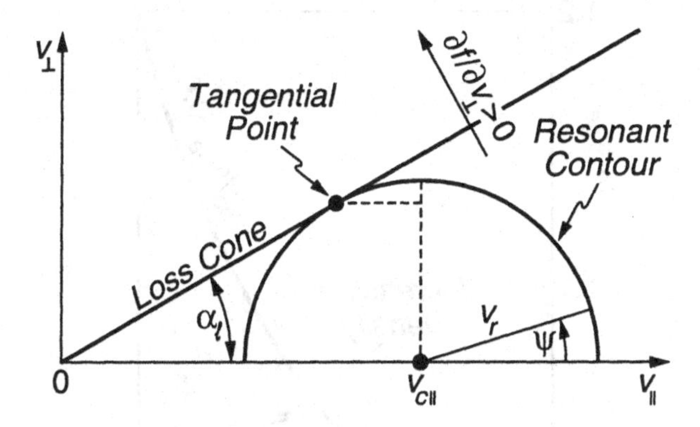

Fig. 5.13. Loss cone as driver of the cyclotron maser instability.

$$\begin{aligned} N_{+1}^2 &= 1 - X \\ K_{+1} &= XY \sin\theta/(1 - X) \\ N_{-1}^2 &= 1 - X(1-X)/(1 - X - Y^2 + XY^2 \cos^2\theta) \\ K_{-1} &= XY \sin\theta/(1 - X - Y^2 + XY^2 \cos^2\theta) \end{aligned} \qquad (5.52)$$

The above growth rate is a very complicated expression, but under favorable conditions it becomes positive. This happens, for instance, when the distribution function of the hot relativistic particles is a loss cone distribution and the resonant circle is entirely inside the loss cone, $\alpha < \alpha_\ell$. Then $f_{e0}(v_\perp)$ has a positive gradient inside the loss cone. It turns out that the contribution of the term $(l\omega_{ge}/v_\perp)\partial f_{eh}/\partial v_\perp$ is destabilizing along the entire contour of integration. This is shown in Fig. 5.13.

Emission Bounds

We can find an approximate estimate of the maximum growth rate of the maser instability by using the value of the distribution function near the point of the resonant circle that is tangential to the loss cone ray (see Fig. 5.13) at $v_\perp = v_r, v_\parallel = v_{c\parallel}$. This value is

$$f_{\max} \approx n_h / 2\pi v_{c\parallel} v_r^2 \qquad (5.53)$$

The X-mode has approximately a refraction index of $N_{-1} \approx 1$. For $l = 1$, we get

$$\boxed{\gamma_{\mathrm{cm,max}} \approx \pi \omega_{pe} \frac{c^2}{v_r v_{c\parallel}} \frac{n_h}{n_0} \frac{\omega_{pe}}{\omega_{ge}}} \qquad (5.54)$$

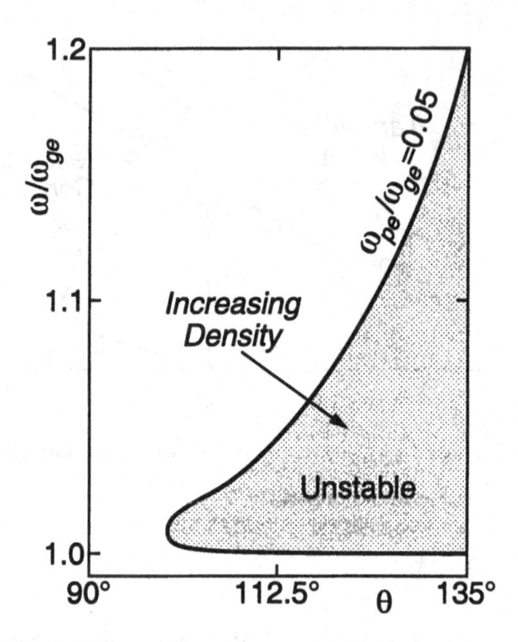

Fig. 5.14. Switching-off boundary $v_r = 0$ for the cyclotron maser.

As is obvious from Fig. 5.13, the range of validity is $\Delta v_{c\parallel} \approx v_r$ and $\Delta v_r \approx v_{c\parallel}\delta\alpha_\ell$, where $(\delta\alpha_\ell)^{-1} \approx |\partial \ln f/\partial\alpha|$ inside the loss cone.

Growth of the instability is also restricted by the simplifying kinematic assumption of small angles. The maximum angle of emission in the X-mode is obtained from its dispersion relation as $\theta_{\max} = \cos^{-1}(v_r/c)$, with an angular spread of the emission cone of $\delta\theta_{\max} \approx \delta v_{c\parallel}/c$. This is a rather narrow angle. The bandwidth of the emission is obtained as

$$\Delta\omega \approx \omega_{ge}\delta\alpha_\ell v_r v_{c\parallel}/c^2 \tag{5.55}$$

Since the emission frequency must be far above the X-mode cut-off, a further restriction arises for the frequency. Using the dispersion relation for the X-mode, one can plot the curve $v_r = 0$ when the maser switches off (Fig. 5.14). This sets an upper limit on the ratio of plasma to cyclotron frequency which is given by

$$\omega_{pe}/\omega_{ge} < v_r/2c < 1 \tag{5.56}$$

Hence, the cyclotron maser instability will work only if the plasma frequency is considerably below the electron-cyclotron frequency. In addition, the growth rates of the higher harmonics decrease as the $2(l-1)$th power of the ratio v_r/c, and the growth rate of the O-mode is very small compared to that of the X-mode.

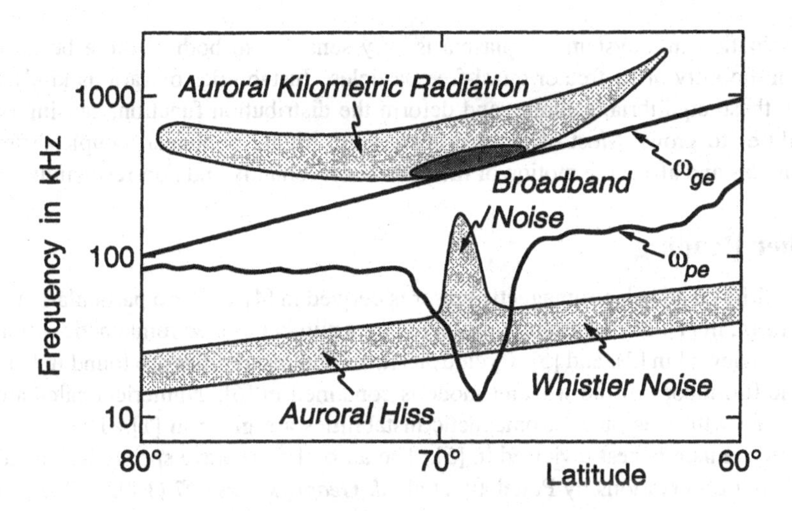

Fig. 5.15. Auroral zone frequency-latitude spectrogram.

Auroral Kilometric Radiation

For the cyclotron maser instability it is important to have very low plasma densities and strong magnetic fields. One favorable situation is encountered in the auroral density decreases during active auroras when the auroral ionospheric plasma is strongly depleted. In this density trough mirroring auroral particles have large loss cones. Moreover, with their energy of several keV the mildly-relativistic approximation applies. This may lead to the emission of X-mode waves at the cyclotron frequency if the cyclotron damping due to the background is overcome by the inverse absorption of X-mode waves induced by the loss cone distribution.

The radiation emitted by loss cone electrons in the auroral region can explain the strong auroral kilometric radiation emitted during substorms. Figure 5.15 sketches the pass of a satellite across the auroral density depletion indicated by the steep drop in ω_{pe}. When crossing the density trough at a few 1000 km altitude, the spacecraft encounters strong auroral kilometric radiation at the local cyclotron frequency. At larger distances it observes propagated emission. At lower frequencies it measures whistler noise and electrostatic hiss, including broadband increases below the cyclotron frequency.

Concluding Remarks

The microinstabilities discussed in the present chapter are caused by deviations from the Maxwellian distribution, a situation nearly never realized in nature. The normal case will be equilibria resulting from competition between energy or momentum inflow and dissipation in a system. A plasma is very sensitive to both of these because of the high mobility of its free or quasi-free particles. Instabilities of various kinds may arise in these equilibria, saturate, and deform the distribution function, causing other instabilities to grow. Most of these processes are highly nonlinear, couple different wave modes and affect the motion of the particles resonantly and non-resonantly.

Further Reading

The full drift kinetic electromagnetic tensor is derived in [4] and, in a particularly transparent form, in [1]. Microinstabilities, including a number of electromagnetic instabilities, are reviewed in [2], and [5]. Cyclotron waves and instabilities are found in [7] and [8]. The full theory of the whistler mode is contained in [6]. Numerical calculations of many growth rates of electromagnetic instabilities are given in [1]. The cyclotron maser mechanism is best reviewed in [5]. The auroral zone wave spectra are modeled after Viking observations by Pottelette et al., *J. Geophys. Res.* **97** (1992) 12029. The dispersion curves and growth rates of the ion-cyclotron instability have been taken from [1].

[1] S. P. Gary, *Theory of Space Plasma Microinstabilities* (Cambridge University Press, Cambridge, 1993).

[2] A. Hasegawa, *Plasma Instabilities and Nonlinear Effects* (Springer Verlag, Heidelberg, 1975).

[3] M. C. Kelley, *The Earth's Ionosphere* (Academic Press, San Diego, 1989).

[4] N. A. Krall and A. W. Trivelpiece, *Principles of Plasma Physics* (McGraw-Hill, New York, 1971).

[5] D. B. Melrose, *Instabilities in Space and Laboratory Plasmas*, (Cambridge University Press, Cambridge, 1991).

[6] S. Sazhin, *Whistler-Mode Waves in a Hot Plasma* (Cambridge University Press, Cambridge, 1993).

[7] T. H. Stix, *Waves in Plasma* (American Institute of Physics, New York, 1992).

[8] A. D. M. Walker, *Plasma Waves in the Magnetosphere* (Springer Verlag, Heidelberg, 1993).

6. Drift Instabilities

The effect of spatial inhomogeneities in the plasma on the instabilities has so far been neglected. However, for longer wavelengths, which are comparable to the natural scales of the plasma, $L_n = |\nabla \ln n|^{-1}$, $L_B = |\nabla \ln B|^{-1}$, and $L_T = |\nabla \ln T|^{-1}$, the change in the plasma parameters with space must be taken into account in the calculation of the wave properties. The result is a new type of waves called *drift modes* which are entirely due to the presence of the plasma inhomogeneity..

Inclusion of inhomogeneity introduces a severe complication. In the fluid approach it implies that the coefficients of the field variables become spatially dependent, and straight forward Fourier transformation of the basic equations is inhibited. In the simplest case, when the inhomogeneity is only in one spatial direction, say on x, the plane wave ansatz is still possible in the two directions y, z transverse to the direction of inhomogeneity. One then reduces the basic system of equations to an ordinary differential equation in x. For weak inhomogeneity one can take advantage of the further approximation of expanding the inhomogeneities around a particular spatial position to first order in x. This approach is the local approximation and is used in the following theory.

6.1. Drift Waves

The fact that all plasmas are inevitably inhomogeneous and thus affected by gradients in the plasma parameters implies that all plasmas are subject to drift wave propagation and drift instability. Accordingly, one calls the drift instability also *universal instability*, meaning all kinds of drift instabilities without further specification. Clearly, drift instabilities can arise for all kinds of waves from electromagnetic Alfvén and magnetosonic waves up to ion-acoustic and plasma waves.

How the drift frequency arises can be understood from a dimensional inspection of the linearized one-dimensional ion continuity equation in an inhomogeneous incompressible plasma

$$\frac{\partial \delta n_i}{\partial t} + \delta v_{ix} \frac{\partial n_0}{\partial x} = 0 \tag{6.1}$$

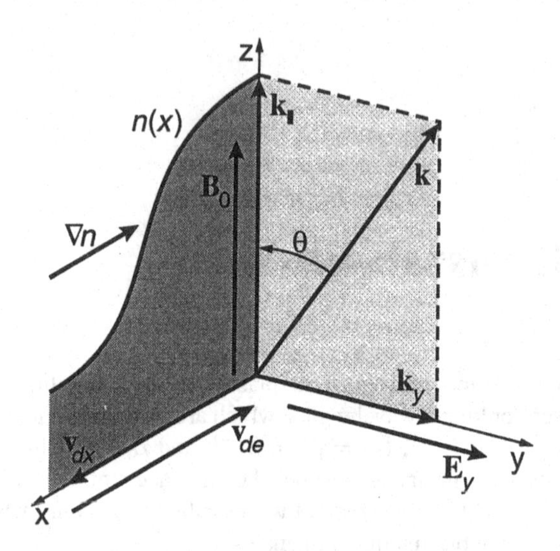

Fig. 6.1. Drift wave geometry.

with $\delta v_{ix} = \delta E_y/B_0 = -ik_y\delta\phi/B_0$ the drift speed. For low frequencies the electrons obey the Boltzmann law

$$\delta n_e = -\frac{en_0}{k_BT_e}\delta\phi \tag{6.2}$$

With the help of the ansatz $\exp(-i\omega t + ik_y y)$ and assuming quasineutrality we find that the oscillation frequency is given by the *electron drift wave frequency*

$$\boxed{\omega_{de}/\omega_{ge} = k_y r_{ge}^2/L_n} \tag{6.3}$$

This slow oscillation of the electrons is caused by the density gradient. It has a frequency much less than the electron gyrofrequency and a wavelength much longer than the gradient scale. Obviously, because the waves are due to the electric field drift, the propagation of the wave is transverse to the density gradient. The system of coordinates is sketched in Fig. 6.1. But the wave current is entirely along the magnetic field because the electric drift does not cause any currents.

Dispersion Relation

In order to obtain the electron response and the wave conductivity, $\sigma = \delta j_\parallel/\delta E_\parallel$, consider the simplified Vlasov equation of the electrons instead of the Boltzmann law

$$\delta f_e = -\delta v_\parallel \frac{\partial f_0}{\partial v_\parallel} - \delta x \frac{\partial f_0}{\partial x} \tag{6.4}$$

with the following expression for the linear excursion, δx, along the density gradient

$$\delta x = i\delta v_x/(\omega - k_\parallel v_\parallel) \tag{6.5}$$

Furthermore, for electrostatic perturbations with vanishing $\nabla \times \delta \mathbf{E} = 0$ one has

$$k_\parallel \delta E_y = k_y \delta E_\parallel \tag{6.6}$$

and, since $\delta v_x = \delta E_y/B_0$, the variation of the parallel electron distribution function is

$$\delta f_e = -\delta E_\parallel \frac{ie}{m_e(\omega - k_\parallel v_\parallel)}\left(\frac{\partial}{\partial v_\parallel} + \frac{k_y}{k_\parallel \omega_{ge}}\frac{\partial}{\partial x}\right) f_0 \tag{6.7}$$

Using Plemelj's formula (see Eq. (I.A.78) in the appendix of our companion book, *Basic Space Plasma Physics*) to replace the resonant denominator this becomes

$$\delta f_e = \frac{\pi e \delta E_\parallel}{m_e}\delta(\omega - k_\parallel v_\parallel)\left(\frac{\partial}{\partial v_\parallel} + \frac{k_y}{k_\parallel \omega_{ge}}\frac{\partial}{\partial x}\right) f_0 \tag{6.8}$$

One can express the parallel current by the integral

$$\delta j_\parallel = \sigma \delta E_\parallel = -e\int\limits_{-\infty}^{\infty} v_\parallel dv_\parallel \delta f_e \tag{6.9}$$

Since this is proportional to the disturbed electric field, $\delta E_\parallel$, the wave conductivity is

$$\sigma(\omega, \mathbf{k}) = -\frac{\pi\epsilon_0\omega_{pe}^2}{n_0|k_\parallel|}\frac{\omega}{k_\parallel}\left(\frac{\partial}{\partial v_\parallel} + \frac{k_y}{k_\parallel \omega_{ge}}\frac{\partial}{\partial x}\right) f_0\bigg|_{v_\parallel=\omega/k_\parallel} \tag{6.10}$$

Let the plasma density be inhomogeneous and otherwise use a Maxwellian undisturbed distribution function, then the conductivity assumes the form

$$\sigma_e = \frac{\pi^{1/2}\epsilon_0\omega^2}{k_\parallel^2\lambda_D^2|k_\parallel|v_{the}}\left(1 - \frac{\omega_{de}}{\omega}\right) \tag{6.11}$$

This expression, inserted into the dielectric function, given in Eq. (I.9.54) of the companion book, and repeated here

$$\epsilon(\omega, \mathbf{k}) = \mathsf{I} + \frac{i\sigma(\omega, \mathbf{k})}{\omega\epsilon_0} \tag{6.12}$$

yields a positive imaginary part of $\epsilon(\omega, \mathbf{k})$ and may thus produce a drift instability.

Frequency and Growth Rate

An example of a drift wave instability due to a density gradient can be obtained by taking a more precise fluid picture for the ions than used above in the dimensional derivation of the drift frequency. If only density gradients exist, L_n is finite while all other scales go to infinity. The ion continuity equation then reads

$$\partial \delta n_i / \partial t + \nabla \cdot (n_0 \delta \mathbf{v}_i) = 0 \tag{6.13}$$

where the density varies in space, and the oscillating ion velocity is given by the linearized and Fourier transformed ion drift velocity

$$\delta \mathbf{v}_i = \delta \mathbf{v}_E + \frac{ie}{m_i \omega}\left(\delta E_\| \hat{\mathbf{e}}_\| - \frac{\omega^2}{\omega_{gi}^2}\delta \mathbf{E}_\perp\right) \tag{6.14}$$

For simplicity the ions are assumed to be cold. Combining these equations and using for electrostatic waves $\delta \mathbf{E} = -\nabla \phi$, one gets for a density gradient along y

$$\delta n_i = \frac{e n_0}{m_i}\left(\frac{k_\|^2}{\omega^2} - \frac{k_\perp^2}{\omega_{gi}^2} - \frac{k_\perp}{L_n \omega \omega_{gi}}\right)\delta \phi \tag{6.15}$$

The perturbed electron density is found from the above disturbed distribution as

$$\delta n_e = -(i k_\|^2 / e\omega)\sigma_e \delta \phi \tag{6.16}$$

Since the plasma is quasineutral, combining the densities yields the dispersion relation

$$\epsilon(\omega, \mathbf{k}) = 1 + \frac{k_\perp^2 c_{ia}^2}{\omega_{gi}^2} - \frac{k_\|^2 c_{ia}^2}{\omega^2} - \frac{\omega_{de}}{\omega} + i\left(\frac{\pi}{2}\right)^{1/2}\frac{\omega}{|k_\||v_{\mathrm{the}}}\left(1 - \frac{\omega_{de}}{\omega}\right) = 0 \tag{6.17}$$

From here the real and imaginary parts of the frequency are obtained under the assumption that $k_\|^2 c_{ia}^2/\omega^2 < k_\perp^2 c_{ia}^2/\omega_{gi}^2 \ll 1$ as

$$\boxed{\omega_{\mathrm{kd}} = \omega_{de}\left(1 - \frac{k_\perp^2 c_{ia}^2}{\omega_{gi}^2}\right)} \tag{6.18}$$

and

$$\boxed{\gamma_{\mathrm{kd}} = \left(\frac{\pi}{2}\right)^{1/2}\frac{k_\perp^2 c_{ia}^2}{\omega_{gi}^2}\frac{\omega_{\mathrm{kd}}^2}{|k_\||v_{\mathrm{the}}}} \tag{6.19}$$

This is the first *kinetic drift instability* we encounter. It is driven by the density gradient and exists only under the above condition on the parallel wavenumber. We note that for finite plasma β this instability is quenched due to electron Landau damping.

6.2. Kinetic Drift Wave Theory

If the influence of the plasma inhomogeneity on wave propagation and instability is small, it will change the frequency only slightly. The natural drift wave frequency will thus fall into the low-frequency range so that the particle magnetic moments are conserved. Hence, the relevant kinetic equation to start from is either the gyrokinetic equation (I.6.25) or the drift kinetic equation (I.6.27) from our companion book. Choosing the latter

$$\frac{\partial f_d}{\partial t} + \frac{\partial}{\partial \mathbf{x}_d} \cdot (\mathbf{v}_d f_d) + \frac{\partial}{\partial v_\parallel}\left(\frac{F_\parallel}{m} f_d\right) = 0 \tag{6.20}$$

with $\mathbf{x}_d$ the guiding center coordinate and

$$\begin{aligned} \mathbf{v}_{d\perp} &= \mathbf{v}_E + \frac{\mathbf{F}\times\mathbf{B}}{qB^2} \approx \mathbf{v}_E + \frac{1}{\omega_g B}\frac{d\mathbf{E}_\perp}{dt} \\ \mathbf{v}_E &= \frac{\mathbf{E}\times\mathbf{B}}{B^2} + \frac{\mathbf{E}_\parallel \times \delta\mathbf{B}}{B^2} \\ \mathbf{F} &= qE_\parallel \frac{\mathbf{B}}{B} - \mu\nabla B - \frac{2k_B T_\parallel}{R_c^2}\mathbf{R}_c - m\frac{d\mathbf{v}_E}{dt} \end{aligned} \tag{6.21}$$

Here μ is the magnetic moment of the particles which is conserved during particle motion in the low-frequency wave field, and $\mathbf{R}_c = -(\mathbf{B}/B\cdot\nabla)\mathbf{B}/B$ is the curvature radius of the magnetic field. In addition we took into account the electric drift in a transverse magnetic field component $\delta\mathbf{B}$ caused in presence of an electromagnetic wave.

Drift Kinetic Equation

In order to obtain the dispersion relation in a linearized theory, we take the following separation of the distribution function

$$f_d(\mathbf{x}_d, \mu, v_\parallel, t) = f_{0d}(\mathbf{x}_d, \mu, v_\parallel) + \delta f_d(\mathbf{x}_d, \mu, v_\parallel, t) \tag{6.22}$$

The undisturbed part does still depend on the spatial inhomogeneity of the plasma. We need to determine the plasma wave conductivity which enters the dielectric response function. Hence, we need to calculate the disturbed plasma current. This current is given as the sum over all species s in the guiding center approximation as

$$\delta\mathbf{j} = \sum_s \int \left[q_s \delta\mathbf{v}_{ds} f_{0ds} - \nabla \times \left(\frac{\mathbf{B}}{B}\mu_s \delta f_{ds}\right)\right] d\mu dv_\parallel \tag{6.23}$$

Note the normalization $\int f_d d\mu dv_\parallel = n_0$. The first term on the right-hand side accounts for the drift current of the component s, while the second term adds the diamagnetic current due to the gradients in the distribution function. For the latter we assume that they are weak and that their characteristic scales are longer than the wavelength. Thus the waves have wavelength short compared to the gradient scales.

We have the freedom to change the system of coordinates and to put the gradient along the y axis and let the wave vector, $\mathbf{k} = (\mathbf{k}_\perp, k_\parallel)$, be in the (x, z) plane. Looking for electrostatic waves with electric field

$$\delta\mathbf{E} = -\nabla\delta\phi = -i(k_\perp, 0, k_\parallel)\delta\phi \tag{6.24}$$

the drift velocity in the inhomogeneous oscillating wave electric field is

$$\delta\mathbf{v}_d = -i\left(\frac{k_\perp}{B}\hat{\mathbf{e}}_y + \frac{\omega - k_\parallel v_\parallel}{\omega_g B}k_\perp\hat{\mathbf{e}}_x\right)\delta\phi \tag{6.25}$$

The next step is to linearize the drift kinetic equation under the condition that the drift is caused in the inhomogeneous and oscillating wave field in the undisturbed inhomogeneous plasma which yields

$$i(k_\parallel v_\parallel - \omega)\delta f_d + \delta v_{dy}\frac{\partial f_{d0}}{\partial y} + i\mathbf{k}\cdot\delta\mathbf{v}_{d\perp}f_{d0} + \frac{q}{m}\delta E_\parallel\frac{\partial f_{d0}}{\partial v_\parallel} = 0 \tag{6.26}$$

The undisturbed drifting distribution function, f_{d0}, contains the inhomogeneity of the plasma, while the wave produces the polarization and $E \times B$ drifts.

Dispersion Relation

Since the waves are electrostatic, we are interested only in the longitudinal part of the dielectric response function (5.2). By using the definition of the dielectric tensor in Eq. (6.12), this part can be expressed through the current. Then one can show that

$$\mathbf{k}\cdot\boldsymbol{\sigma}\cdot\mathbf{k} = i\mathbf{k}\cdot\delta\mathbf{j}/\delta\phi \tag{6.27}$$

where we used Ohm's law for waves and the above representation of the wave electric field through the electrostatic potential. Inserting into Eqs. (6.12) and (5.2) yields

$$\epsilon_\mathrm{L}(\omega, \mathbf{k}, \mathbf{x}) = 1 - \frac{\mathbf{k}\cdot\delta\mathbf{j}}{\epsilon_0\omega k^2\delta\phi} \tag{6.28}$$

Using an inhomogeneous but isotropic Maxwellian distribution as the undisturbed distribution and evaluating the current, one gets the dielectric response function of drift waves

$$\epsilon_\mathrm{L}(\omega, \mathbf{k}, \mathbf{x}) = 1 + \sum_s\left\{\frac{\omega_{ps}^2}{\omega_{gs}^2}\frac{k_\perp^2}{k^2} + \frac{1}{k^2\lambda_{Ds}^2}\left[1 - \left(1 - \frac{\omega_{ds}}{\omega}\right)Z(\zeta_s)\right]\right\} \tag{6.29}$$

The argument of the dispersion function is $\zeta_s = \omega/k_\parallel v_{\text{th}s}$ and the drift wave frequency ω_{ds} which appears in the dispersion function is determined by

$$\omega_{ds} = \mathbf{k} \cdot \mathbf{v}_{ds0} \tag{6.30}$$

The zero-order drift velocity entering this expression is given by

$$\mathbf{v}_{ds0} = \text{sgn}(q_s) v_{\text{th}s} \mathbf{B} \times \left[(r_{rs} \nabla) \ln f_{ds0} \right] / B \tag{6.31}$$

This *drift frequency* depends on the inhomogeneity of the plasma which is contained in the spatial dependence of the zero-order distribution function. The drift velocity is the diamagnetic drift velocity for the species s. It is thus interesting that it is only the diamagnetic drift which gives rise to drift waves in a plasma. In other words, it is the various gradients which the pressure variation contributes that give rise to drift waves.

Electromagnetic Corrections

The simplified analysis presented here is based on two assumptions, the assumption of small particle gyroradii and the electrostatic or longitudinal assumption. A more precise and more sophisticated approach starts from the full Vlasov equation and expands the distribution function up to first order in y with the expansion coefficients L_n^{-1}, L_B^{-1}, and L_T^{-1}, yielding for the longitudinal dielectric function

$$\epsilon_{\mathrm{L}}(\omega, \mathbf{k}, \mathbf{x}) = 1 + \sum_s \chi_{\mathrm{L}s}(\omega, \mathbf{k}, \mathbf{x}) \tag{6.32}$$

with the susceptibilities, $\chi_{\mathrm{L}s}$, defined as

$$\chi_{\mathrm{L}s}(\omega, \mathbf{k}, \mathbf{x}) = \frac{1}{k^2 \lambda_{Ds}^2} \left\{ 1 - \sum_{l=-\infty}^{\infty} \left[1 - \mathcal{O}_s \left(1 + \frac{l\omega}{\eta_s \omega_{gs}} \right) \right] \frac{\omega}{\omega - l\omega_{gs}} \Lambda_l(\eta_s) Z(\zeta_s) \right\} \tag{6.33}$$

where the differential operator, $\mathcal{O}_s$, is defined as

$$\mathcal{O}_s = \text{sgn}(q_s) k_\perp r_{gs} \cos\theta \frac{\omega_{gs}}{\omega} \left[(r_{gs} \nabla_\perp) \ln n_s + (r_{gs} \nabla_\perp T_s) \frac{\partial}{\partial T_s} \right] \tag{6.34}$$

with θ the angle between $\mathbf{k}_\perp$ and the x axis. In many cases this operator can be approximated by $\mathcal{O}_s \approx \omega_{ds}/\omega$.

The expression for the dielectric tensor becomes more complicated if one takes into account electromagnetic interactions. In order to obtain it one follows the procedure used to derive the magnetized dielectric tensor in Sec. 10.4 of our companion book. The resulting dispersion relation becomes a function of frequency, wavenumber, and space coordinate

$$D(\omega, \mathbf{k}, x) = 0 \tag{6.35}$$

Again, as has implicitly been done in the electrostatic case, this dispersion relation is evaluated in the *local approximation*, at $x = 0$, under the assumption that the gradients are weak and the gradient length is longer than the wavelength and the gyroradius

$$r_{gs} \ll L_n, L_B, L_T \tag{6.36}$$

a condition which is a prerequisite for the validity of the theory of adiabatic particle motion. The most important change in the dispersion relation introduced by a density gradient is that the resonant denominator in Eq. (1.35) is replaced according to

$$(k_\parallel v_\parallel + l\omega_{gs} - \omega) \rightarrow [k_\parallel v_\parallel + (k_\perp v_\perp^2/2L_n\omega_{gs}) + l\omega_{gs} - \omega] \tag{6.37}$$

This change of the resonance leads to the appearance of drift modes.

6.3. Drift Modes

Spatial inhomogeneities and the associated drift modes interact best with low-frequency waves. Hence, the drift instability is most important for the excitation of ion-cyclotron, Alfvén, and lower-hybrid waves.

Drift-Cyclotron Instability

Cyclotron modes can become unstable in density gradients and excite the *drift-cyclotron instability*. From the electrostatic magnetized dispersion relation (1.35) one can derive the following response function

$$\epsilon(\omega, \mathbf{k}) = 1 + \frac{\omega_{pe}^2}{\omega_{ge}^2} + \frac{1}{k^2\lambda_{Di}^2}\left(1 - \frac{\omega_{di}}{\omega}\right)\left(1 - \frac{\omega}{\sqrt{2\pi}k_\perp r_{gi}(\omega - l\omega_{gi})}\right) = 0 \tag{6.38}$$

Clearly, this equation has cyclotron harmonic solutions near the harmonics of the ion cyclotron frequency $\omega = l\omega_{gi}$. Instability is obtained under the condition that

$$\frac{r_{gi}}{L_n} > 2l\left(\frac{m_e}{m_i}\right)^{1/2}\left(1 + \frac{\omega_{pe}^2}{\omega_{ge}^2}\right)^{1/2} \tag{6.39}$$

The right-hand side of the last condition reduces to the root of the mass ratio times $2l$ whenever the plasma density is low. It is interesting to note that the condition for instability sets an upper limit on the number of cyclotron harmonics. This property differs from the current-, beam- or loss cone-driven cyclotron instabilities discussed above, where the limit was set by the decrease of the growth rate. Here it appears as a cut-off of the growth rate. The number of allowed unstable modes is found from the full expression for the growth rate as

$$l_{\rm m} \approx (r_{gi}/2L_n)(m_i/m_e)^{1/4} \approx 3r_{gi}/L_n \tag{6.40}$$

and the growth rate of the highest excited harmonic is obtained as

$$\gamma_{l\rm m} \approx (8\pi)^{1/4}(m_i/m_e)^{1/4}(r_{gi}/L_n)\omega_{gi} \tag{6.41}$$

Sometimes gradients in the electron distribution in the magnetosphere may be responsible for the observed upper harmonic cut-offs.

Drift-Alfvén Instability

In the extremely-low frequency limit we are in the domain for Alfvén or magnetosonic modes and must take into account the electromagnetic correction mentioned above. The approximate dispersion relation, given here without proof, is

$$D(\omega,\mathbf{k}) = N^2\epsilon_{\rm L}(\omega,\mathbf{k}) - \epsilon_{\|}(\omega,\mathbf{k})\epsilon_{\perp}(\omega,\mathbf{k}) = 0 \tag{6.42}$$

where the components of the dielectric tensor are modified to include the inhomogeneity

$$\begin{aligned} \epsilon_{\rm L}(\omega,\mathbf{k}) &= 1+\sum_s \chi_{s{\rm L}}(\omega,\mathbf{k}) \\ \epsilon_{\|}(\omega,\mathbf{k}) &= 1+\sum_s \chi_{s\|}(\omega,\mathbf{k}) \\ \epsilon_{\perp}(\omega,\mathbf{k}) &= 1+\sum_s \chi_{s\perp}(\omega,\mathbf{k}) \end{aligned} \tag{6.43}$$

and the approximate susceptibilities are given by

$$\begin{aligned} \chi_{s{\rm L}} &= (k_\perp^2\chi_{s\perp} + k_\|^2\chi_{s\|})/k^2 \\ \chi_{s\perp} &= \frac{1}{k_\perp^2\lambda_{Ds}^2}\left(1-\frac{\omega_{ds}}{\omega}\right)[1-\Lambda_0(\eta_s)] \\ \chi_{s\|} &= \frac{1}{k_\|^2\lambda_{Ds}^2}\left(1-\frac{\omega_{ds}}{\omega}\right)[1-Z(\zeta_{s0})]\,\Lambda_0(\eta_s) \end{aligned} \tag{6.44}$$

In the limit of small electron gyroradius, $kr_{ge} \ll 1$, and with $k_\| v_{{\rm th}i} \ll k_\| v_{{\rm th}e}$, the dispersion relation assumes a tractable form

$$\begin{aligned} \left(1-\frac{\omega_{gi}}{\omega}\right)\frac{T_e}{T_i}\frac{k_\|^2 v_A^2}{\omega^2}\eta_i &= \left[\left(1-\frac{\omega_{de}}{\omega}\right) - \frac{k_\|^2 c_{ia}^2}{\omega^2}\left(1-\frac{\omega_{di}}{\omega}\right)\Lambda_0(\eta_i)\right] \\ &\quad \cdot \left[\left(1-\frac{\omega_{di}}{\omega}\right) - \frac{\eta_i}{1-\Lambda_0(\eta_i)}\frac{k_\|^2 v_A^2}{\omega^2}\right] \end{aligned} \tag{6.45}$$

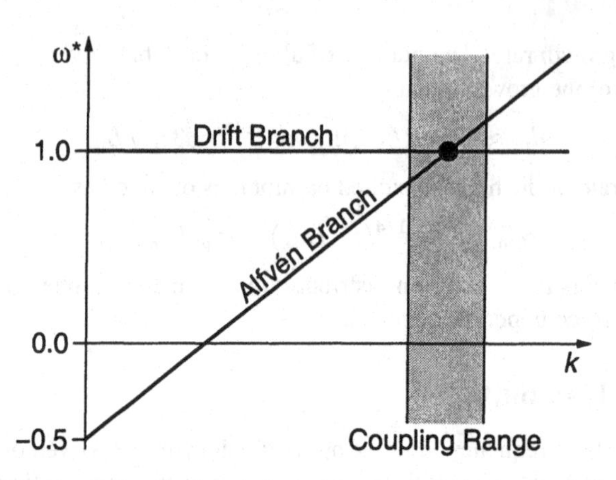

Fig. 6.2. Instability due to crossing of drift and Alfvén wave branches.

For small η_i this dispersion relation yields with vanishing left-hand side Alfvén and ion sound waves. On the other hand it simplifies considerably for an isothermal plasma with $T_e = T_i$ to become

$$\frac{k_\parallel^2 v_A^2}{\omega^2} = \frac{1}{\eta_i}\frac{\omega^* - 1}{\omega^* + 1}\left[\omega^*(\omega^* + 1) - \frac{k_\parallel^2 v_A^2}{\omega_{de}^2}\frac{\eta_i}{1 - \Lambda_0(\eta_i)}\right] \tag{6.46}$$

where we defined $\omega^* = \omega/\omega_{de}$. The three real solutions of this equation for small η_i are ω_{de}, and the conjugate pair

$$\omega = -\tfrac{1}{2}\omega_{de}\left[1 \pm \left(1 + 4k_\parallel^2 v_A^2/\omega_{de}^2\right)^{1/2}\right] \tag{6.47}$$

Instability arises because one of the pair branches crosses the ω_{de} branch, as sketched in Fig. 6.2. In the vicinity of the crossing point

$$k_\parallel v_A = \sqrt{5}\omega_{de}/4 \tag{6.48}$$

instability arises due to the coupling of two wave modes. The maximum growth rate is found for $\eta_e \ll 1, \zeta_{0e} \ll 1$ from $\gamma \propto \mathrm{Im}\,\epsilon_L/(\partial D/\partial\omega)$. It occurs at

$$k_{\parallel\max}^2 \approx 2\omega_{de}^2/v_A^2 \tag{6.49}$$

Only the imaginary part of ϵ_L comes into play in this approximation. One finds

$$\boxed{\gamma_{dA,\max} \approx \frac{(\pi\eta_i)^{1/2}}{3}\frac{v_A}{v_{the}}\omega_{de} \approx \left(\frac{m_e}{m_i\beta}\right)^{1/2}\frac{v_{thi}}{L_n}} \tag{6.50}$$

as the maximum growth rate of the *drift-Alfvén instability* where the right-hand side applies for plasma beta in the range $\beta \gg m_e/m_i$.

The mechanism of the drift Alfvén instability is the coupling of the Alfvén mode to the drift mode of constant frequency. The drift wave feeds the Alfvén wave with the energy required for wave growth. This is another mechanism of excitation of Alfvén waves which adds to the mechanism we already know as the firehose instability. There the wave is driven by macroscopic pressure differences, while here it is driven by the kinetic effect of the plasma inhomogeneity. As long as the inhomogeneity is maintained the Alfvén wave can be excited. But when the inhomogeneity is depleted, this kind of instability shuts off. In the magnetosphere one may expect excitation of Alfvén waves by this mechanism in all places, where a large amount of plasma is freshly injected and produces strong density gradients.

Lower-Hybrid Drift Instability

One of the most important instabilities is the *lower-hybrid drift instability*. The reason for its importance is that it excites waves near the lower-hybrid frequency which is a natural resonance. Hence, the instability can reach large growth rates. The energy needed to excite the instability is taken from the diamagnetic drift of the plasma in a density gradient. This is similar to the modified two stream instability insofar that the diamagnetic drift gives rise to a transverse current in the plasma which acts in a way corresponding to the current drift velocity of the modified two stream instability.

In general, the lower-hybrid drift instability is an electromagnetic instability causing whistler waves near the resonance cone to grow. But in a good approximation one can treat it as an electrostatic instability, in which case we must refer only to the longitudinal inhomogeneous dielectric response function, $\epsilon_{\rm L}$. Because the wavelength should be much larger than the gyroradius of the electrons, we have $\zeta_e \gg 1$. Moreover, in the electrostatic limit we assume that the ions are unmagnetized, drifting with v_{di}. Then the real part of the longitudinal response function yields

$$\epsilon_r(\omega, \mathbf{k}) = 1 + \frac{\omega_{pe}^2}{\omega_{ge}^2}\frac{k_\perp^2}{k^2} + \frac{1}{k^2\lambda_D^2}\frac{\omega_{de}}{\omega} + \frac{1}{k^2\lambda_{Di}^2}[1 - Z(\zeta_i)] = 0 \tag{6.51}$$

Here we have

$$\zeta_i = (\omega - \mathbf{k}\cdot\mathbf{v}_{di})/kv_{\rm thi} \tag{6.52}$$

It is easy to see that for nearly perpendicular propagation, in which case the wave is actually electrostatic, $k_\perp^2 \approx k^2$, and for wave phase velocities much larger than the ion thermal speed, $\omega/k \gg v_{\rm thi}$, the real part of the solution for the frequency, ignoring the imaginary part of the response function, becomes simply the lower-hybrid frequency

$$\omega_{\rm lhd}^2 \approx \omega_{lh}^2 \tag{6.53}$$

To find the more precise expressions for the wave frequency and the growth rate of the instability, one makes use of the Galilean invariance and transforms to a system in which the ions are at rest. In this particular system the electrons drift with velocity $-\mathbf{u} = \mathbf{v}_{di}$, such that the real part of the dispersion relation takes the form

$$1 + \frac{\omega_{pe}^2}{\omega_{ge}^2} + \frac{1}{k^2\lambda_D^2}\frac{\omega_{de}}{(\omega - \mathbf{k}\cdot\mathbf{v}_{di})} + \frac{1}{k^2\lambda_{Di}^2} = 0 \tag{6.54}$$

This is a dispersion relation which corresponds to a two-stream instability. We know that it has solutions for $\omega < \mathbf{k}\cdot\mathbf{v}_{di}$. Further assuming weak instability, the growth rate can be calculated exactly as for the case of the modified two-stream instability, maintaining the imaginary part of the ion plasma dispersion function in Eq. (6.51). Thus it is the inverse ion Landau damping at the positive gradient of the ion distribution functions which feeds the instability. For $v_{di} < v_{\mathrm{th}i}$ and weak growth rate the frequency obtained from (6.51) is

$$\omega_{\mathrm{lhd}} = -\frac{\Lambda_0(\eta_e)k_y v_{di}}{1 + T_e/T_i - \Lambda_0(\eta_e)} \tag{6.55}$$

and the growth rate obtained under the assumption $\gamma \ll \omega$ is

$$\gamma_{\mathrm{lhd}} = \omega_{\mathrm{lhd}}\left(\frac{\pi}{2}\right)^{1/2}\frac{T_e}{T_i}\frac{v_{di}}{v_{\mathrm{th}i}}\frac{\Lambda_0(\eta_e)}{[1 + T_e/T_i - \Lambda_0(\eta_e)]^2} \tag{6.56}$$

A numerical example of the dependence of frequency and growth rate on the wavenumber is shown in Fig. 6.3.

When $T_e \ll T_i$, the maximum of the growth rate occurs at a frequency

$$\boxed{\omega_{\mathrm{lhd,max}} \approx \omega_{lh} \approx \mathbf{k}\cdot\mathbf{v}_{di}/2} \tag{6.57}$$

The maximum growth rate is given by

$$\boxed{\gamma_{\mathrm{lhd,max}} \approx 0.6\,\omega_{lh}v_{di}^2/v_{\mathrm{th}i}^2} \tag{6.58}$$

Since the ion drift velocity can be quite high, this growth rate can become very large.

Extensions of the theory to larger ratios T_e/T_i are possible and show that the growth rate is still large for increasing electron temperatures. Moreover, extensions to oblique propagation and inclusion of electromagnetic corrections show that the lower-hybrid drift instability exists also at larger angles in the whistler band. It may thus play an important role in many applications to space plasmas. One should, however, mention that the conventional lower-hybrid drift instability, as presented here, heavily

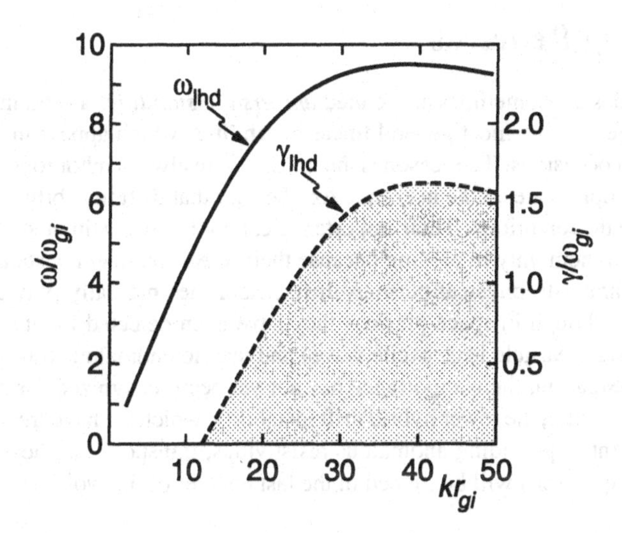

Fig. 6.3. Lower-hybrid drift instability for $r_{gi}/L_n = 0.5$.

depends on plasma β. When β increases to $\beta > 1$, the instability is quenched. On the other hand, temperature gradients help to destabilize the instability. These are important facts which must be taken into account when applying the lower-hybrid drift instability to real space plasma conditions.

The lower-hybrid drift instability has been applied to low-frequency electrostatic waves excited in the shock front region of the Earth's bow shock, the density gradient region at the magnetopause and in the low latitude boundary layer, the density gradient region in the Earth's magnetotail plasma sheet, and the density gradients in the upper ionosphere. Excitation of lower-hybrid drift waves at the bow shock is probably suppressed by the high beta of the oncoming solar wind. More important is the excitation of lower-hybrid drift waves in the gradients at the magnetopause, both the transition region and the low latitude boundary layer. Relatively high wave intensities have been reported here close to the lower-hybrid frequency. Lower-hybrid waves may add to transport of cold plasma across the magnetopause if the wave intensities are high.

Lower-hybrid waves can be excited in the tail plasma sheet as long as one stays outside of the neutral sheet current layer proper. The increase in the measured electric wave spectra toward low frequencies in Fig. 4.17 is an indication of the presence of lower-hybrid waves in the tail (for plasma sheet conditions the lower-hybrid frequency is close to 10 Hz). In the ionosphere, in spite of the steep density gradients appearing in the E- and F-layers, the lower-hybrid drift instability is stabilized by the presence of frequent collisions. Nevertheless, it can evolve under special circumstances at occasional very steep gradients.

Concluding Remarks

Drift instabilities are sometimes also called *universal instabilities* a term indicating that drift instabilities are the most general linear instabilities which appear in almost every place and at all occasions. The reason is that plasmas are always inhomogeneous at least on the microscopic scales. One may therefore be sure that drift instabilities will be met under all realistic conditions. Their importance cannot be overestimated. Nevertheless we have treated them only in passing because their linear treatment is standard.

In space and astrophysical plasmas drift instabilities naturally play an important role. However, though in space plasmas they have been detected in situ and are well known to generate particle loss, excite waves and lead to anomalous transport, heating and energy dissipation, their importance has not yet been recognized for astrophysical plasmas. In particular, the lower hybrid drift instability, which we have treated in length, is very important in providing anomalous resistivities, transport, and heat conduction. Some of these questions will be treated in the last chapter of this volume.

Further Reading

The full drift kinetic electromagnetic tensor is derived in [2], in [3], and in a particularly transparent form in [1]. A short but informative article on instabilities in inhomogeneous plasmas is found in [4].

[1] S. P. Gary, *Theory of Space Plasma Microinstabilities* (Cambridge University Press, Cambridge, 1993).

[2] N. A. Krall and A. W. Trivelpiece, *Principles of Plasma Physics* (McGraw-Hill, New York, 1971).

[3] A. B. Mikhailovskii, *Theory of Plasma Instabilities II* (Consultants Bureau, New York, 1976).

[4] A. B. Mikhailovskii, in *Handbook of Plasma Physics, Vol. 1*, eds. A. A. Galeev and R. N. Sudan (North-Holland Publ. Co., Amsterdam, 1983), p. 586

7. Reconnection

The magnetohydrodynamic instabilities discussed in Chap. 3 are ideal instabilities. They lead to bending and deformation of magnetic flux tubes, but the frozen-in condition remains valid. This condition breaks down when the plasma becomes non-ideal. Whenever this happens, the magnetic field starts diffusing across the plasma, and magnetic flux can be exchanged between different plasmas in mutual contact. The process of magnetic flux exchange is a diffusive process with magnetic diffusivity resulting from either collisions or being the consequence of nonlinear interactions of the kind reviewed in Chap. 12. Any such diffusion process implies that the magnetic field lines, which in an ideal plasma can be considered as unbreakable strings, are re-ordered such that the field line configuration after re-ordering looks different from the initial one.

Physically the magnetic flux is rearranged during diffusion, a process which usually is slow. However, in some model cases this kind of re-ordering can become comparably fast when the diffusion process is restricted to a region of small spatial extent. This is the important case when one speaks of *magnetic reconnection*. Magnetic reconnection or merging plays a key role in the processes at the magnetopause, in the magnetotail, in solar active regions, and in a number of astrophysical applications like accretion disks. It has experienced enormous attention in the space plasma community during the last forty years. Both, remote and in situ observations of magnetic diffusion processes in space have revealed that re-ordering of magnetic fields may proceed at very high speeds when restricted to narrow spatial regions.

It must be emphasized that from the physical point of view there is nothing particular about reconnection insofar as any process which causes violation of the ideal conditions in a plasma and thus leads to collisional effects, always results in diffusion of magnetic fields and thus reconnection and rearranged magnetic field topology. The real question is how and where such diffusivities are generated in an otherwise ideal or collisionless plasma. But once they exist, it is quite natural that oppositely directed field components will cancel each other and cause reconnection. In the following we review the currently accepted and competing reconnection models for both collisional and collisionless plasmas. Since in the collisional case the diffusivity is natural and its origin must not be explained, we start with the collisional reconnection models before discussing some versions of collisionless reconnection.

7.1. Reconnection Rates

Consider two mutually approaching ideal magnetohydrodynamic flows containing oppositely directed magnetic fields until they meet in a plane at $x = 0$, as shown in Fig. 7.1. In ideal magnetohydrodynamics the frozen-in condition requires that each flow element is fixed to its particular field line. Hence, when the flows meet they will be unable to mix and pressure balance will slow them down and force them to rest. But if the magnetic fields in the two flows are oppositely directed and of the same strength, they may annihilate each other at the plane of contact, allowing the plasmas to mix.

Magnetic Diffusion

Mixing and annihilation is permitted only when the plasma is non-ideal, because only then the magnetic field can diffuse across the flow and annihilate. This is described by the general induction equation (I.5.2) in our companion book

$$\frac{\partial \mathbf{B}}{\partial t} = \nabla \times (\mathbf{v} \times \mathbf{B}) + \frac{1}{\mu_0 \sigma} \nabla^2 \mathbf{B} \tag{7.1}$$

where σ is the constant electrical conductivity. The diffusion time of the magnetic field is given in Eq. (I.5.6) of our companion book

$$\tau_d = L^2 \sigma \mu_0 \tag{7.2}$$

where L is the global length scale of the change in $\mathbf{B}$, and the diffusion velocity is

$$v_d = L/\tau_d = 1/L\sigma\mu_0 \tag{7.3}$$

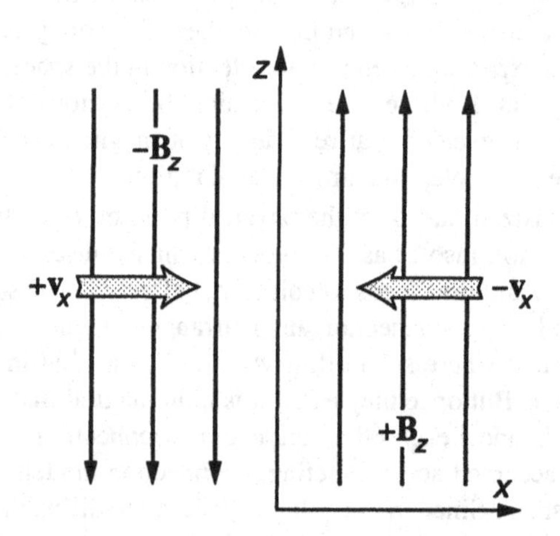

Fig. 7.1. Counterstreaming flows leading to reconnection.

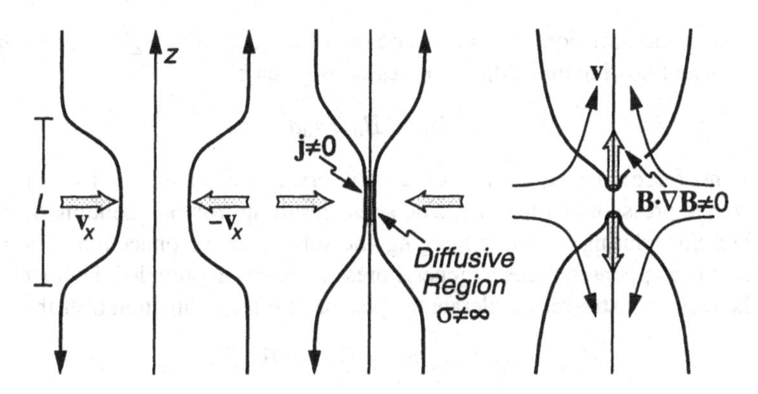

Fig. 7.2. Formation of current sheet and neutral point in reconnection.

Assume that the two approaching but oppositely directed field lines are locally bent over the length, L, along the field. At the nose of the bended field lines they will contact first. Diffusion and annihilation will set in here and the field lines will be broken and reconnected in another topology. A magnetic neutral point is created spontaneously at the former point of contact, with zero magnetic field strength in its center. There is no need for the flow to stop, because part of it can flow along the magnetic field lines into the other region, part of it will deviate and start transporting the newly merged field lines away from the neutral point. This motion is driven by the magnetic tension force stored in the highly bent reconnected field lines. Hence, the direction of the deviated flow will be perpendicular to that of the incoming flow. Figure 7.2 visualizes this kind of process.

During the initial approach of two magnetized flows a current layer is created, with the current flowing in the y direction. When the magnetic field is bent locally, the current increases. Hence, the initial instability is a transverse current instability, leading to diffusion of magnetic flux and current concentration in the neutral point, which in three dimensions will extend into a neutral line or X-line. However, for stationary flow this initial phase must be small. For *stationary reconnection* diffusion becomes stationary and an equilibrium between the inflowing mass and magnetic flux, the magnetic diffusion, and the outflowing mass and magnetic flux is reached soon.

Sweet-Parker Reconnection

Historically the first stationary model of reconnection is the *Sweet-Parker reconnection* model. It permits for steady flow and diffusion of the magnetic field across a long *diffusion region* of length, $2L$, and narrow width, $2d$. The model is based on the conservation of mass, momentum, energy, and magnetic flux in the flow, from the ideal plasma to both sides of the diffusion region into and out of the diffusion region. From Ampères law the current flowing in the diffusion region is given by

$$j_{dy} \approx B_{0z}/\mu_0 d \tag{7.4}$$

The Lorentz force entering the momentum conservation equation consists of two terms, the magnetic pressure and the magnetic stress resulting from the bending of the magnetic field lines during magnetic merging and subsequent reconnection. The pressure force term is compensated due to general pressure balance, provided the boundary is at rest. The magnetic stresses accelerate the plasma into the z direction over the length L

$$(\mathbf{j}_{dy} \times \mathbf{B}_{dx}) \cdot \hat{\mathbf{e}}_z \approx j_{dy} B_{dx} = B_{0z} B_{dx}/\mu_0 d \tag{7.5}$$

In the stationary case, this force balances the nonlinear velocity term in the momentum equation

$$nm_i(\mathbf{v} \cdot \nabla)v_{dz} \approx nm_i v_{dz}^2/L \approx B_{0z} B_{dx}/\mu_0 d \tag{7.6}$$

Since the magnetic field is divergence-free, we have

$$B_{0z}/L \approx B_{dx}/d \tag{7.7}$$

so that the equation for the outflow velocity becomes

$$v_{dz}^2 \approx B_{0z}^2/\mu_0 nm_i = v_{A0}^2 \tag{7.8}$$

which is the Alfvén speed in the inflowing plasma. For a plasma moving into the diffusion region with inflow speed v_{0x} we can estimate the efficiency of the reconnection process by defining a *reconnection rate*, $\mathcal{R} = v_{0x}/v_{dz} = M_{A0}$, which is equal to the incident Mach number. Under stationary conditions the plasma leaves the neutral point region with the Alfvén speed of the inflowing plasma. Hence, the reconnection proceeds via a large-amplitude Alfvén wave. Actually, it is a rotational discontinuity, where the normal component of the magnetic field is produced by the reconnection process.

The magnetic field is incompressible and carried into and out of the diffusion region at the speed of the flow. Hence, we have in analogy to Eq. (7.3)

$$v_{0x} = 1/d\sigma\mu_0 \tag{7.9}$$

with $2d$ the width of the diffusion region. Conservation of mass requires that the rate of mass inflow equals that of the outflow. Because of the incompressibility of the medium

$$v_{0x} L \approx v_{A0} d \tag{7.10}$$

Now we can eliminate the width of the diffusion layer, d, obtaining

$$v_{0x}^2 \approx v_{A0}/L\sigma\mu_0 \tag{7.11}$$

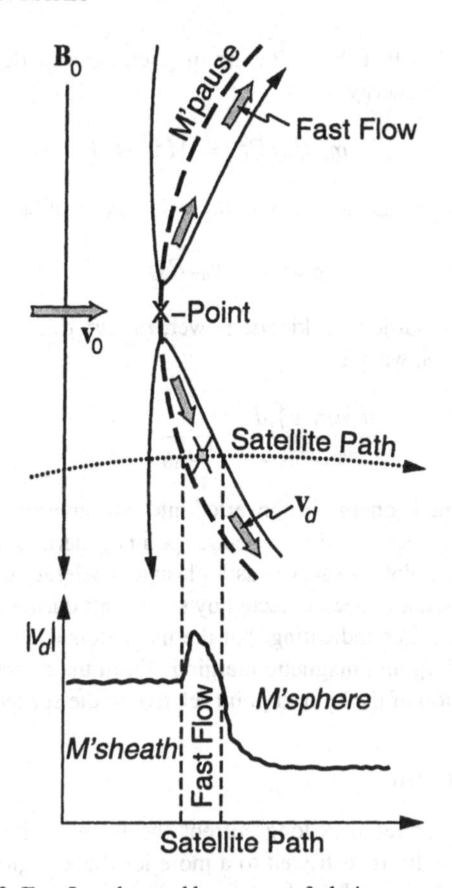

Fig. 7.3. Fast flow detected by spacecraft during reconnection.

or, if we divide by the Alfvén velocity and define the Alfvén Mach number in the usual way as the ratio of the velocity to the Alfvén speed and the magnetic Reynolds number defined in Sec. 5.1 of our companion book, $R_{m0} = L\sigma v_{0A}\mu_0$

$$\boxed{\mathcal{R}_{\mathrm{SP}} = \mathcal{M}_{\mathcal{A}\prime} = \mathcal{R}_{\Updownarrow\prime}^{-\infty/\epsilon}} \tag{7.12}$$

This is the *Sweet-Parker reconnection rate*. It shows that the speed at which the magnetic field can diffuse in and through the diffusion region is determined by the magnetic Reynolds number and is low for high numbers. Thus in space plasmas as for instance the plasma at the magnetopause which are practically collisionfree the reconnection process must be very slow unless anomalous effects generate much lowered conductivities.

Reconnection implies that the inflowing magnetic energy density is much larger than the inflowing kinetic energy

$$\mu_0 n m_i v_{0x}^2 / B_{0z}^2 = M_{A0}^2 \ll 1 \tag{7.13}$$

Concerning the outflow, we recognize that magnetic flux must be conserved

$$v_{0x} B_{0z} = v_{dz} B_{dx} \tag{7.14}$$

Forming the ratio of the outflowing kinetic power, $nm_i v_{dz}^3 d/2$, to the inflowing magnetic power, $v_{0x} B_{0z}^2 L/\mu_0$, we get

$$\frac{\mu_0 n m_i v_{dz}^3 d}{2\, v_{0x} B_{0z}^2 L} = \frac{v_{dz}^3}{2 v_{A0}^3} = \frac{1}{2} \tag{7.15}$$

Half the inflowing magnetic energy is converted into flow energy and thus transformed into acceleration. Hence, reconnection at neutral points generates high-speed flows escaping from the neutral point and serves as a plasma acceleration mechanism. These high-speed flows have actually been detected by spacecraft during high magnetic shear magnetopause crossings, thus indicating that the magnetopause can behave as a rotational discontinuity undergoing magnetic merging. From the direction of the flow one can determine the position of the neutral point relative to the spacecraft (see Fig. 7.3).

Petschek Reconnection

Sweet-Parker reconnection requires long current sheets and is a rather slow process. When the finite conductivity is restricted to a more localized region, the reconnection process becomes faster because L is shorter. This mechanism does, however, require more complicated processes to work at the magnetopause near the reconnection site. The outer region has no current flow and, since reconnection proceeds only in a small diffusion region around the neutral point, not all plasma can cross the diffusion region. Most of the plasma must turn around before reaching the interface between the two counterstreaming flows. Because this change in flow direction and speed will be abrupt, it must occur at a shock. Since the magnetic field lines are refracted toward the shock normal, the shocks involved are *slow mode shocks* (cf. Fig. 8.11 in our companion book and Fig. 7.4 in this volume). Here 3/5 of the inflowing magnetic energy are converted into kinetic energy behind the shock, the remaining 2/5 are used to heat the plasma.

Although there is no reconnection at the shocks, a normal component, B_n, exists. The slow shocks are large-amplitude slow magnetosonic waves, which in a hot plasma travel with the Alfvén speed based on its normal field

$$v_{ss}^2 = B_n^2 / n m_i \mu_0 \tag{7.16}$$

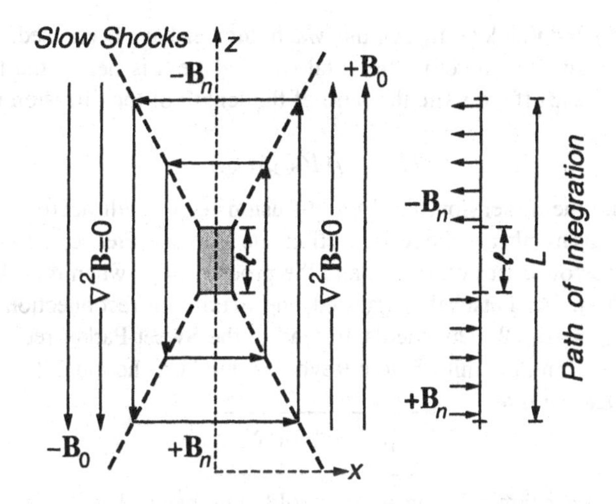

Fig. 7.4. Mechanism of Petschek Reconnection.

Figure 7.4 shows the idealized geometry of the Petschek model with the narrow and short diffusion region of length ℓ along z and the four slow shocks connected to it. These shocks extend into the current-free environment, where $\nabla^2\mathbf{B} = 0$. The shocks change the direction of the magnetic field by 90°, as is required for reconnection. These fields are the reconnected fields which leave, as in the Sweet-Parker model, from the diffusion region and are driven by magnetic stresses.

The field change at the shock is a disturbance of the initial external field, $\mathbf{B}_0$, in the outer region. This disturbance can be calculated assuming that each element of the shock causes a small magnetic disturbance, $\delta B = b/z$, which decays with distance of the element from the diffusion region along the shock surface, z. The flux caused by it is the product of the disturbance field and the surface element in one dimension, πz, yielding $\pi z b/z = b\pi$. But because the element has the length dz along the shock, the flux is also equal to $2B_n dz$ and thus $b = (2B_n/\pi)dz$. Integrating along the shock length, L, over all contributions of the elements excluding the diffusion region of length, ℓ, where no shocks exist, gives

$$\frac{2B_n}{\pi}\left(\int_{-L/2}^{-\ell/2}\frac{dz}{z} - \int_{\ell/2}^{L/2}\frac{dz}{z}\right) = -\frac{4B_n}{\pi}\ln\left(\frac{L}{\ell}\right) \tag{7.17}$$

This field must be added to the external field at large distances. The result is

$$B_{0z} = B_0\left[1 - \frac{4M_{A0}}{\pi}\ln\left(\frac{L}{\ell}\right)\right] \tag{7.18}$$

where Eq. (7.16) and the definition of the Mach number have been used. The external Mach number is small. Therefore the total external field is nearly the field at large distances in this case. If we write the ratio of the length of the diffusion region to the lateral scale as

$$\ell/L = 1/(R_{m0}M_{A0}) \tag{7.19}$$

it is obvious that the dimension, ℓ, of the diffusion region will decrease if either the magnetic Reynolds number or the reconnection rate increase. Hence, high reconnection rates require a narrow diffusion region and the process stops when their lateral extent becomes too short. As Petschek suggested, the maximum reconnection rate will be achieved for $B_{0z} = B_0/2$. This yields, instead of the Sweet-Parker reconnection rate given in Eq. (7.12), in the limit of large Reynolds numbers the much faster maximum *Petschek reconnection rate*

$$\boxed{\mathcal{R}_{\mathrm{P}} \approx \pi/\forall \ln \mathcal{R}_{\Updownarrow},} \tag{7.20}$$

It depends only logarithmically on the Reynolds number and varies much less with conductivity. Petschek reconnection is therefore a very efficient mechanism to merge magnetic fields and to provide magnetic diffusion through narrow diffusion regions.

Because the conductivities in space plasmas are so high, Sweet-Parker reconnection is usually a very inefficient process which will not lead to violent mixing of magnetic fields and the related energy releases. Processes which locally decrease the conductivity will enhance reconnection. But Petschek reconnection, whenever it occurs, will always be much faster providing fast dissipation and effective mixing as well as violent acceleration of plasma from the reconnection region. However, it is at present not clear that Petschek reconnection can evolve in natural systems.

7.2. Steady Collisionless Reconnection

The reconnection rates given in the previous section all refer to the collisional regime with at least a small resistivity. In space plasmas such resistivities do not exist a priori. Hence, the magnetohydrodynamic approach becomes invalid, and one should switch to a steady kinetic approach where particle dynamics initiates reconnection. In the next sections we will consider an instability, the *tearing mode*, which may work in both the resistive and collisionless regimes. Here we briefly discuss the possibility of a steady collisionless state.

Linear Regime

Consider a two-dimensional configuration with all quantities independent of y. This configuration is shown schematically in Fig. 7.5. Then the magnetic field can be represented as $\mathbf{B} = -\hat{\mathbf{e}}_y \times \nabla A(x, z)$, where A is the only surviving vector potential com-

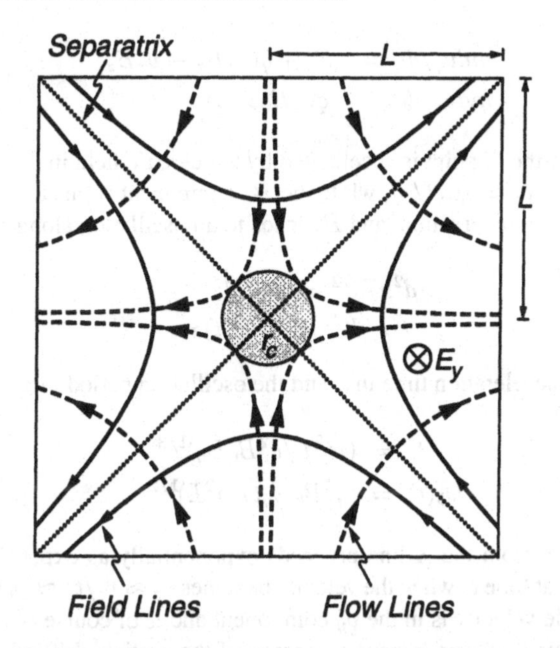

Fig. 7.5. Two-dimensional scheme of collisionless reconnection.

ponent, and because of the assumption of steadiness, the electric field is $\mathbf{E} = E_y\hat{\mathbf{e}}_y$. Initially $A_0 = B_0/(2L)(x^2 - z^2)$, so that

$$\mathbf{B}_0 = (B_{0x}, B_{0y}, B_{0z}) = B_0(z/L, 0, x/L) \tag{7.21}$$

In such a configuration the particles drift adiabatically with $E\times B$-drift velocity. The plasma flows in along the z axis and out along x. Only in the vicinity of the neutral point, in the region of radius

$$r_c \approx L(mE_y/qB_0^2L)^{1/3} \tag{7.22}$$

this flow is violated. Here conversion of magnetic energy into flow energy occurs. The size of this region can be estimated realizing that the inertial term in the equation of motion comes into play near the neutral point, where $B = B_0r/L$, when the convective derivative of the flow velocity, $\mathbf{v}_E = \mathbf{E}\times\mathbf{B}/B^2$, equals the inertial term, $md\mathbf{v}/dt$, or

$$v_E dv_E/dr \approx E_y^2L^2/B_0^2r^3 \approx qE_y/m \tag{7.23}$$

which immediately leads to the above expression (7.22). Inside the dissipation region one must solve the set of equations

$$mdv_x/dt \;=\; qv_yB_0x/L$$

$$\begin{aligned} m dv_y/dt &= qE_y + q(v_z B_x - v_x B_z) \\ m dv_z/dt &= -q v_y B_0 z/L \end{aligned} \tag{7.24}$$

A particle entering the X-point is accelerated by the electric field in the y direction and receives a speed $v_y(t) \approx qE_y t/m$, while the B_z component wants to eject the particle from the X-line in the x direction, and B_x leads to an oscillation along z, so that

$$\begin{aligned} d^2x/dt^2 &= tx/t_c^3 \\ d^2z/dt^2 &= -\omega_b^2(t) z \end{aligned} \tag{7.25}$$

where the critical acceleration time in x and the oscillation period in z are given by

$$\begin{aligned} t_c &= (m^2 L/q^2 B_0 E_y)^{1/3} \\ \omega_b(t) &= (q^2 B_0 E_y t/m^2 L)^{1/2} \end{aligned} \tag{7.26}$$

Hence, the ejection from the X-line proceeds exponentially as $\exp(t/t_c)^{3/2}$, and the particle reaches r_c at time t_c when the velocity becomes $v_c \approx r_c/t_c \approx (qE_y^2 L/mB)^{1/3}$. The gain in particle velocity is in the v_y component and is of course of the order of v_c. It is thus transformed into the transverse energy of the particle drifting away from the X-line. The energy turns out to be proportional to $m^{1/3}$ such that the main dissipation is due to ions rather than electrons. Moreover, the net electric current is carried by the ions, because it is basically a polarization current, $I \approx en_i v_c r_c^2$. It is therefore sufficient to consider the ion motion. In other words, the collisionless reconnection in this approximation is driven by ion inertia.

Outside the dissipation region ion dynamics is simple electric drift with streamlines

$$dx/v_{Ex} = dz/v_{Ez} \tag{7.27}$$

which are hyperbolic with $xz =$ const. The ion flux entering the dissipation region is bounded by the streamline with $x_c \approx z_c \approx r_c$ and, for steady flow,

$$n_0 x_L v_{Ez}(L) = n_i x_c v_{Ez}(r_c) \tag{7.28}$$

where n_e is the undisturbed electron background density, and the index L as well as the dependence on L indicate that the corresponding quantities must be taken at distance L (or at infinity). Taking into account the electric field drift expressions, one finds for ion number and current densities

$$\begin{aligned} n_i &\approx n_0(r_c/L)^2 \approx n_0(m_i E_y/eLB_0^2)^{2/3} \\ j &\approx en_i v_c \approx en_0(E_y/B_0)(m_i E_y/eLB_0^2)^{1/3} \end{aligned} \tag{7.29}$$

Thus the total magnetic energy dissipation is proportional to the third power of the electric field, $Q_B = E_y I \propto E_y^3$.

When the magnetic field generated by the electric current becomes comparable to the initial field, B_0, this linear approximation will break down. This takes place at a field strength $\delta B \propto \mu_0 jr/2$. Replacing the current density and radius in this expression and using the external Mach number, $M_{A0} = v_0/v_{A0}$, with $v_0 = E_y/B_0$ gives the following limit for the linear regime

$$M_{A0} \leq M_c = \tilde{R}_{m0}^{-1/4} \tag{7.30}$$

where the Reynolds number has been defined as the ratio of the total length to the external ion inertial length, $\tilde{R}_{m0} = (L\omega_{pi}/c)^{1/2}$. This indicates that the reconnection is determined by the ion inertia. The collisionless Reynolds number is very large for large L and small inertial lengths. But the Reynolds number enters at the fourth root of this ratio. For larger reconnection rates one must extend the calculation to the nonlinear regime.

We now simplify to the incompressible case outside the reconnection region far away from the X-line. Then the external ion density is equal to n_0, and

$$j \approx en_0 v_0 \approx en_0(eLE_y^2/m_i B_0)^{1/3} \tag{7.31}$$

The linear approximation holds under the restriction on the electric field

$$E_y \leq E_{yc} = m_i v_{A0}^2 c/eL^2\omega_{pi} \tag{7.32}$$

which may be written as $M_{A0} \leq M^* \approx \tilde{R}_{m0}^{-1}$. This condition is more restrictive than Eq. (7.30) and readily violated. Then reconnection enters the nonlinear regime.

Nonlinear Regime

We now sketch the nonlinear estimates for the assumed large reconnection rates. The electric field is $E_y > E_{yc}$. Let us use dimensionless quantities with the electric field normalized to E_{yc}, length to c/ω_{pi}, magnetic field to $B_0 c/L\omega_{pi}$, and time to L/v_{A0}. This yields the normalized vector potential $A_0 = (x^2 - z^2)/2$. The geometry is shown in Fig. 7.6. It is an elongated reconnection site of length $2L$ and width $2d$ for the normalized electric field $E_y \gg 1$, far above the critical field, E_{yc}. Then $dB_x/dz \approx j$ for the normalized magnetic field and current in the reconnection site, under the assumption that we can take the current as constant. From the condition $\nabla \cdot \mathbf{B} = 0$ we obtain

$$B_z(x) \approx \frac{jd}{\pi} \int_{L-x}^{L+x} \frac{d\xi}{\xi} \approx \frac{2jd}{\pi}\frac{x}{L} \tag{7.33}$$

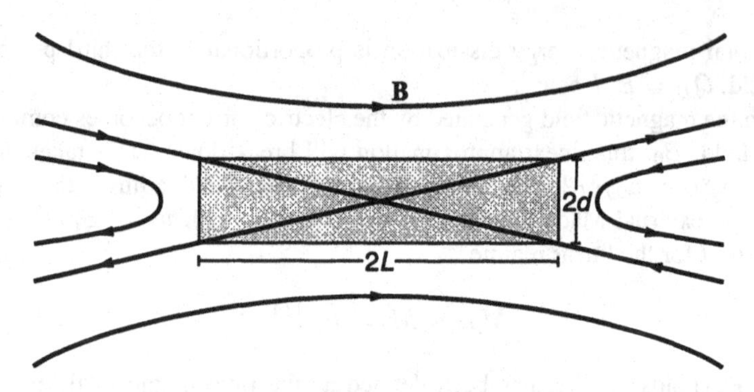

Fig. 7.6. Elongated reconnection site.

which yields for the derivative of the small normal magnetic field component $dB_z/dx \approx 2jd/\pi L \ll 1$, which is nearly zero. The main variation in the field is in the x component, and the reconnection site is a nearly one-dimensional current sheet.

Following the same procedure as in the linear regime we find that the time scale, t_c, of deviation from the x axis and the oscillation frequency, ω_b, are now given by

$$
\begin{aligned}
t_c &\approx \left(E_y \frac{dB_z}{dx}\right)^{-1/3} \\
\omega_b(t) &\approx \left(tE_y \frac{dB_x}{dz}\right)^{1/2}
\end{aligned}
\tag{7.34}
$$

These scales are very different from the linear scales. In particular, t_c is now very long since the derivative of the normal component B_z, is small, and the bounce-averaged motion of the ions is decoupled from the oscillation along the z axis. This motion is described by the following equations

$$
\begin{aligned}
d\bar{v}_x/dt &= \bar{v}_y \bar{x} \\
d\bar{v}_y/dt &= 1 - \bar{v}_x \bar{x}
\end{aligned}
\tag{7.35}
$$

where we normalized the time to the new t_c, length to $x_c = E_y t_c^2$, and velocities to $v_c = E_y t_c$. Ions entering the dissipation region with $\bar{v}_y = 0$ will experience free acceleration along y and exponential expulsion from the neutral line for short times, $t < t_c$. For $t > t_c$ their motion is an electric drift. This indicates that the length x_c defines the edge of the dissipation region in x direction

$$x_c \approx E_y t_c^2 \approx L \tag{7.36}$$

The width, d, of the dissipation region is estimated from the amplitude of the ion bounce motion at time t_c. We do not go through this somewhat complicated scaling argument, but present the final result which is obtained. This scaling can be reduced to a dependence of all interesting quantities on the electric field (or the reconnection rate). Then current, width and density in the dissipation region scale as powers of E_y

$$d \approx E_y^{-4/3} \approx n_0^{-1} \approx j^{-7/4} \tag{7.37}$$

indicating that the width becomes narrow for large fields, and the density and current increase. It can also be shown that the reconnection rate in this nonlinear case scales approximately as

$$\boxed{\mathcal{R}_{CL} \approx \tilde{R}_{A0}^{-1/2}} \tag{7.38}$$

depending on the ion inertial Reynolds number. Formally, this scaling is about the same as the Sweet-Parker scaling, but the ion Reynolds number entering here may help to increase the reconnection rate.

7.3. Resistive Tearing Mode

So far we have avoided to touch the problem of how reconnection sets on. The instability envisaged must work in a plasma where at least the magnetic field is inhomogeneous, changing its direction across the current sheet. This inhomogeneity implies that the currents flowing across the field are also inhomogeneous. Thus the instability is by nature a macroinstability, but it is not necessarily a magnetohydrodynamic instability. Since, in addition, a dissipative region is required around the neutral point, the instability must be either resistive or must give rise to some kind of collisionless dissipation in the diffusion region to permit the magnetic field to diffuse. In the latter case the instability will be collisionless but dissipative. We will discuss this type of instability in the next section, but here briefly review the idea of a collisional fluid plasma instability. Its name is *resistive tearing instability*, because it will cause the magnetic field to form a succession of magnetic islands and neutral points along the current layer.

Mechanism

Let us assume a long current sheet, like the neutral sheet current in the center of the Earth's plasma sheet. Figure 7.2 in our companion book shows the schematic change in the magnetic field across the current layer. Small disturbances in the current flow will cause the current to pinch off due to the self-generated magnetic field of the current. Instability occurs when the plasma has a finite resistivity. This effect is restricted to the region of weak magnetic field, the neutral sheet, as one realizes from inspection of Ohm's law

$$\delta\mathbf{E} + \delta\mathbf{v} \times \mathbf{B}_0 = \delta\mathbf{j}/\sigma \tag{7.39}$$

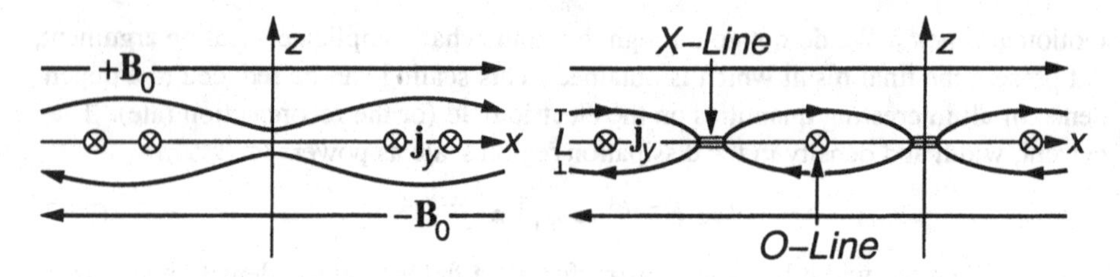

Fig. 7.7. Formation of magnetic islands in a current sheet.

In the center of the neutral sheet where $B_0 = 0$ the electric field and current are supported by the finite conductivity only, while outside the convection term dominates and conductivity can be neglected as long as it is large enough. We assume that the region where the conductivity is important is $-(d/2) < z < (d/2)$. Here the induction law given in Eq. (7.1) becomes

$$\frac{\partial \delta \mathbf{B}}{\partial t} = \frac{1}{\mu_0 \sigma} \nabla^2 \delta \mathbf{B} \tag{7.40}$$

while the exterior equation has the particularly simple form

$$\frac{\partial \delta \mathbf{B}}{\partial t} = \nabla \times (\delta \mathbf{v} \times \mathbf{B}_0) \tag{7.41}$$

The problem consists in solving these two equations together with the inhomogeneous linearized equation of motion which determines the velocity distortion, $\delta\mathbf{v}$, under the boundary conditions across the neutral current sheet. A simple dimensional estimate for a thin current layer with scale length, $L_B = |B_0/\nabla B_0|$, of the background magnetic field, $\mathbf{B}_0$, suggests a growth rate

$$\gamma_{\text{tea}} = 4/\mu_0 \sigma d^2 \tag{7.42}$$

under the condition that the wavelength in the x direction is longer than the thickness of the current layer, $k_x < 1/L_B \approx 1/d$. The growth rate still contains the undefined width, *d*, of the current layer, but narrow layers will have much larger growth rates than thick layers. Moreover, when the conductivity decreases the growth rate increases. The result of this instability will be that long wavelength magnetic islands form in the current layer containing a number of neutral points where the magnetic field reconnects (see Fig. 7.7). These islands contain closed toroidal magnetic field lines converging onto the current in their center. They bear the name magnetic O-points or, in three dimensions, *O-lines*, in contrast to the neutral points or lines which separate the islands, and the last field line belonging to the islands which goes through all the neutral points is the *separatrix*.

The resistive tearing mode may explain the mechanism of how a current layer may become unstable and cause reconnection. The resistive tearing mode is believed to play a decisive role in solar physics, while it is the collisionless tearing mode discussed in the next section, which is of importance in magnetospheric applications.

Growth Rate

The evolution of the total magnetic field is determined by Eq. (7.1)

$$\frac{\partial(\mathbf{B}+\delta\mathbf{B})}{\partial t} = \nabla\times[\delta\mathbf{v}\times(\mathbf{B}+\delta\mathbf{B})] + \frac{1}{\mu_0\sigma}\nabla^2(\mathbf{B}+\delta\mathbf{B}) \tag{7.43}$$

Here $\delta\mathbf{v}$ is the velocity perturbation field. The unperturbed state is assumed static, such that we are dealing with a non-driven two-dimensional case, $\partial/\partial y = 0$. Let us introduce stream functions ψ, $\delta\psi$, and $\delta\phi$ for $\mathbf{B}$, $\delta\mathbf{B}$, and $\delta\mathbf{v}$, respectively. Then any of these vectors assumes the form

$$\mathbf{B} = (-\nabla_z\psi, B_y, \nabla_x\psi) \tag{7.44}$$

and so on. The component B_y is a constant, which is non-zero only for the undisturbed magnetic field. Moreover, one finds $\mathbf{B}\cdot\nabla\psi = 0$, i.e., the stream function ψ is constant along a field line. The same holds for the other quantities. Ampère's law yields for the perturbed current density in terms of the stream function

$$\mu_0\nabla^2\delta\psi = -\delta j_y \tag{7.45}$$

which is the only non-zero component of the current disturbance. With the help of Ohm's law the induction equation is transformed into

$$\frac{\partial\delta\psi}{\partial t} = B_x\nabla_x\delta\phi + \frac{1}{\mu_0\sigma}\nabla^2\delta\psi \tag{7.46}$$

We now assume that the initial field is of the form $B_{x0} = \delta BG(z)$, with $\delta B = \text{const}$, and $B_{z0} = 0$. We also allow for a non-zero B_y. The function $G(z)$ determines the change in the magnetic field across the current layer, for instance a tanh-function. It is formally given by $G(z) = \mathbf{k}\cdot\mathbf{B}_0/kB_0$, which close to $z = 0$ is a straight line. Moreover, $G(z)$ must be an odd function in order to allow for tearing to develop. Because of this definition we have

$$\nabla_z\psi = -\delta BG(z) \tag{7.47}$$

This equation is used in the following to eliminate the stream function ψ. The velocity is determined by Eq. (3.71)

$$nm_i d\delta\mathbf{v}/dt = \delta\mathbf{j}\times\mathbf{B} - \nabla p \tag{7.48}$$

where p is the pressure which does not play a role in the incompressible case we consider. Linearizing and introducing the following ansatz for the stream functions

$$\begin{bmatrix} \delta\psi \\ \delta\phi \end{bmatrix} = \begin{bmatrix} \delta\psi_0(z)\cos(k_x x) \\ \delta g(z)\gamma(k_x\delta B)^{-1}\sin(k_x x) \end{bmatrix} \exp(\gamma t) \tag{7.49}$$

the whole system of linearized equations for the stream functions, resulting from the incompressible equation of motion, the induction law, and Ohm's law, can be reduced after some tedious algebra to two coupled linear equations

$$\begin{aligned} \delta\psi_0 - G\delta g &= \frac{1}{\gamma\tau_d}\left(\frac{d^2\delta\psi_0}{dz^2} - \kappa^2\delta\psi_0\right) \\ -\gamma^2\kappa^{-2}\tau_A^2\left(\frac{d^2\delta g}{dz^2} - \kappa^2\delta g\right) &= G\left(\frac{d^2\delta\psi_0}{dz^2} - \kappa^2\delta\psi_0\right) - \delta\psi_0\frac{d^2G}{dz^2} \end{aligned} \tag{7.50}$$

for $\delta\psi$ and δg, where we defined $\kappa = k_x d$, the Alfvén time

$$\tau_A = d/v_A \tag{7.51}$$

and used the diffusion time given in Eq. (7.2). Moreover, the coordinates are made dimensionless by defining $z \to z/d$. In the external region these equations, neglecting resistivity, yield $\tau_d \to \infty$ and $\delta\psi_0 = G\delta g$ and thus

$$G\left(\frac{d^2\delta\psi_0}{dz^2} - \kappa^2\delta\psi_0\right) = \delta\psi_0\frac{d^2G}{dz^2} \tag{7.52}$$

We specify the external field by choosing $G(z) = \tanh z$ and find for the solutions below and above the current sheet

$$\begin{aligned} \delta\psi_{0+}(+|z|) &= (1 + \kappa^{-1}\tanh z)\exp(-\kappa z) \\ \delta\psi_{0-}(-|z|) &= (1 - \kappa^{-1}\tanh z)\exp(\kappa z) \end{aligned} \tag{7.53}$$

The boundary conditions are $\delta\psi_{0\pm} \to 1$ at the center of the current sheet, $z \to 0$. The derivative, $d\delta\psi_0/dz = \delta\psi_0'$, of $\delta\psi_0$ stays discontinuous here. If we use $\Delta = [\delta\psi_{0+}' - \delta\psi_{0-}']_{z=0}/\delta\psi_0(0)$, we obtain for the jump in the derivative of $\delta\psi_0$

$$\Delta = 2(1 - \kappa^2)/\kappa \tag{7.54}$$

Now solve for the interior of the dissipation region. Since this region is very narrow, $\delta\psi$ will be constant inside. Moreover, $G = \tanh z \approx z$ here. Furthermore, the wavelength of the tearing mode is much larger than the width of the diffusion region, such that $\kappa^2|\delta\psi_0| \ll |d^2\delta\psi_0/dz^2|$ and $\kappa^2|\delta g| \ll |d^2\delta g/dz^2|$. This requirement reduces the differential equations for $\delta\psi_0$ and δg to

$$\begin{aligned} \frac{d^2\delta\psi_0}{dz^2} &= \gamma\tau_d\left[\delta\psi_0(0) - z\delta g\right] \\ \frac{d^2\delta\psi_0}{dz^2} &= \frac{\gamma^2\tau_A^2}{\kappa^2}\frac{1}{z}\frac{d^2\delta g}{dz^2} \end{aligned} \tag{7.55}$$

The matching condition demands that the derivatives of $\delta\psi_0$ account for the above jump Δ at the boundary of the dissipation region, $z = 1$. We integrate the second of the above equations with respect to z over the dissipation layer to obtain the jump in the derivative expressed in terms of $g(z)$, divide by the constant $\delta\psi_0(0)$ in the inner region and compare with Δ to obtain

$$\Delta = -\frac{\gamma^2\tau_A^2}{\delta\psi(0)\kappa^2}\int_{-\infty}^{\infty}\frac{dz}{z}\frac{d^2\delta g}{dz^2} \tag{7.56}$$

This is the matching condition. In order to find its explicit form one must solve for $g(z)$. This is achieved in the following way. One eliminates the second derivative of $\delta\psi$ from the two differential equations (7.55) and transforms the resulting differential equation into the simple form

$$y'' + q^2 y = q \tag{7.57}$$

where the prime indicates differentiation with respect to $q = z(\kappa^2\tau_d/\gamma\tau_A^2)^{1/4}$, and $y = -\delta g q/\delta\psi_0(0)$. As $q \to \infty$ the function $y \to -1/q$.

The solution of the above differential equation can be expressed as a definite integral. The matching condition (7.56) requires integration over y''. Hence, we get

$$\Delta = \gamma^{5/4}\tau_d^{3/4}\tau_A^{1/2}\kappa^{-1/2}\mathcal{I} \tag{7.58}$$

where $\mathcal{I} = \int_{-\infty}^{\infty} y'' dq/q$ is a numerical value. The equation for the growth rate γ, i.e., the dispersion relation, is obtained by combining this formula with the expression for Δ obtained from Eq. (7.54). Numerical solutions have shown that it is sufficient to put $q = 2$ for the halfwidth of the diffusion region. Combining all the variables one ultimately obtains for the growth rate

$$\gamma_{\rm tea} = \frac{1}{\tau_d}\left[\frac{2(1-\kappa^2)}{\mathcal{I}\kappa^{1/2}}\right]^{4/5}\left(\frac{\tau_d}{\tau_A}\right)^{2/5} \tag{7.59}$$

This is the general expression for the resistive tearing mode growth rate in a narrow plane current sheet of the particular shape given by the function $G(z)$. Due to constancy inside the layer, it diverges for $\kappa = 0$. A more precise calculation leads to $\kappa_{\max} =$

$(4\tau_A/\tau_d)^{1/4}$ and a maximum growth rate

$$\boxed{\gamma_{\mathrm{tea,max}} \approx (2\tau_A \tau_d)^{-1/2}} \tag{7.60}$$

This growth rate depends on the inverse of the diffusive and Alfvén times in the current layer. Large diffusive and Alfvén times will decrease it. In the magnetosphere and solar wind the diffusive times are very long. Hence, resistive tearing will not play an important role. One must consider collisionless wave-particle interactions as the relevant processes for initiating fast reconnection here.

7.4. Collisionless Tearing Mode

When the plasma is collisionless, reconnection can evolve only as kinetic instability with ion inertia as the force which mixes the magnetic fields. The resulting instability is the *collisionless tearing instability*.

Configuration

As in the case of resistive tearing consider a planar current sheet (see Fig. 7.7), but no resistivity. A small disturbance in the position of the distribution of the current filaments along x will cause the filaments to attract each other and magnetic islands will start to form spontaneously. In a two-dimensional current sheet the magnetic field field is fully described by the only non-vanishing component of the vector potential, A_y, as

$$\mathbf{B}(x, z) = \left(-\frac{\partial A_y}{\partial z}, 0, \frac{\partial A_y}{\partial x} \right) \tag{7.61}$$

Initially the magnetic field is a Harris-type field with no B_z component (see Fig. 7.2 in our companion book). The tearing mode will produce a periodic variation along x. Hence, the distorted electric and vector potentials can be represented as plane waves

$$\begin{aligned} \delta A_y(x, z, t) &= \delta A(z) \exp(-i\omega t + ikx) \\ \delta \phi(x, z, t) &= \delta \Phi(z) \exp(-i\omega t + ikx) \end{aligned} \tag{7.62}$$

It is most convenient to use the energy principle, derived in Sec. 3.6, to the evolution of the instability. In this simple two-dimensional case, it can be written as the energy balance between the magnetic fluctuation energy and the energy dissipated by the current in the Harris sheet

$$\frac{1}{2\mu_0} \frac{\partial}{\partial t} \int |\delta \mathbf{B}|^2 \, dz = - \int \delta \mathbf{j} \cdot \delta \mathbf{E}^* dz \tag{7.63}$$

where the asterisk indicates the conjugate complex of the electric wave field. The right-hand side is the Ohmic loss in the reconnecting current sheet. It accounts for the growth or damping of the magnetic field.

Electron Tearing Instability

Let us use a slightly modified Harris equilibrium with the magnetic field given as

$$\mathbf{B}_0 = \hat{\mathbf{e}}_x B_0 \tanh(z/d) + B_{0z}\hat{\mathbf{e}}_z \tag{7.64}$$

where the zero-order normal component of the magnetic field is small such that the ions behave unmagnetized in this component while the electrons are magnetized, implying

$$\frac{r_{ge}}{d} < \left(\frac{B_{0z}}{B_0}\right)^2 < \frac{r_{gi}}{d} \tag{7.65}$$

Intuitively one may argue that the strong magnetization of the electrons in the field B_{0z} will introduce some stiffness of the magnetic field and thus a stabilizing effect on the electron-driven tearing mode, the *electron tearing instability*. In addition, the half width of the current layer, Δ, is assumed to be larger than the ion gyroradius

$$\Delta^2 < r_{gi}d \tag{7.66}$$

Electrons will thus be strongly magnetized while the ions are weakly magnetized.

With the help of the above definitions the zero-order undisturbed but inhomogeneous particle distribution functions are

$$f_{s0}(\mathbf{v}, z) = \frac{n_0}{\pi^{3/2}v_{\mathrm{th}s}^3} \exp\left[-\frac{v^2 + v_{ds}^2}{v_{\mathrm{th}s}^2} + \frac{2m_s v_{ds}(v_y + q_s A_{0y})}{k_B T_s} - \frac{q_s\phi_0}{k_B T_s}\right] \tag{7.67}$$

where the diamagnetic drift velocity in the density gradient of the Harris current layer of halfwidth $d = L_n$ has been introduced as

$$v_{ds} = -k_B T_s / q_s B_0 L_n \tag{7.68}$$

Solving the linearized Vlasov equation for the disturbance δf_s of the particle distribution functions using the representations of the electric and magnetic fields through the potentials yields

$$\delta f_s(\mathbf{v}, z, t) = \frac{q_s f_{0s}}{k_B T_s}\left[v_{ds}\delta A_y - \delta\phi + i\omega \int_{-\infty}^{t} (v_y \delta A_y - \delta\phi)dt'\right] \tag{7.69}$$

As usual, the integration over t' must be performed along the undisturbed particle orbits. This requires the solution of the undisturbed particle motion inside and outside the current layer. The term outside the integral describes the adiabatic particle response. Hence, the integral term is the non-adiabatic correction on the first order distribution function.

In calculating the adiabatic response we apply the energy principle. Using the adiabatic part of the disturbed distribution function

$$\delta f_{\mathrm{ad}s} = \frac{q_s v_{ds}}{k_B T_s} \delta A_y f_{0s} \tag{7.70}$$

the adiabatic current can be calculated from the velocity integral over $v \delta f_{\mathrm{ad}s}$. Inserting into the right-hand side of Eq. (7.63), one obtains

$$\int dz\, \delta j_{y\mathrm{ad}} \delta E_y^* = -\sum_s \frac{e^2 v_{ds}^2}{2k_B T_s} \frac{\partial}{\partial t} \int dz\, n(z)\, |\delta A(z)|^2 \tag{7.71}$$

Now, expressing the magnetic field distortion in Eq. (7.63) through the vector potential, the energy principle assumes the form

$$\frac{\partial}{\partial t}\left\{\int dz \left[\left|\frac{d\delta A}{dz}\right|^2 + \left(k^2 - \frac{2}{d^2 \cosh^2(z/d)}\right)|\delta A|^2\right]\right\} = -2\mu_0 \mathrm{Re} \int dz\, \delta j_{y\mathrm{ad}} \delta E_y^* \tag{7.72}$$

For sufficiently narrow current sheets, $k^2 d^2 \ll 1$, the tearing mode energy is negative, and any dissipation will lead to instability.

Growth Rate

One can obtain a very rough estimate of the electron tearing growth rate when restricting to electrons and assuming that a simple Ohm's law holds for the current inside the current layer and that outside the layer current is zero

$$\delta j_y = \sigma \delta E_y \qquad |z| \leq \Delta_e \tag{7.73}$$

Here the conductivity is modeled as $\sigma \approx ne^2/m_e v_{\mathrm{the}} k$, implying that the collision time equals the free flight time of the electrons over one wavelength. Then to order of magnitude the energy principle (7.72) yields as a first step the approximate growth rate of the one-dimensional electron tearing instability

$$\gamma_{\mathrm{etea}} \approx \left(\frac{r_{ge}}{d}\right)^{3/2} \left(1 + \frac{T_i}{T_e}\right) k v_{\mathrm{the}} \tag{7.74}$$

which holds only for long wavelengths. Use has been made of $1/k > d$ and of the expression for the electron inertial length, $c^2/\omega_{pe}^2 = r_{ge}^2(1 + T_i/T_e)$, which is a consequence of overall pressure balance across the layer.

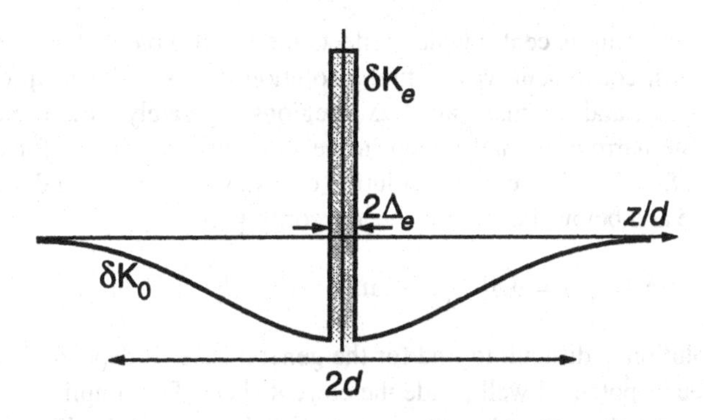

Fig. 7.8. Adiabatic and non-adiabatic pseudo-potentials in the tearing mode.

The above growth rate does not yet include the non-adiabatic electron contribution contained in Eq. (7.69). Assuming low frequency disturbances, $\omega/k \ll c$, the electrostatic potential, $\delta\phi$, can be neglected in (7.69). Moreover, inside the current layer the particles can be assumed to be unmagnetized. They will thus move along straight undisturbed orbits, $x = v_x t$ at $|z| \leq \Delta_s$, and the vector potential, δA_y, can be assumed as constant in the narrow current layer. Integrating over these orbits, we find

$$\delta f_s = \left[v_{ds} - \Theta(|z| - \Delta_s) \frac{\omega v_y}{\omega - k v_x} \right] \frac{q_s \delta A_y}{k_B T_s} \tag{7.75}$$

Here $\Theta(x)$ is the Heaviside step function. Again, this expression is used to calculate the oscillating current as function of the vector potential. From Maxwell's first equation $\nabla \times \delta\mathbf{B} = \mu_0 \delta\mathbf{j}$, one derives the following equation for the amplitude, $A(z)$, of δA_y

$$\frac{d^2 A(z)}{dz^2} - A(z) \left[k^2 + \delta K_0^2(z) + \sum_s \delta K_s^2(\omega, \mathbf{k}, z) \right] = 0 \tag{7.76}$$

where $\delta K_0^2(z)$ and the sum over δK_s^2 are the adiabatic and non-adiabatic contributions of the particle motion and currents to the square of the wavenumber. These pseudo-potentials are given by

$$\begin{aligned} \delta K_0^2(z) &= -2d^{-2} \cosh^{-2}(z/d) \\ \delta K_s^2(\omega, \mathbf{k}, z) &= -i\sqrt{\pi}(\omega_{ps}^2 \omega / c^2 |k| v_{\mathrm{th}s}) \Theta(|z| - \Delta_s) \end{aligned} \tag{7.77}$$

Equation (7.76) is a one-dimensional Schrödinger-like equation for a particle moving in an external potential. The form of this pseudo-potential is schematically shown in Fig. 7.8. The adiabatic potential forms a shallow sink containing a narrow non-adiabatic electron contribution in its center which reflects the vector potential acting as pseudo-particle. The most convenient way to find a solution is to solve the Eq. (7.76) in the external, $|z| > \Delta_s$, and internal, $|z| < \Delta_s$, regions separately, and to match the solutions across the narrow internal region in the same way as we did for the resistive tearing mode in Sec. 7.3. The external solution decays with increasing distance in both directions above and below the current sheet according to

$$\delta A_{\rm ex}(z) = \delta A(0)[1 + \tanh(z/d)/kd] \exp(-k|z|) \tag{7.78}$$

The internal solution is difficult to find for the general case. It depends heavily on the form of the pseudo-potential well inside the current sheet. For simplicity one neglects the adiabatic contribution as well as the ion non-adiabatic term. In this case the internal pseudo-potential is simply of rectangular form and provided by the electrons only, yielding

$$\delta A_{\rm in}(z) = \delta A(0) \cosh(z \delta K_e) \tag{7.79}$$

at $|z| < \Delta_e$. The matching condition is the same as in the resistive tearing case, requiring continuity of the logarithmic derivatives of the two solutions at the boundaries of the inner and outer parts. This condition yields

$$\delta K_e \tanh(\Delta_e \delta K_e) = (1 - k^2 d^2)/(kd^2 + \Delta_e) \tag{7.80}$$

Let us consider the limit $kd^2 > \Delta_e$ and $\delta K_e^2 \Delta_e^2 \ll 1$. Then

$$\frac{1 - k^2 d^2}{kd^2 + \Delta_e} = \frac{1}{2} \int_{-\infty}^{\infty} \delta K_e^2 \, dz \tag{7.81}$$

and the more precise growth rate of the one-dimensional electron tearing instability is obtained from the last matching condition, inserting the expression for δK_e^2

$$\boxed{\gamma_{\rm etea} = \sqrt{\pi}\left(1 + \frac{T_i}{T_e}\right)\left(\frac{r_{ge}}{d}\right)^{5/2} (1 - k^2 d^2) \omega_{ge}} \tag{7.82}$$

This growth rate is proportional to the electron gyrofrequency and to the ratio of electron gyroradius to the width of the current layer to the 5/2-th power. On the other hand it is reduced at short wavelengths, becoming comparable to the width of the current layer. One thus expects that long wavelength tearing islands will have the largest growth rates. On the other hand, any mechanism shortening the wavelength will drive the instability toward marginal stability with wavelength of the order of the width of the current layer. Marginally stable island will have a nearly circular shape.

Ion Tearing Instability

So far we have neglected the effect of the small normal magnetic field component, B_{0z}, on the evolution of the one-dimensional tearing instability. Realistic current sheets and equilibria are always two-dimensional with non-vanishing normal magnetic fields and the magnetic field across the sheet will never turn exactly zero in any place. As we already mentioned, the bulk low-energy electron component will in such a case become strongly magnetized in the B_{0z} field component in the center of the magnetic neutral sheet. It will stay frozen-in and not allow the electron tearing mode to become unstable. In such cases the ions drive the instability due to their larger inertia and unmagnetized nature, if their gyroradius is large enough across the current sheet. A calculation very similar to that presented in the previous section leads to the growth rate of the one-dimensional *ion tearing instability*

$$\boxed{\gamma_{\text{itea}} \approx \sqrt{\pi}\left(\frac{r_{gi}}{d}\right)^{5/2}\left(1+\frac{T_e}{T_i}\right)(1-k^2d^2)\omega_{gi}} \tag{7.83}$$

This growth rate scales as the ion-cyclotron frequency. It is positive as long as the tearing wavelength is longer than the halfwidth of the current sheet, $k^2d^2 < 1$. The corresponding mode grows much slower than the electron tearing mode and is of longer wavelength. Similarly, the marginally stable width of the islands will be of the order of the ion gyroradius. Since the electron tearing mode is stabilized much earlier, it is reasonable to assume that the dominant tearing mode under collisionless conditions is ion tearing driven by inertial effects. When kd increases beyond one, the mode is damped. However, strongest damping occurs for $k^2d^2 = 5$. For current sheets of larger width the damping goes to zero again.

Two-Dimensional Tearing

However, because of the two-dimensionality of the problem, this is not the full story. One must take into account the energy contribution of the frozen-in magnetized electrons. This case corresponds to the range of ratios given in Eq. (7.65). The magnetized electrons perform a gyration about the normal magnetic field component, B_{0z}, in the current neutral sheet and a bounce motion between the two mirror points of the magnetic field at the boundaries. Both motions are fast oscillations, which do not contribute to instability and can be averaged out when determining the electron motion. We are thus left with an effective average drift motion

$$\delta v_{ex} = \delta E_y / B_{oz} \tag{7.84}$$

of the electrons in the bounce- and gyro-averaged electric wave field of the tearing mode. This value is inserted in the averaged electron continuity equation, and the wave electric field is replaced by the average vector potential to obtain

$$\delta n_e/n_0 = ik\delta A_y/B_{0z} \tag{7.85}$$

In this two-dimensional model the magnetic field lines in the vicinity of the neutral sheet can be approximated by a simple parabolic shape

$$x(z) = x_0 + B_0 z^2/2B_{0z}d \tag{7.86}$$

which yields for the phase of the tearing mode along the magnetic field

$$kx(z) = kx_0 + R[Q^2(z) - 1] \tag{7.87}$$

where we have used the Taylor expansion of the magnetic field near the neutral plane, $B(z) \approx B_{0z} + (\partial^2 B/\partial z^2)z^2/2$ and introduced the abbreviations $R = B_{0z}kd/2B_0 \ll 1$ and $Q(z) = B(z)/B_{0z}$. Since this is the phase of the tearing mode, one can express the density variation as

$$\delta n_e \approx \langle \delta A(z) \exp\left[iRQ^2(z)\right]\rangle (ikn_0/B_{0z}) \exp\left[-i\omega t + ikx - iRQ^2(z)\right] \tag{7.88}$$

where the angular brackets indicate averaging over the electron bounce motion in the neutral layer. Again, this averaging can be circumvented assuming that the vector potential is constant throughout the current layer. This allows to replace the term in angular brackets by the value of the vector potential, $\delta A(\Delta_e)$, at the boundaries of the internal region, $\pm\Delta_e$. Since at the boundary the phase is $\pi/2$, the size of this region is

$$\Delta_e/d \approx (\pi B_{0z}/B_0 kd)^{1/2} \tag{7.89}$$

The electron current of the tearing mode perturbation is the diamagnetic current flowing in the neutral sheet. This current is given by

$$\delta j_e = ik_B(T_e + T_i)k\delta n_e/B_{0z} \tag{7.90}$$

where ik is the inverse oscillating gradient scale of the density variation. Thus the total work done by this current over the width of the sheet is

$$\frac{1}{2}\int dz\, \delta j_e(z)\delta A_y^*(z) \approx -\Delta_e \frac{|\delta A(\Delta_e)|^2}{2\mu_0}\frac{k^2 B_0^2}{2B_{0z}^2} \tag{7.91}$$

Since this energy tries to stabilize the instability, one must compare the free energy with this expression. We can use Eq. (7.78) for $\delta A(\Delta)$ and the matching condition

$$\frac{1 - k^2 d^2}{\Delta_e + kd^2} = \frac{1}{2}\sum_s \int_{-\infty}^{\infty} |\delta A(z)|^2 \delta K_s^2(z) dz \tag{7.92}$$

to obtain the condition for instability in the particular form

$$\frac{1-k^2d^2}{k^2\Delta_e d(kd+\Delta_e/d)} > \frac{1}{2}\frac{B_0^2}{B_{0z}^2} \tag{7.93}$$

This condition implies that the compression energy of the plasma in the narrow current layer is not large enough to exceed the free energy, if

$$r_{ge}/d < B_{0z}/B_0 < (r_{ge}/d)^{1/2} \tag{7.94}$$

In this case most of the electrons drift along the x direction inside the neutral sheet of size $\Delta_e \approx \sqrt{r_{ge}d}$, while exhibiting a bounce oscillation between the boundaries of the neutral sheet and performing a gyratory motion in the normal field, B_{0z}. The small thickness of the electron current layer allows to neglect $\Delta_e \ll d$ against d to obtain

$$\frac{1}{k^2d^2} - 1 > \left(\frac{B_0}{B_{0z}}\right)^2 \left(\frac{r_{ge}}{d}\right)^{1/2} \frac{kd}{2} \tag{7.95}$$

as condition of instability. Under these favorable conditions, the ion tearing instability will grow at the growth rate in Eq. (7.83) whenever the growth rate exceeds the value

$$\boxed{\gamma_{\mathrm{itea}} = \gamma'\omega_{gi}(1-k^2d^2) > \omega_{gi}B_{0z}/B_0} \tag{7.96}$$

with $\gamma' = \sqrt{\pi}\,(r_{gi}/d)^{5/2}(1+T_e/T_i)$. Eqs. (7.95) and (7.96) set thresholds on the ion tearing instability: a large normal component of the magnetic field in the center of the neutral sheet tends to stabilize the collisionless tearing mode. To visualize the unstable range we combine the conditions into one, defining $b = (B_{0z}/B_0)$. Then instability arises for

$$\gamma'(1-k^2d^2) > b > (kd)^{3/2}(1-k^2d^2)^{-1/2}(r_{ge}/d)^{1/4} \tag{7.97}$$

The graphic representation of this condition is shown in Fig. 7.9. For given b and γ' instability arises in the shaded region. Hence, there is a restricted domain of wavenumbers and of normal magnetic field components, depending on the electron and ion gyroradii, where the ion tearing mode can develop.

Magnetospheric Substorms

The collisionless ion tearing mode is part of the most successful substorm model (see Sec. 5.6 of our companion book). Under quiet conditions the magnetic field in the center of the Earth's plasma sheet is dipolar, with a relatively large non-negligible normal component, B_{0z}, and the tearing mode is stabilized by the pressure of the electrons trapped in the center of the neutral sheet. In the substorm growth phase new open

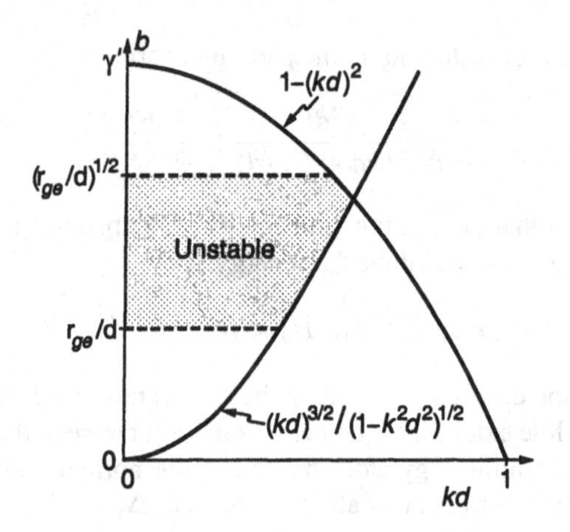

Fig. 7.9. Region of collisionless ion tearing instability.

magnetic flux, created by dayside magnetic merging, is convected into the tail and increases the tail magnetic pressure. The field becomes stretched and the tail current sheet becomes narrower with increased current flow in the sheet. This process decreases the normal magnetic field component, B_{0z}, while it increases B_0 until the right-hand side of Eq. (7.96) becomes small enough for the ion tearing growth rate to exceed the threshold, leading to reconnection and substorm onset.

7.5. Percolation

Observations of the magnetic field structure and plasma flow at the magnetopause often reveal that reconnection does not proceed in a simple way, where oppositely directed magnetic fields merge at one single point at the magnetopause close to the stagnation point. Intuitively, it is easy to imagine that remote points on the magnetopause surface are not correlated. In order to be correlated they should be connected by flow of information. The speed of information flow is of the order of the Alfvén speed. Thus two points separated by a distance such that the Alfvén travel time is longer than the growth time of tearing instability will under similar conditions decouple and reconnect separately. This leads to the picture that the magnetopause at a certain time may be the site of many independent merging events going on. Different flux tubes may reconnect and open up the magnetosphere at different positions.

Tearing Island Overlap

Actually, the tearing instability generates magnetic islands which are separated by X-points. For thin sheets these islands are ordered spatially into long chains. Since there is no reason for the islands to have the same size, two different effects may arise in such chains. The first is related to the evolution of an instability which causes the island to merge into the largest possible island size, where the dimension of the island is restricted by the width of the current layer. This instability can be interpreted as an instability in which the largest island eats up all smaller islands. In magnetohydrodynamic theory this *coalescence instability* is rather fast and has no threshold.

However, tearing instability may occur independently in several adjacent parallel current layers and the tearing instability may generate islands of different scales. Since these islands are formed by magnetic fields lines, the formation of a tearing mode spectrum implies that islands may overlap and magnetic field lines may become intermingled. In such a case one may speak of turbulent diffusion of magnetic field lines. Regions will be generated where the magnetic flux from both sides of the current layer is screened, while in other regions field lines or flux tubes penetrate the boundary current layer. This process is called *percolation* and is believed to produce localized and temporally varying reconnection which leads to the interconnection of flux tubes from the magnetosheath and the magnetosphere. Such flux tubes are known to form the so-called *flux transfer events* frequently detected in the vicinity of the magnetopause.

Field Line Dynamics

The basic equation for magnetic field line dynamics can be derived from a consideration of the magnetic field line of a single tearing mode with wavenumber, **k**, in a model where the magnetic field rotates from one side to the other of the current layer, including a guide field magnetic component parallel to the current. At the magnetopause such a configuration corresponds to a magnetic field model with magnetic shear as shown in Fig. 7.10. In such a model the main magnetic field points in the z direction, but existence of the shear implies that a continuous tangential component exists parallel to the diamagnetic current flowing in the magnetopause in y direction

$$\mathbf{B} = B_0 \tanh(\xi/d)\hat{\mathbf{e}}_z + B_{0y}\hat{\mathbf{e}}_y \tag{7.98}$$

with ξ the distance measured from the magnetopause. Since $\nabla \cdot \mathbf{B} = 0$, the components of the magnetic field satisfy

$$\frac{d\xi}{\delta B(\mathbf{k}) \sin(k_z z + k_y y)} = \frac{dy}{B_{0y}} = \frac{dz}{B_z(\xi)} = \frac{ds}{B(s)} \tag{7.99}$$

where $B(\mathbf{k})$ is the perturbation amplitude of the magnetic field caused by the tearing mode, s is the coordinate along the magnetic field line, x, y, z the coordinates in the

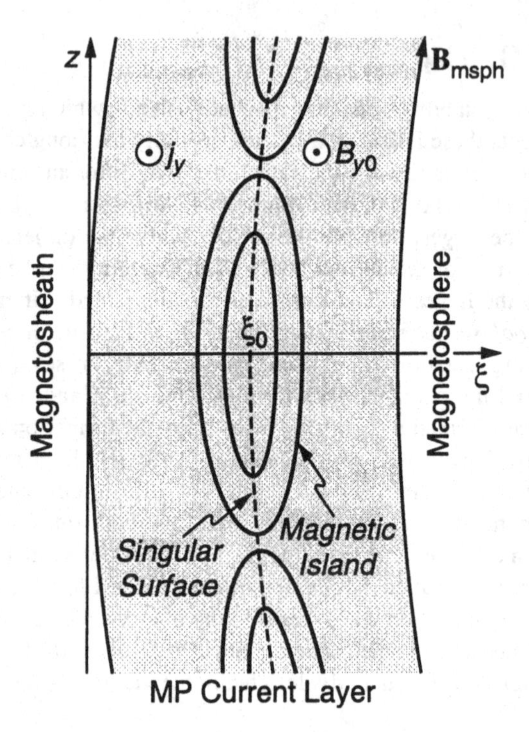

Fig. 7.10. Geometry for magnetopause tearing and percolation.

undistorted magnetic field, ξ_0 the position of a given magnetic surface (see Fig. 7.10). The magnetic surface is defined as the singular surface where the wave vector component parallel to the actual local magnetic field, including the tearing mode disturbance, vanishes, $k_\parallel(\xi_0) = 0$, corresponding to the condition for tearing mode instability to appear locally at $\mathbf{k} \cdot \mathbf{B}_0 = 0$. Any such magnetic surface can be described by a Harris sheet as $B = B_0 \tanh(\xi/d)$, with ξ the coordinate in the frame of this surface and d the total width of the magnetic transition. Note that the parallel index refers to the local parallel wavenumber in the superposition of the ambient and the distorted fields.

In the vicinity of this surface the phase of the tearing mode may be expanded into a Taylor series. We define the new coordinate $x = \xi - \xi_0$, $dx = d\xi$ and write

$$k_y y + k_z z = \left.\frac{dk_\parallel(x)}{dx}\right|_{\xi=\xi_0} \int x(s)ds \tag{7.100}$$

The integration is taken over the coordinate, s, along the magnetic field of the island. The extensions of the resonant regions are defined separately for ions and electrons as $\Delta_j = |\omega/k_\parallel' v_{\mathrm{th}j}|$ (the prime indicates the derivative of $k_\parallel$ with respect to ξ or x.)

Integrating the above equation yields the result

$$x = \pm \left\{ \frac{2\delta B(\mathbf{k})}{k_{\parallel}' B_0} \left[2\lambda^2 - 1 - \cos\left(k_{\parallel}' \int x(s)ds \right) \right] \right\}^{1/2} \tag{7.101}$$

The integration constant λ^2 determines whether the magnetic surfaces are open, $\lambda > 1$, or closed, $\lambda < 1$. The separatrix between open and closed island field lines is determined by the condition $\lambda = 1$. This yields the halfwidth of the islands of a given wavenumber, $\mathbf{k}$, at the particular resonant surface position

$$w(\xi_0, \mathbf{k}) = 2[\delta B(\mathbf{k})/B_0 k_{\parallel}'(\xi_0, \mathbf{k})]^{1/2} \tag{7.102}$$

This width is a function of the position, ξ_0, of the singular resonant layer, the center of the magnetic surface. In addition it depends on the wavenumber. Hence, it can be different for different layers and wavenumbers. The widths form an entire spectrum of different lengths and magnetic islands belonging to different layers may overlap. For two different islands at the resonant surfaces, ξ_{01} and ξ_{02}, the overlap condition is

$$w_1(\xi_{01}, \mathbf{k}_1) + w_2(\xi_{02}, \mathbf{k}_2) > |\xi_{01} - \xi_{02}| \tag{7.103}$$

Migration Coefficient

Whenever this condition is satisfied, the two magnetic surfaces cease to exist separately but belong to both islands at the same time. The magnetic field lines of these two islands mix, the common surface and the field lines start migrating across the islands and the layers. This process is comparable to some kind of Brownian motion with a spatial step, $\Delta x \approx w$, performed during a characteristic stepping time, $\tau_w = s/v$, where v is a characteristic velocity of a magnetic perturbation. The length of the field line corresponding to one Brownian step is estimated by

$$s_w \approx (k_{\parallel}' w)^{-1} \tag{7.104}$$

Using this expression for the width of the islands allows to define a geometric field line migration coefficient. This coefficient has the dimension of a length

$$D_{\rm geo} \approx \frac{w^2}{s_w} \approx \left| \frac{\delta B(\mathbf{k})}{|k_{\parallel}'|^{1/3} B_0} \right|^{3/2} \tag{7.105}$$

In this expression the magnetic fluctuation spectrum, $[\delta B(\mathbf{k})]^2$, appears.

The geometric migration coefficient can be rewritten with the help of the total turbulent magnetic energy, $W_{\rm tea} = \sum_{\mathbf{k}} |\delta B(\mathbf{k})|^2 / B_0^2$, normalized to the undisturbed magnetic energy. The number of islands in the total layer is $N \approx d/\langle w \rangle$, where $\langle w \rangle \approx$

w is the average width of the islands. This yields for the normalized turbulent magnetic energy density in the shear tearing mode:

$$W_{\mathrm{tea}} \approx N(\delta B/B_0)^2 \approx w^3 k_{\parallel}'^2 d \tag{7.106}$$

which is approximately the normalized energy released during the shear tearing mode instability under the condition that many modes are excited. The geometric field line diffusion coefficient can now be expressed as

$$D_{\mathrm{f,geo}} \approx \frac{W_{\mathrm{tea}}}{|k_{\parallel}'|d} \approx \pi \sum_{\mathbf{k}} \left[\frac{|\delta B(\mathbf{k})|^2}{B_0^2} \right] \delta[k_{\parallel}(x)] \tag{7.107}$$

This version of the migration coefficient is the quasilinear resonant diffusion coefficient of migrating field lines, valid for random phase interactions between the overlapping modes. Hence, the overlap of the islands can be interpreted as a randomization of field lines whenever the tearing mode islands overlap.

Some of the magnetic flux tubes will start migrating across the current layer, leading to interconnected magnetosheath-magnetosphere flux tubes which appear as flux transfer events in magnetic recordings. Figure 7.11 illustrates such an overlap and how a magnetic flux tube may locally penetrate across the magnetopause current layer forming a flux transfer event. The number of singular surfaces and X-points along the ξ axis across the magnetopause boundary layer is always an odd number and in the overlap regions the local magnetic field is amplified. Because of the finite shear component, $B_{0y} \neq 0$, the magnetic field in the X-points is not zero. Only the z components of the magnetic field cancel each other. A flux tube will usually enter as a B_{0y}-tube through one of the X-points on one side and turn around the overlapping islands until it leaves from an X-point on the other side of the current layer. Hence, any migrating flux tube will constitute a three-dimensional entity.

The geometric field line diffusion coefficient does not have the dimension of a diffusion coefficient. In order to get a real diffusion coefficient, one has to multiply $D_{\mathrm{f,geo}}$ by the characteristic velocity, v, of a magnetic perturbation. The appropriate speed is the Alfvén velocity. Hence, overlap of magnetic island will cause migration of magnetic field field lines corresponding to a percolation diffusion coefficient

$$D_{\mathrm{per}} \approx v_A W_{\mathrm{tea}}/|k_{\parallel}'|d \tag{7.108}$$

This diffusion coefficient can be applied only to magnetic flux tube diffusion and only to those regions where islands overlap. Particle diffusion is not described by it.

As already mentioned earlier, stabilization of the tearing mode will occur when the island width becomes comparable to the ion gyroradius, $w \approx r_{gi}$. Thus the diffusion coefficient at stabilization is proportional to the third power of the ion gyroradius. If we

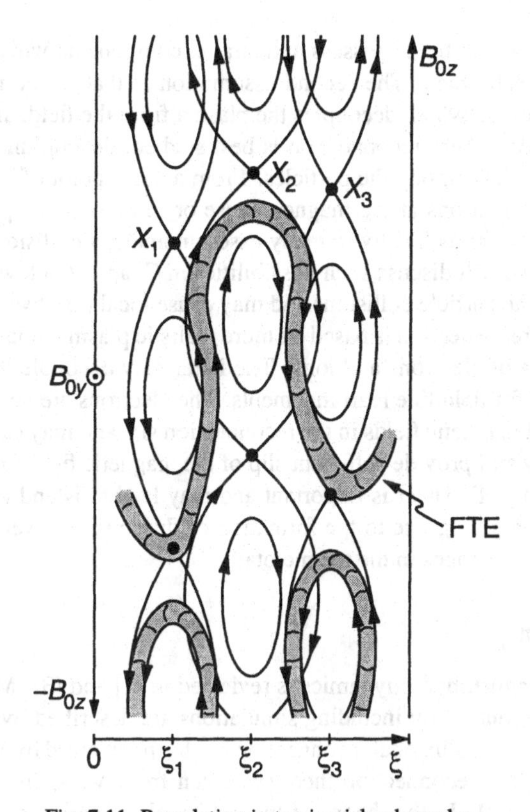

Fig. 7.11. Percolation in tearing island overlap.

replace the derivative of the singular wavenumber by $k'_{\|} \approx kB_0/B_{0y}d$, the percolation diffusion coefficient at saturation

$$\boxed{D_{\text{per,sat}} \approx \frac{v_A k r_{gi}^3}{d} \frac{B_0}{B_{0y}}} \tag{7.109}$$

assumes a particularly simple form which is valid as long as the shear component of the magnetic field has a finite value, B_{0y}. For vanishing shear, percolation ceases to exist and the character of the tearing mode changes.

Concluding Remarks

Reconnection is one particular and very fruitful concept of fast and localized magnetic field line diffusion. It is based on two assumptions. The first assumption is that the two interacting magnetic fields possess antiparallel components which may annihilate when contacting each other. The second assumption is that in the region of contact some mechanism exists which decouples the plasma from the field. In magnetohydrodynamics, the model where reconnection is best studied, decoupling can be achieved by resistive interactions among the particles. From a naive point of view they are thus not applicable to conditions at the magnetopause or in most space plasmas. But this view is incorrect insofar as locally in many cases anomalous collisions may be generated nonlinearly. We will discuss such possibilities in Chap. 12 below. The anomalous collisions replace the particle collisions and may cause local resistive reconnection.

Collisionless reconnection is based on more realistic plasma models which include the different inertia of electrons and ions. The latter may decouple the particles from the field and allow for field line rearrangements. The electrons are very sensitive to the presence of normal magnetic fields in the reconnection site and may cause stabilization. But ion inertia may still provide sufficient slip of the magnetic field for reconnection to be maintained. Magnetic shear is important and may lead to island overlap and magnetic field migration giving rise to the formation of flux transfer events at the dayside magnetopause and flux ropes in the magnetotail.

Further Reading

Reconnection in magnetohydrodynamics is reviewed in [1] and [3]. Magnetohydrodynamic aspects of reconnection including simulations are described by Biskamp, *Phys. Reps.*, **237** (1994) 179. Collisional reconnection has been reviewed by White in [4]. The stationary collisionless reconnection theory is taken from Vekstein and Priest, *Phys. Plasmas* **2** (1995) 3169. In the theory of the collisionless tearing mode we followed [2]. The section about percolation theory is based on Galeev et al., *Space Sci. Rev.* **44** (1986) 1.

[1] D. Biskamp, *Nonlinear Magnetohydrodynamics* (Cambridge University Press, Cambridge, 1993).

[2] A. A. Galeev, in *Magnetospheric Plasma Physics*, edited by A. Nishida (D. Reidel Publ. Co., Dordrecht, 1982), p. 143.

[3] E. Priest, *Solar Magnetohydrodynamics* (D. Reidel Publ. Co., Dordrecht, 1984).

[4] M. N. Rosenbluth and R. Z. Sagdeev, *Handbook of Plasma Physics, Vol. 1: Basic Plasma Physics I*, edited by A. A. Galeev and R. N. Sudan (North-Holland Publ. Co., Amsterdam, 1983), p. 611.

8. Wave-Particle Interaction

With this chapter we enter the field of *nonlinear effects* in space plasma physics. We have chosen to split the discussion of the nonlinear effects into two parts. The first part (Chaps. 8–9) discusses weakly turbulent effects, which are usually dealt with by using *perturbation theory*, while the second part (Chaps. 10–11) concentrates on strongly turbulent phenomena. Chapter 12 concludes this volume with a number of selected applications of nonlinear space plasma theory.

The presentation of the nonlinear theory differs from that of linear theory in that it is more review-like than tutorial. The reason for this difference is twofold. Plasma physics is fundamentally nonlinear. Linear theory illuminates only a narrow window out of the wealth of all effects which can proceed in a plasma. Mathematically, linear theory uses a well-developed apparatus of algorithms. Naturally, for the much wider field of nonlinear theory no such general algorithms are available.

While discussing the linear aspects of space plasma physics has filled our companion book *Basic Space Plasma Physics* and the first half of the present volume, describing the nonlinearities in comparable depth would expand this presentation to unreasonable size. Even if we wanted, it would be impossible, because nonlinear plasma physics is still, after 35 years of intensive research, in its infancy. Few analytical methods are known to treat nonlinear effects, and most of these methods rely on approximations and lowest-order perturbation theory. This methodological lack of a general nonlinear theory and the difficulties and restrictions of nonlinear analytical methods have driven most researchers into the partly qualitative field of numerical simulations.

The present chapter starts with describing the primary effect of large amplitude waves on the particle orbits, leading to *particle trapping* in a wave, followed by a discussion of *exact nonlinear waves*, which can be calculated in a one-dimensional electrostatic plasma in the kinetic approximation. Then we present the theory of *weak turbulence* or weak *wave-particle interaction*, and a discussion of *quasilinear theory* for different types of waves, the today still most important part in space plasma applications. We continue with an application of this theory to *pitch angle diffusion*. The final sections deal with a brief introduction into *wave-wave interaction*, in the two approximations of *coherent* and *incoherent* interactions, and its application to solar radio bursts.

8.1. Trapping in Single Waves

The simplest nonlinear effect is particle trapping. When a plasma wave reaches large amplitudes, either because it has been injected at large amplitude into the plasma by external means or because it has grown to large amplitude due to instability, several entirely new effects set in. The first and simplest one is particle trapping in the potential field of the wave.

Particles Trapped in a Well

Consider, for instance, a potential well created by an ion in a plasma. In the fluid description electrons in the vicinity of the ion will readily deplete this well outside the Debye radius, λ_D. But if one takes into account the different velocities of the electrons, say, in a Maxwellian distribution of particles, it becomes clear that only low energy electrons will be trapped in the vicinity of the ion in the potential trough. The condition for trapping is that the particle energy is less than the potential energy of the electric field, $W_e < e|\phi|$. The particle distribution then splits into a trapped and a non-trapped distribution. In a shallow potential well it is reasonable to assume that the trapped distribution is constant, $f_{tr}(W_e) = f_e(0)$, where $f_e(0)$ is the distribution at the center of the well. The particle density is calculated from the moment of the distribution function as

$$n_e(\phi) = 2n_0 \left[\int_\xi^\infty f_e(W_e, \phi) dv + \int_0^\xi f_e(0) dv \right] \tag{8.1}$$

and $\xi = (2e\phi/m_e)^{1/2}$. The factor 2 takes care of the symmetry of distribution function and potential well. It is reasonable to choose a Boltzmann distribution for the electron distribution function

$$f_e(W, \phi) = \frac{(m_e}{2\pi m_e k_B T_e)^{1/2}} \exp\left(-\frac{W_e - e\phi}{k_B T_e} \right) \tag{8.2}$$

Calculating the integrals, one finds that the density can be represented as a sum of three terms which depend on $\psi = e\phi/k_B T_e$

$$n_e(\psi) = n_0 \left\{ \left[1 - \mathrm{erf}(\psi^{1/2}) \right] \mathrm{e}^\psi + 2(\psi/\pi)^{1/2} \right\} \tag{8.3}$$

Here the function $\mathrm{erf}(x)$ is Gauss' error integral

$$\mathrm{erf}(x) = \frac{2}{\pi^{1/2}} \int_0^x \exp(-t^2) dt \tag{8.4}$$

For very shallow potentials and potential energies much less than the electron thermal energy, the integral can be expanded in the small-amplitude limit to obtain

$$n_e(\phi) = n_0 \left[1 + \frac{e\phi}{k_B T_e} - \frac{4}{3\pi^{1/2}}\left(\frac{e\phi}{k_B T_e}\right)^{3/2}\right]. \tag{8.5}$$

This expression shows that to lowest order the dependence of the density on the potential is a linear function. However, due to the trapping of particles, another nonlinear term comes into play.

For large argument, $\xi \propto \psi \gg 1$, the first two terms in the full expression of the density cancel because the error function becomes unity. The density is determined entirely by the trapped particle component

$$n_e = 2n_0 \left(\frac{e\phi}{\pi k_B T_e}\right)^{1/2} \tag{8.6}$$

The density increases only as the half power of the potential, yielding a weaker dependence on the potential than in the Boltzmann case.

Particle Trapping in a Monochromatic Wave

Similarly, particles can become trapped in a wave potential if the particle kinetic energy in the wave frame is less than the potential energy of the wave. It is immediately clear that trapping will be largest for resonant particles moving at approximately the same velocity as the wave and experiencing a nearly stationary electric wave potential

$$\phi(x,t) = \phi_0 \cos(kx - \omega t) = \phi_0 \cos(kx') \tag{8.7}$$

where the wave coordinates have been transformed into the wave frame by

$$x' = x - (\omega/k)t \tag{8.8}$$

The particle velocity also transforms and becomes

$$v' = v - \omega/k \tag{8.9}$$

such that the total energy of the particle in the wave frame of reference is

$$W_e = \tfrac{1}{2}mv'^2 - e\phi_0 \cos kx' \tag{8.10}$$

When considering the motion of the electron in the two-dimensional particle phase space (x', v'), the particle moves along the lines of constant energy W_e. Figure 8.1 shows the schematic form of these lines. There are two families of curves in this phase

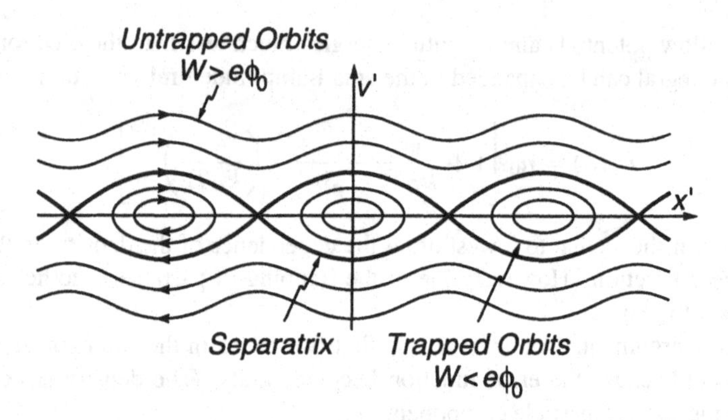

Fig. 8.1. Phase space trajectories of particles moving in the electrostatic potential of a wave.

space. One of the families for low particle velocities exhibits closed lines corresponding to closed particle trajectories. These curves describe trajectories of the particles trapped in the wave potential. The second family of curves at higher particle speeds consists of open lines corresponding to untrapped electrons.

Trapped trajectories have negative total energy, $W_e < 0$. Hence, the particles on such orbits bounce back and forth between the walls of the wave potential exhibiting an oscillatory motion which is periodic in phase space. One can easily estimate the frequency of this bounce motion considering small-amplitude oscillations of the particles near the bottom of the wave potential well, i.e., for particles on trajectories close to the center of the O-type trapped orbits. For those particles the cosine-potential can be expanded to lowest order, yielding the equation of motion

$$md^2x'/dt^2 = -e\phi_0 k^2 x' \tag{8.11}$$

which is the equation of a harmonic oscillator of frequency

$$\omega_b = |e\phi_0 k^2/m|^{1/2} \tag{8.12}$$

This is the *particle trapping frequency*. Since $\delta E = -ik\phi$ it is proportional to the root of the electric wave field amplitude and increases with growing wave field, the larger the wave field amplitude, the faster the trapped resonant electrons will oscillate in the field. In addition, for larger field amplitudes more particles will become trapped by the deepening of the potential trough and the widening of the wave resonance. Because it is just the resonant particles, which are easiest trapped in the wave potential wells, it is clear that Landau damping may become strongly affected by particle trapping. Indeed, one immediately deduces that trapping can be neglected only as long as the trapping frequency is much smaller than the Landau damping or growth rate

$$\omega_b \ll \gamma_l \tag{8.13}$$

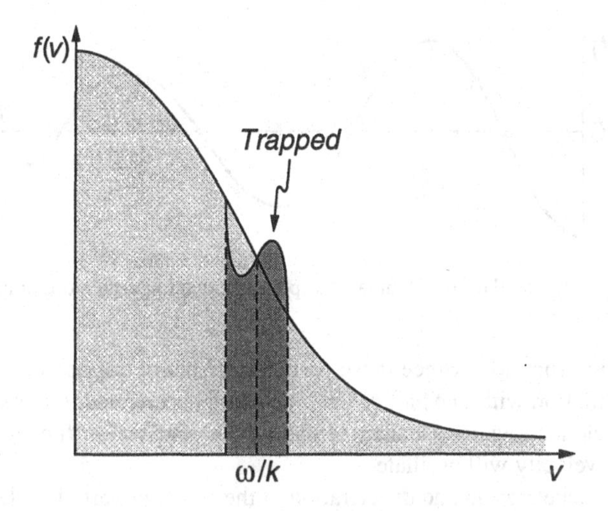

Fig. 8.2. Modulation of the resonant part of the distribution function by trapping.

i.e., for very small amplitudes. During instability, when the amplitude increases, the state when trapping becomes important is readily reached. Intuitively one can argue, however, that this violation of the usual concept of Landau damping and Landau growth appears only for nearly monochromatic waves of narrow bandwidth. When the bandwidth of the excited waves is large, as may be the case for hot gentle beam-excited Langmuir waves, the particle trajectories for the different waves in phase space will mix and the closed trajectories will possibly become distorted, thereby decreasing trapping. Under these conditions the usual concept of Landau damping and growth remains valid. For this to happen the phase velocity spread in phase space must be larger than the width of the closed particle orbits, $m[\Delta(\omega/k)]^2 > 2e\phi_0$,. This requirement corresponds, for constant frequency, to a bandwidth of the spectrum exceeding

$$\Delta k/k > \sqrt{2}\omega_b/\omega \tag{8.14}$$

Trapping Oscillations

In the opposite case, when trapping becomes important in a monochromatic sufficiently large-amplitude wave, the Landau damping or growth rate in Eq. (4.3), rewritten as

$$\gamma_l = \frac{\pi\omega_{pe}^2}{k^2(\partial\epsilon/\partial\omega)} \left.\frac{\partial f(v)}{\partial v}\right|_{v=\omega/k} \tag{8.15}$$

will become modulated by the bounce frequency of the particles. This version of the equation shows explicitly that the value of γ_l depends on the few resonant particles

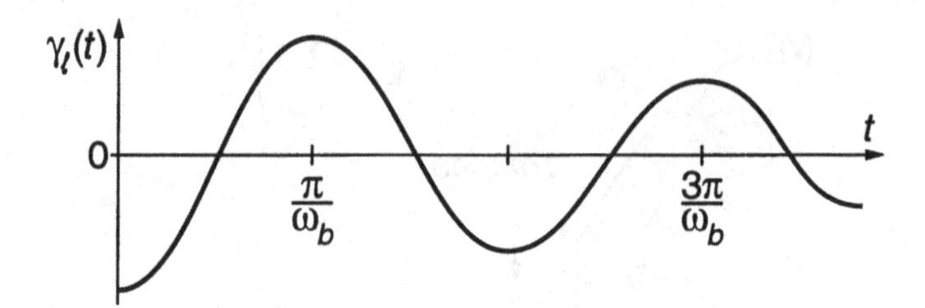

Fig. 8.3. Oscillation of Landau damping rate due to particle trapping.

in the distribution function. Since these particles are being trapped in the wave, they perform an oscillation with one half of the time being accelerated, the other half of the time being decelerated again. Hence, the part of the distribution function centered at the wave phase velocity will oscillate.

Due to this acceleration and deceleration of the resonant particles, the distribution function may periodically develop positive slopes (see Fig. 8.2), and reduce the Landau damping rate temporarily as suggested by Eq. (8.15). Hence, the Landau damping rate will oscillate at the trapping frequency, ω_b, of the particles in the large-amplitude wave potential field. During this oscillation the damping rate turns positive, an effect leading to further amplification of the already large wave amplitude and to further amplification of the trapping. Since this oscillation will dissipate the wave energy in the long time run, the oscillation will damp out (see Fig. 8.3) and tend to become zero for $t \to \infty$. We can calculate the temporal evolution of the wave energy from

$$\frac{d|\phi_0(t)|^2}{dt} = 2\gamma_l(t)|\phi_0(t)|^2 \tag{8.16}$$

which yields the result that the wave energy is exponentially modulated according to

$$|\phi_0(t)|^2 = |\phi_0(0)|^2 \exp\left[2\int_0^t \gamma_l(\tau)d\tau\right] \tag{8.17}$$

Due to the exponential dependence of the wave energy on the growth rate, the variation of the growth rate by trapping of resonant particles causes large-amplitude oscillations at the trapping frequency, ω_b, in the amplitude of the large-amplitude Langmuir wave.

Similar arguments apply to instability-generated Langmuir waves. For instance, in the presence of a gentle beam the trapping of the low energy part of the beam at speeds just below the wave phase velocity causes an increase in the steepness of the beam and an increase in the growth rate during half of the bounce period, but a decrease during the other half. Hence, the amplitude of the wave starts oscillating at the trapping frequency in a large-amplitude Langmuir wave after the first short linear growth phase of the wave, with sharp switch-off periods of the growth rate of the order of half the trapping frequency. In addition, because resonance is slightly shifted, it also becomes possible that the frequency of the Langmuir wave starts oscillating. However, these fine structure effects are difficult to detect in space plasmas.

8.2. Exact Nonlinear Waves

The former sections have made clear that trapping plays an important role in the nonlinear evolution of plasma waves. In principle, when investigating the dynamics of such wave from a kinetic point of view one should distinguish between two components of the particle distribution function, the trapped and the transmitted particles. Both distributions are not independent neither mutually nor of the wave amplitude. They are determined by the Vlasov equation and the Poisson equation. The question arises then, under which conditions one can find exact solutions of this system of equations which determine in a self-consistent way all three quantities.

The complex nonlinear equations describing the evolution of plasma waves can usually not be solved exactly. There is a very small number of occasions when such solutions can be constructed. The simplest one is in the magnetohydrodynamic approximation, where it can be demonstrated that incompressible Alfvén waves in ideal magnetohydrodynamics also exist at large amplitudes. The more interesting question is, however, what kind of exact nonlinear waves can exist in a Vlasov plasma. Such waves exist only in one-dimensional plasmas and are known as *Bernstein-Greene-Kruskal waves* or simply BGK modes.

BGK Modes

BGK modes are exact one-dimensional electrostatic solutions of the unmagnetized Vlasov-Poisson system of equations. In the wave frame they are non-oscillating entities which move at a certain velocity, $v_0 = \text{const}$, across the plasma laboratory frame and can be considered as large-amplitude stationary structures. Inside these structures one encounters an equilibrium between the particles and the waves, the stability of which against external disturbances can be considered separately. It is of considerable interest that the plasma allows such structures to exist for comparably long times.

If a one-dimensional system of Vlasov-Poisson equations is transformed by replacing $x \to x - v_0 t$ and $v \to v - v_0$ to a stationary moving system, $\partial/\partial t \to 0$, the now stationary distribution function, $f_{s0}(x - v_0 t, v - v_0)$, and stationary electric field, $E(x - v_0 t)$, obey

$$
\begin{aligned}
v\frac{\partial f_{s0}(x,v)}{\partial x} &= -\frac{q_s}{m_s}E(x)\frac{\partial f_{s0}(x,v)}{\partial v} \\
\frac{\partial E(x)}{\partial x} &= \sum_s \frac{q_s}{\epsilon_0}\int dv\, f_{s0}(x,v)
\end{aligned} \tag{8.18}
$$

Introducing the new energy variable

$$
W_s = \tfrac{1}{2}m_s v_s^2 + q_s\phi(x) \tag{8.19}
$$

where $E = -\partial\phi/\partial x$, and ϕ the electrostatic potential, it is possible to express the solution of the Vlasov equation as a superposition of solutions for the negative and positive energy domains, $f_{s<}$ and $f_{s>}$, respectively

$$
f_{s0}[W_s, \mathrm{sgn}(v)] = f_{s<}(W_s)\Theta(-v) + f_{s>}(W_s)\Theta(v) \tag{8.20}
$$

where $\Theta(v)$ is the Heaviside step function. Substituting the distribution (8.20) into the stationary Poisson equation and replacing $dv = \pm dW_s[2m_s(W_s - q_s\phi)]^{-1/2}$, we find

$$
\frac{\partial^2\phi}{\partial x^2} = -\sum_s \int_{q_s\phi}^{\infty} \frac{f_{s<}(W_s) + f_{s>}(W_s)}{\epsilon_0[2m_s(W_s - q_s\phi)]^{1/2}} dW_s \tag{8.21}
$$

Formally this equation can be regarded as the equation of motion of a pseudo-particle at 'position ϕ' in 'time x'. For such a particle one can derive a 'first integral of motion' via multiplying by $\partial\phi/\partial x$ and integrating once with respect to x

$$
\frac{1}{2}\left(\frac{\partial\phi}{\partial x}\right)^2 + V(\phi) = \text{const} \tag{8.22}
$$

The pseudo-potential $V(\phi)$ is obtained from the right-hand side of Eq. (8.21), denoted by $G(\phi)$, as

$$
V(\phi) = -\int_{\phi_0}^{\phi} d\phi\, G(\phi) \tag{8.23}
$$

Depending on the geometrical form of this potential function, one obtains periodic and aperiodic solutions for $\phi(x)$ by a simple quadrature of Eq. (8.22)

$$
x - x_0 = \frac{\pm 1}{\sqrt{2}}\int_{\phi_0}^{\phi} \frac{d\phi}{|V(\phi) - V(\phi_0)|^{1/2}} \tag{8.24}
$$

If the pseudo-potential $V(\phi)$ for given particle distributions in the trapped and untrapped regions has a well in the region of interest, one gets a discrete spectrum of periodic waves. These periodic solutions for $\phi(x)$ represent spatial modulations of the wave potential caused by the interaction of the large-amplitude wave with the trapped particles.

Particle Distribution Functions

Equation (8.21) can be used to determine the distribution function of the trapped particle component. The trapped electrons have energies $W_{e<} < -e\phi_{min}$. Their distribution is defined as

$$f_{\rm etr} = f_{e>} + f_{e<} = 2f_{e>} = 2f_{e<} \tag{8.25}$$

If we define their number density as

$$g(e\phi) = \int\limits_{-e\phi}^{-e\phi_{min}} \frac{f_{\rm etr}\, dW_e}{[2m_e(W_e + e\phi)]^{1/2}} \tag{8.26}$$

which vanishes at the minimum potential, $g(e\phi_{min}) = 0$, the solution of Eq. (8.21) for the trapped electron component can be expressed as

$$f_{\rm etr}(W_e) = \frac{(2m_e)^{1/2}}{\pi} \int\limits_{e\phi_{min}}^{-W_e} \frac{dg(u)}{du} \frac{du}{(-u - W_e)^{1/2}} \tag{8.27}$$

This solution holds as long as $W_e < -e\phi_{min}$, so that the expression in the root term in the denominator is real. The expression for $g(u)$ entering this solution is a complicated expression of the untrapped particle distributions

$$g(u) - \frac{\epsilon_0}{e}\frac{\partial^2\phi}{\partial x^2} = \int\limits_{u}^{\infty} \frac{(f_{i<} + f_{i>})dW_i}{[2m_i(W_i - u)]^{1/2}} - \int\limits_{-u_{min}}^{\infty} \frac{(f_{e<} + f_{e>})dW_i}{[2m_e(W_e + u)]^{1/2}} \tag{8.28}$$

The untrapped distribution can be freely chosen. For simple sinusoidal potential function, $\phi = \phi_0 \cos kx$, with the ion distribution function being a monoenergetic ion beam, $f_{i>} = 0$, and

$$f_{i<} = a_i[2m_i(W_+ + u)]^{1/2}\delta(W_i - W_+) \quad \text{for} \quad W_+ > u \tag{8.29}$$

to make sure that no ions are trapped, with $f_{e>} = 0$ for the untrapped electrons, and

$$f_{e<} = a_e[2m_e(W_- - u)]^{1/2}\delta(W_e - W_-) \quad \text{for} \quad W_- > u \tag{8.30}$$

one obtains simple expressions for the densities of ions and untrapped electrons

$$n_s = a_s \left[\frac{W_\pm \pm u_0}{W_\pm \mp u(x)} \right]^{1/2} \tag{8.31}$$

where the upper sign corresponds to ions, the lower to electrons. These definitions permit to determine the function g

$$g(u) = -\frac{\epsilon_0 k^2 u}{e^2} + \sum_s n_s(u) \tag{8.32}$$

and with its help to find the trapped electron distribution. This distribution is given for

$$\epsilon_0 k^2 u / e^2 = a_e - a_i \tag{8.33}$$

which serves as the relation between the so far undetermined constants, a_e and a_i, the amplitude of the wave, u, and the wavenumber, k. The trapped solution is given by

$$\pi f_{\mathrm{etr}}(W_e) = [2m_e(e\phi_0 - W_e)]^{1/2} \left(-\frac{2\epsilon_0 k^2}{e^2} + \frac{a_i}{W_+ + W_e} + \frac{a_e}{W_- - W_e} \right) \tag{8.34}$$

This distribution function must be positive, a condition which can be expressed as

$$a_e \geq \frac{\epsilon_0 k^2}{e^2} \frac{(2W_+ - e\phi_0)(W_- + e\phi_0)}{(W_+ + W_-)} \tag{8.35}$$

where we put $W_e = -e\phi_0$, since this maximizes the number of trapped particles.

8.3. Weak Particle Turbulence

The previous section has convinced us that nonlinear interactions between waves and particles can lead under certain circumstances to stationary states consisting of large-amplitude waves and particle distributions separated into trapped and untrapped populations. Since it is in most cases impossible to search for such exact stationary states, a different approach for treating nonlinear problems must be looked for. The simplest such an approach is based on a perturbation expansion. This is the approach chosen in the treatment of weakly turbulent wave-particle interactions in a plasma. The corresponding theory is also known under the name of *weak plasma turbulence*.

Quasilinear Theory

Quasilinear theory is the first and zero-order step in weak particle interaction theory. Its assumptions are that the particle distribution in a plasma is weakly affected by the presence of the self-excited wave spectrum in a random-phase uncorrelated way, and this effect on the particle distribution will self-consistently quench the instability. In this section we derive the basic equations of quasilinear theory. In spite of the restrictions imposed, these equations are, however, of more general use when discussing particle acceleration and diffusion later in a generalized context.

The starting point is the full Maxwell-Vlasov system of equations, of which we write down only the Vlasov equation

$$\frac{\partial f_s}{\partial t} + \mathbf{v}\cdot\nabla f_s + \frac{q_s}{m_s}(\mathbf{E}+\mathbf{v}\times\mathbf{B})\cdot\frac{\partial f_s}{\partial \mathbf{v}} = 0 \tag{8.36}$$

for the s component of the plasma. Splitting the distribution function and fields into average slowly varying parts, f_{s0}, $\mathbf{E}_0 = 0$ and $\mathbf{B}_0$, and oscillating parts δf_s, $\delta\mathbf{E}$ and $\delta\mathbf{B}$, and assuming that the long-time, large-volume averages over the oscillating fields and distribution vanish

$$\langle \delta f_s\rangle = \langle \delta\mathbf{E}\rangle = \langle \delta\mathbf{B}\rangle = 0 \tag{8.37}$$

the average of the Vlasov equation (8.36) gives

$$\boxed{\frac{\partial f_{s0}}{\partial t} + \mathbf{v}\cdot\nabla f_{s0} + \frac{q_s}{m_s}(\mathbf{v}\times\mathbf{B})\cdot\frac{\partial f_{s0}}{\partial \mathbf{v}} = -\frac{q_s}{m_s}\left\langle(\delta\mathbf{E}+\mathbf{v}\times\delta\mathbf{B})\cdot\frac{\partial \delta f_s}{\partial \mathbf{v}}\right\rangle} \tag{8.38}$$

This equation describes the phase space evolution of the slowly varying ensemble averaged part of the distribution function under the action of the fields and the interaction between the fluctuations of the fields and those of the probability distribution. The latter are contained in the non-vanishing averaged term in angular brackets on the right-hand side of Eq. (8.38). Since we have not made any assumption about the smallness of the fluctuations, Eq. (8.38) is generally valid for any fluctuation amplitude, as long as there is a clear separation between the fluctuations and the average behavior of the plasma. This assumption is usually well satisfied.

Equation (8.38) is the fundamental equation describing the nonlinear dynamics of the plasma. Solving it is a formidable task, because it requires a priori knowledge of the fluctuation fields in order to calculate the average term on the right-hand side. This term has the nature of a Boltzmann collision term. Hence, even in a collisionless plasma, where all particle correlations can be neglected, there are still some kind of collisions, which affect the evolution of the average distribution function and presumably drive the plasma ultimately towards thermal equilibrium. These collisions are of anomalous nature. They are entirely due to the nonlinear coupling between the particles and the fluctuating wave fields. Actually, they are the result of scattering the particle motion in the short-wavelength high-frequency field fluctuations.

Weak Gentle Beam Turbulence

Equation (8.38) is valid under the random-phase approximation. Only then when all fluctuating phases mix in an irregular way, such that their effects cancel in the average statistical ensemble, the contribution of the correlation between the wave and particle fluctuations can be condensed into an average pseudo-collision integral of the kind shown on the right-hand side of Eq. (8.38). If this is the case, the structure of this correlation term allows for a very efficient formulation of the average interaction between the particles and fields in terms of a *weakly turbulent wave-particle interaction* theory.

Assuming that the fluctuations are small, they can be calculated from linear theory, and the pseudo-collisional correlation term is calculated as the random-phase averaged correlation of the linear fluctuations. Introducing a smallness parameter, λ, chosen as the ratio of the average fluctuation energy to the average thermal energy of the average distribution function,

$$\lambda = \frac{\langle \epsilon_0 \delta |\mathbf{E}(\mathbf{x},t)|^2 \rangle}{2\langle n \rangle k_B \langle T \rangle} \ll 1 \tag{8.39}$$

the perturbation expansions for distribution function and electric field are given as

$$\begin{aligned} f_s &= f_{s0} + \lambda \delta f_{s1} + \lambda^2 \delta f_{s2} + \dots \\ \delta \mathbf{E} &= \lambda \delta \mathbf{E}_1 + \lambda^2 \delta \mathbf{E}_2 + \dots \end{aligned} \tag{8.40}$$

where the indices, $1, 2, \dots$ indicate the different orders.

Turbulence theory derives equations for each of these orders in ascending succession. Let us discuss here the purely electrostatic case and the first order only. We also restrict ourselves to Langmuir waves with the ions as immobile neutralizing background, as in Sec. 10.1 of our companion book, *Basic Space Plasma Physics*. But this time we do not specify the undisturbed distribution function to be a Maxwellian, since we are interested in the modification of the zero-order distribution due to the unstable wave spectrum. However, because such a reaction can be expected only at large wave amplitudes, we suppose that a gentle beam of low density crosses the plasma and leads to instability. In such a case the wave frequency and growth rate of the broadband Langmuir waves are given by Eqs. (4.2) and (4.3)

$$\omega(k) = \omega_{pe}\left(1 + \tfrac{3}{2}k^2\lambda_D^2\right) \tag{8.41}$$

$$\gamma(k,t) = \omega(k)\frac{\pi\omega_{pe}^2}{2k^2}\left.\frac{\partial f_{0b}(v,t)}{\partial v}\right|_{v=\omega/k} \tag{8.42}$$

The waves propagate on the cool background plasma, and the instability is driven by the positive slope of the zero-order beam distribution, $f_{0b}(v,t)$, at the resonance. This distribution function evolves with time and will thus determine the evolution of the growth rate of the Langmuir waves. As a consequence, the electrostatic wave field,

$\delta E(t) = -ik\delta\phi(k,t)$, will also evolve with time according to

$$\delta E(k,t) = \delta E(k,0) \exp\left\{-\int_0^t [i\omega(k) - \gamma(k,\tau)]d\tau\right\} \tag{8.43}$$

Intuitively, one expects that the change in the distribution function due to the growing wave field will deplete the positive slope of the distribution function, decrease the growth rate, and saturate the instability.

To demonstrate this behavior we need to derive the equation for the zero-order distribution function and solve it together with the above equation for the wave amplitude. We start from the one-dimensional unmagnetized electrostatic version of Eq. (8.38), assuming, for simplicity, that the zero-order state is homogeneous for all times. This permits to drop the second term on the left-hand side, which contains the gradient of the zero-order distribution function. This yields the simplified equation for the time evolution of the zero-order electron beam distribution function

$$\frac{\partial f_{0b}(v,t)}{\partial t} = \frac{e}{m_e}\left\langle \delta E \frac{\partial \delta f}{\delta v}\right\rangle \tag{8.44}$$

The correlation term on the right-hand side of this expression can be calculated introducing the linear solutions for the Fourier components of the electric wave field, $\delta E(k)$, and the disturbance of the distribution function

$$\delta f(k) = i\frac{e}{m_e}\frac{\delta E(k)}{\omega - kv}\frac{\partial f_{0b}(v,t)}{\partial v} \tag{8.45}$$

and integrating over space and time. The last expression makes the distortion of the distribution function a function of the zero-order distribution.

The integrals over the product of the two Fourier series can be performed when observing that the single terms are orthogonal. This leads to the replacement of $k \to -k, \omega \to -\omega$ in the terms of one of the series, while all cross-multiplied terms are zero. The two sums thereby reduce to one single sum over the product $\delta E(k,\omega)\delta E(-k,-\omega)$. Since the wave field must be real, $\delta E(-k,-\omega) = \delta E^*(k,\omega)$. Hence, we arrive at

$$\boxed{\frac{\partial f_{0b}(v,t)}{\partial t} = \frac{\partial}{\partial v}\left[D(v,t)\frac{\partial f_{0b}(v,t)}{\partial v}\right]} \tag{8.46}$$

This equation has the form of a diffusion equation for the zero-order beam distribution function, $f_{0b}(v,t)$, in velocity space with diffusion coefficient

$$D(v,t) = \mathrm{Re}\left\{\frac{ie^2}{m_e^2}\sum_k \frac{|\delta E(k)|^2}{kv - \omega(k) + i\gamma(k,t)} \exp\left[2\int_0^t \gamma(k,\tau)d\tau\right]\right\} \tag{8.47}$$

It is a special case of the much more general *Fokker-Planck equation*. The beam distribution will thus spread with time in the velocity space under the action of the unstable Langmuir wave field. This process does not proceed in real space and can thus only be observed in the measured wave field or in a direct measurement of the particle distribution function. The diffusion coefficient is proportional to the spectral density of the electric wave field, $|\delta E(k)|^2$. For large average wave amplitudes it will be high. The actual evolution of the wave field is described by Eq. (8.43), with the growth rate given by Eq. (8.42).

Since the diffusion coefficient is a velocity and time dependent quantity, the full system of nonlinear equations for the beam distribution function is not easy to solve. However, one realizes that the interaction between the waves and the beam distribution takes place basically in the resonant region, with particles from outside the resonant region to a large extent being insensitive to the presence of the wave spectrum. This allows to replace the resonant denominator in the diffusion coefficient by a delta function

$$D(v,t) = \frac{\pi e^2}{m_e^2} \int W_w(k,t) \delta(\omega - kv)\, dk \tag{8.48}$$

where we replaced the sum over the various field amplitudes by an integral

$$\int W_w(k,t)\, dk = \sum_k |\delta E(k)|^2 \exp\left[2 \int_0^t \gamma(k,\tau) d\tau\right] \tag{8.49}$$

over the temporally variable spectral density function of the wave continuum, $W_w(k,t)$, and write instead of Eq. (8.43)

$$\frac{\partial W_w(k,t)}{\partial t} = 2\gamma(k,t) W_w(k,t) \tag{8.50}$$

Equations (8.46), (8.48) and (8.50) together with the expression for the growth rate form the basic system of equations of the *quasilinear theory* of beam excited Langmuir waves. These equations are also called the *quasilinear equations* in order to express that they are the lowest order state of the nonlinear theory, where only the reaction of the entire resonant broadband wave spectrum on the zero-order distribution function is considered while all other kinds of wave-wave and wave-particle interactions are dropped. Neglecting these other interactions, which are proportional to the square of the wave intensity, $W_w^2(k)$, is justified as long as the wave intensity is small such that

$$\gamma/\omega \gg W_w / n_0 k_B T_e \tag{8.51}$$

Fortunately the structure of the quasilinear equations is such that it is possible to rearrange them into a form, which allows to get insight into the physics of the quasilinear relaxation process. If we multiply the quasilinear diffusion coefficient given in Eq. (8.48) by $\partial f_{0b}/\partial v$, we can replace $dk = (dk/dv)dv$ and take advantage of the δ-function to replace $k = \omega_{pe}/v$. With Eqs. (8.42) and Eq. (8.50), we get

$$D(v,t)\frac{\partial f_{0b}(v,t)}{\partial v} = -\frac{e^2}{m_e^2\omega_{pe}}\frac{1}{v^3}\frac{\partial}{\partial t}W_w\left(\frac{\omega_{pe}}{v}\right) \tag{8.52}$$

After substitution into Eq. (8.46), we obtain the following conservation law

$$\frac{\partial}{\partial t}\left[f_{0b}(v,t) + \frac{e^2}{m_e^2\omega_{pe}}\frac{1}{v^3}\frac{\partial}{\partial t}W_w\left(\frac{\omega_{pe}}{v}\right)\right] = 0 \tag{8.53}$$

Assuming that the initial energy in the turbulence is negligibly small, of the order of the thermal fluctuation energy, integration over the region of the positive slope of the distribution function, from $v_{\min}$ to v, gives

$$\frac{e^2}{m_e^2\omega_{pe}v^3}\lim_{t\to\infty}W_w\left(\frac{\omega_{pe}}{v}\right) = -\int_{v_{\min}}^{v} dv\,\left[f_{0b}(v,\infty) - f_{0b}(v,0)\right] = A \tag{8.54}$$

From this expression one can estimate the quasilinear saturation level of the Langmuir turbulence excited by the gentle beam in the resonant region. The constant, A, on the right-hand side is approximately proportional to the area under the positive slope of the beam distribution function. The saturation energy can be defined as

$$W_{w\text{sat}} = \frac{\epsilon_0}{2}\sum_k |\delta E(k,\infty)|^2 \approx W_w(k,\infty)\Delta k \tag{8.55}$$

where Δk is the spectral width. Hence, the saturation level is

$$\frac{W_{w\text{sat}}}{n_0 k_B T_e} \approx \frac{\Delta k}{2\omega_{pe}}\frac{v_b^3}{v_{\text{the}}^2}A \tag{8.56}$$

Because $\Delta k \approx \omega_{pe}/v_b$ and for instability $v_b > \sqrt{3}v_{\text{the}}$, the saturation level turns out to be of the order of $W_{w\text{sat}}/n_0k_BT_e > 3A$. We thus find that quasilinear velocity space diffusion produces a final stationary wave spectrum, the total intensity of which is proportional to the resonant area under the beam distribution function and increases with beam velocity, v_b. Of course, this is only approximately correct, because we neglected all interactions between the waves themselves. In the long-time limit these interactions become important and will lead to additional modulation of the wave saturation level.

The most important result obtained from quasilinear theory can be drawn from an inspection of Eq. (8.52). In the final state the right-hand side of this equation vanishes, since the wave spectrum saturates at a constant time-independent level

$$D(v,\infty)\frac{\partial f_{0b}(v,\infty)}{\partial v} = 0 \tag{8.57}$$

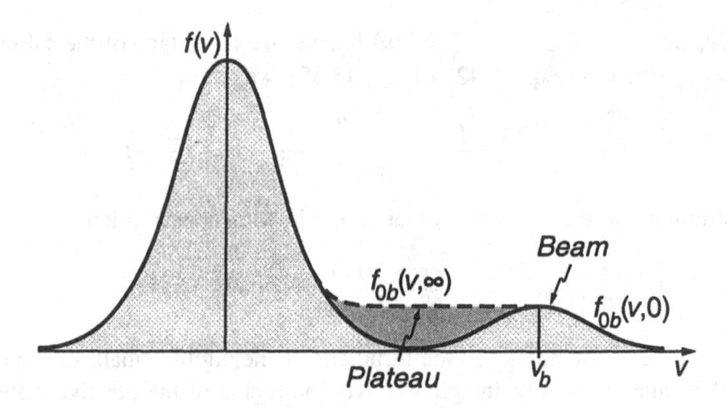

Fig. 8.4. Plateau formation in quasilinear diffusion.

Because the diffusion coefficient cannot vanish unless the wave spectrum vanishes, this implies that the resonant part of the gentle beam distribution function evolves towards a plateau with vanishing velocity derivative, $\partial f_{0b}(v,\infty)/\partial v = 0$. In quasilinear theory the positive slope of the distribution function tends to depleted, as shown schematically in Fig. 8.4. As long as a positive velocity gradient exists, the particles in the positive slope diffuse from the region of the slope down into the valley towards the thermal distribution. This is the physical meaning of the quasilinear velocity space diffusion process described by the quasilinear equations.

It is clear that the wave spectrum does not persist forever after quasilinear diffusion has ceased. Interactions of higher order will destroy the spectrum. Moreover, a weak positive slope will always remain such that in a real equilibrium this slope just compensates for the loss of waves from the resonant region. In addition, propagation effects of the particle distribution will also modify the conditions of quasilinear saturation.

Weak Lower-Hybrid Drift Turbulence

Out of the large number of waves undergoing quasilinear relaxation we pick out here one particularly interesting electrostatic mode in a magnetized plasma, the lower-hybrid drift mode discussed in Sec. 6.3. Since quasilinear theory describes the time evolution of the equilibrium distribution function, it is necessary to retain the dependence on the distribution function in the expression for the growth rate of the lower-hybrid instability, without assuming it to be a Maxwellian. For the sake of simplicity, we restrict ourselves to purely perpendicular propagation. Then the dispersion relation can be written as

$$D(k_\perp,\omega) = 1 + \chi_e(k_\perp,\omega) + \chi_i(k_\perp,\omega) = 0 \tag{8.58}$$

where we introduced the abbreviations

$$\begin{aligned}
\chi_e(k_\perp,\omega) &= \frac{\omega_{pe}^2}{\omega_{ge}^2}\left[1-\frac{1}{k_\perp L_n}\frac{\omega_{ge}}{\omega-k_\perp v_{de}}\right] \\
\chi_i(k_\perp,\omega) &= \frac{\omega_{pi}^2}{k_\perp}\int\frac{dv_\perp}{\omega-k_\perp v_{de}}\frac{\partial f_{0i}(v_\perp,t)}{\partial v_\perp}
\end{aligned} \tag{8.59}$$

This time it is the ion distribution function which carries the free energy, and therefore the ion distribution function will evolve quasilinearly in time. The growth rate in the low drift velocity regime, $v_{de} < v_{\mathrm{th}i}$, is obtained from the linear instability theory and given by

$$\gamma = -\frac{\pi v_{\mathrm{th}i}^2 k_\perp^2 v_{de}}{2|k_\perp|(1+k_\perp^2/k_{\max}^2)^2}\left.\frac{\partial f_{0i}}{\partial v_\perp}\right|_{v_\perp=\omega/k_\perp} \tag{8.60}$$

where $v_{de} = -v_{di}$, and $k_{\max}$ is defined as

$$k_{\max}^2\lambda_{Di}^2 = 2/(1+\omega_{pe}^2/\omega_{ge}^2) \tag{8.61}$$

the value of the perpendicular wavenumber where the growth rate maximizes. Hence, quasilinear saturation of lower-hybrid turbulence can proceed in two ways, either by depleting the resonant part of the distribution function, such that the velocity gradient tends to zero as for Langmuir turbulence, or by reduction of the drift speed, $v_{de} \to 0$.

The evolution of the spectral density follows Eq. (8.50), which we now write as

$$\partial W_w(k_\perp,t)/\partial t = 2\gamma(k_\perp,t)W_w(k_\perp,t) \tag{8.62}$$

Remembering that $W_E(k_\perp,t) = \epsilon_0|\delta E(k_\perp,t)|^2/2$ is the electric field energy, and that the wave energy contains a contribution from the displacement of the electrons, i.e., a contribution from the polarization of the medium, $\partial(\omega\epsilon)/\partial\omega$, we can write the wave energy as

$$W_w(k_\perp,t) = \left(1+\omega_{pe}^2/\omega_{ge}^2\right) W_E(k_\perp,t) \tag{8.63}$$

The resonantly unstable region covers the range of ion velocities $0 < v < v_{de} = v_E$. In this resonant range the ion velocity evolves according to the quasilinear equation

$$\frac{\partial f_{0i}(\mathbf{v},t)}{\partial t} = \frac{\partial}{\partial \mathbf{v}}\cdot\left[\mathsf{D}_i(\mathbf{v},t)\cdot\frac{\partial f_{0i}(\mathbf{v},t)}{\partial \mathbf{v}}\right] \tag{8.64}$$

which is the three-dimensional generalization of the electrostatic quasilinear diffusion equation (8.46). Here the diffusion coefficient is defined as

$$\mathsf{D}_i(\mathbf{v},t) = \frac{2i\omega_{pi}^2}{n_0 m_i}\int\frac{d^3k}{k^2}\frac{\mathbf{kk}\,W_E(\mathbf{k},t)}{\omega(\mathbf{k})-\mathbf{k}\cdot\mathbf{v}} \tag{8.65}$$

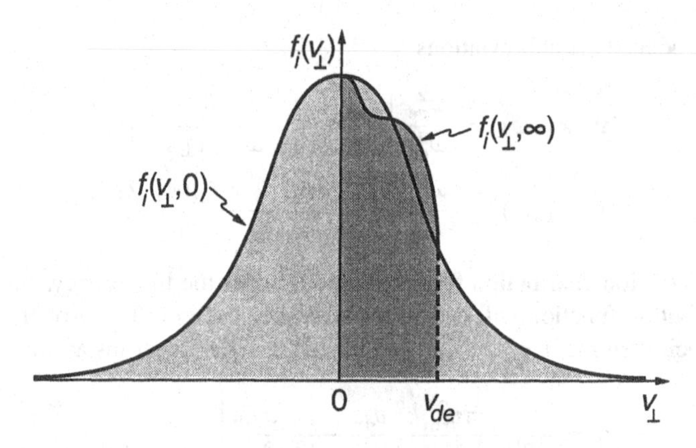

Fig. 8.5. Plateau formation in a lower-hybrid resonant ion distribution.

In our perpendicular case the quasilinear diffusion equation (8.64) and the diffusion coefficient (8.65) reduce to one dimension

$$\frac{\partial f_{0i}(v_\perp,t)}{\partial t} = \frac{\partial}{\partial v_\perp}\left[D_\perp(v_\perp,t)\frac{\partial f_{0i}(v_\perp,t)}{\partial v_\perp}\right] \tag{8.66}$$

$$D_\perp(v_\perp,t) = \frac{2\pi\omega_{pi}^2}{n_0 m_i}\int dk_\perp W_E(k_\perp,t)\,\delta(\omega - k_\perp v_\perp) \tag{8.67}$$

With the resonant wavenumber, $k_0^2(v_\perp) = k_{\max}^2 v_\perp/|v_{de} - v_\perp|$, and the group velocity, $v_{gr} = \partial\omega/\partial k_\perp$, the expression for the diffusion coefficient can be integrated to yield

$$D_\perp(v_\perp,t) = \frac{2\pi\omega_{pi}^2 W_E(k_0,t)}{n_0 m_i\,|v_{gr}(k_0) - v_\perp|} \tag{8.68}$$

Multiplying Eq. (8.66) with $f_{0i}(v_\perp,t)$ and integrating over the resonant region in velocity space yields

$$\frac{\partial}{\partial t}\int dv_\perp f_{0i}^2(v_\perp,t) = -2\int dv_\perp D_\perp(v_\perp,t)\left[\frac{\partial f_{0i}(v_\perp,t)}{\partial v_\perp}\right]^2 \leq 0 \tag{8.69}$$

Hence, the lower-hybrid drift instability also saturates by forming a plateau in the resonant region of velocity space, $0 < v_\perp < v_{de}$, leading to $\partial f_{0i}(v_\perp,\infty)/\partial v_\perp = 0$ for asymptotically vanishing lower-hybrid drift growth rate, $\gamma(\infty) = 0$. This behavior is shown schematically in Fig. 8.5.

To obtain information about the mode of energy transport during this relaxation and plateau formation process, we multiply the quasilinear diffusion equation (8.66) by

$v_\perp^2$ and integrate. Inserting the diffusion coefficient from Eq. (8.67) and performing the integration over the δ-function yields

$$\frac{n_0 m_i}{2}\frac{\partial}{\partial t}\int v_\perp^2 f_{0i} dv_\perp = \int 2\gamma(k_\perp,t)W_w(k_\perp,t)dk_\perp \tag{8.70}$$

Since $\partial W_w/\partial t = 2\gamma W_w$ from Eq. (8.62), this equation becomes a conservation law containing the resonant ion kinetic energy density, $W_{ires} = (n_0 m_i/2)\int dv_\perp v_\perp^2 f_{0i}$

$$\partial(W_{ires} - W_w)/\partial t = 0 \tag{8.71}$$

Thus the quasilinear saturation level can be estimated by calculating the final resonant particle kinetic energy level in the plateau region. Fig. 8.5 shows that there is a particular velocity, v_*, inside the resonant region, for which during plateau formation the number of particles is conserved. This implies that the value of the distribution function at $v_\perp = v_*$ is unchanged for all times, and in particular $f_{0i}(v_*,0) = f_{0i}(v_*,\infty)$. Assuming that the initial distribution is a Maxwellian, $f_{0i,\mathrm{m}}$, we get

$$v_{de} f_{0i,\mathrm{m}}(v_*) = \int_0^{v_{de}} dv_\perp f_{0i,\mathrm{m}}(v_\perp) \tag{8.72}$$

For $v_{de}^2 \ll v_{\mathrm{th}i}^2$ this equation simply yields $v_* = v_{de}/\sqrt{3}$. Calculating for this low drift velocity regime the net energy change

$$\Delta W_{ires} = W_{ires}(\infty) - W_{ires}(0) \approx 0.05\, n_0 k_B T_i\, (v_{de}/v_{\mathrm{th}i})^5 \tag{8.73}$$

we find that it goes as the fifth power of the drift-to-thermal velocity ratio. Since the total energy is conserved, this is also the change in wave energy between initial and saturated state, ΔW_w. Since the initial wave energy is at the thermal fluctuation level and can be neglected, we can integrate the wave spectral energy, $W_w(k_\perp)$, from Eq. (8.63), over the wavenumber space and obtain the wave energy

$$\left(1+\frac{\omega_{pe}^2}{\omega_{ge}^2}\right)\int dk_\perp \left(1+\frac{k_\perp^2}{k_{\mathrm{max}}^2}\right) W_E(k_\perp,\infty) = \Delta W_{ires} \tag{8.74}$$

which can be used to estimate the saturated electric field fluctuation level, $W_E(k_\perp,\infty)$. This becomes particularly easy when one assumes that the spectrum is peaked at the maximum wavenumber. Then the bracket containing the ratio of wavenumber to maximum wavenumber under the integral becomes 2, and one finds that

$$W_E(k_\perp,\infty) = 0.1\frac{m_i}{m_e}\left(\frac{v_{de}}{v_{\mathrm{th}i}}\right)^3 W_{\mathrm{eq}} \tag{8.75}$$

The quantity W_{eq} introduced here is the electric field fluctuation level, which is obtained when the final state can be characterized by equipartition between the kinetic energy and the wave energy. In this final state one half of the energy is in the waves

$$\boxed{W_{\text{eq}} = \frac{n_0 m_e v_{de}^2}{8(1+\omega_{pe}^2/\omega_{ge}^2)}} \tag{8.76}$$

This would be the predicted saturation level if current relaxation instead of quasilinear plateau formation would be the mechanism which saturates the lower-hybrid drift instability. Equipartition of particle and wave energy is frequently assumed in situations where the saturation mechanism is unknown. However, the equipartition saturation level may differ strongly from the quasilinearly reached saturation level.

The quasilinear theory gives the possibility to estimate the quasilinear relaxation time of the ion distribution function until plateau formation is nearly completed. This asymptotic time interval, taken to be infinite in the above calculation, is a long but finite time span, given approximately by the ratio of the square of the ion thermal velocity to the quasilinear diffusion coefficient

$$\tau_{ql} \approx v_{\text{th}i}^2/2D_\perp \tag{8.77}$$

To order of magnitude, we can estimate the quasilinear diffusion coefficient

$$D_\perp \approx \frac{4\pi}{n_0 m_i}\frac{\omega_{pi}^2}{k_{\max} v_{de}} W_E \tag{8.78}$$

where we used the maximum growing wavenumber and $v = v_{de}/2$. Since we already know the saturation level, the quasilinear relaxation time is estimated to

$$\tau_{ql} \approx \frac{k_{\max} v_{de}}{4\pi\omega_{pi}^2}\frac{n_0 k_B T_i}{W_E} \tag{8.79}$$

Hence, knowing the spectral energy of the electric fluctuations in the lower-hybrid drift instability, one can estimate the relaxation time. Using the above expressions and the expression for the maximum linear growth rate, $\gamma_{\max} = \omega_{lh}(\sqrt{2\pi}/8)(v_{de}^2/v_{\text{th}i}^2)$, one finds

$$\tau_{ql}\gamma_{\max} \approx 1.4(v_{\text{th}i}/v_{de})^2 \tag{8.80}$$

For completeness we investigate the behavior of the non-resonant ions and electrons. Both are described by the same quasilinear equation, but with the non-resonant part of the diffusion coefficient, i.e., the principal-value part and not the δ-function resonant part. Formally this equation can be written for the ions as

$$\frac{\partial F_{0i}}{\partial t} = \frac{\partial}{\partial v_\perp}\left[D_{nr}(v_\perp, t)\frac{\partial F_{0i}}{\partial v_\perp}\right] \tag{8.81}$$

where we used a capital F to distinguish the non-resonant from the resonant distribution functions, and the non-resonant diffusion coefficient is obtained assuming that the frequency is low enough to allow to write

$$D_{nr}(v_\perp, t) = \frac{2\omega_{pi}^2}{m_i n_0} \int dk_\perp \frac{\gamma(k_\perp, t) W_E(k_\perp, t)}{k_\perp^2 v_\perp^2} \tag{8.82}$$

With the help of this equation the energy conservation assumes the form

$$\frac{n_0 m_i}{2} \frac{\partial}{\partial t} \int v_\perp^2 dv_\perp F_{0i} = -2\omega_{pi}^2 \int \frac{dk_\perp}{k_\perp^2} \gamma W_E \int \frac{dv_\perp}{v_\perp} \frac{\partial F_{0i}}{\partial v_\perp} \tag{8.83}$$

Because in the non-resonant region the distribution function remains approximately unchanged, it can be replaced by a Maxwellian on the right-hand side. After carrying out the differentiation this removes the singularity and produces

$$\frac{\partial}{\partial t} \left[W_{i,nr} - 2 \int \frac{dk_\perp}{k_\perp^2 \lambda_{Di}^2} W_E(k_\perp, t) \right] = 0 \tag{8.84}$$

where $W_{i,nr} = (n_0 m_i/2) \int v_\perp^2 dv_\perp F_{0i}$ is the non-resonant kinetic ion energy and the spectral field energy is replaced by its time variation, $2\gamma W_E = \partial W_E/\partial t$. At maximum growth rate we have $k_\perp^2 = k_{\text{max}}^2$, and the $k_\perp$-integral over the spectral density gives the total energy density of the wave field, $W_E = 2(1 + \omega_{pe}^2/\omega_{ge}^2) \int dk_\perp W_E(k_\perp)$ and thus

$$\partial (W_{i,nr} - \tfrac{1}{2} W_w)/\partial t = 0 \tag{8.85}$$

The non-resonant ion interaction is thus non-negligible in the total energy budget. This small distortion of the non-resonant bulk distribution is best accounted for as a non-resonant heating effect of the bulk ion plasma by the non-resonant absorption of the lower-hybrid wave energy. We can estimate this kind of heating from the last equation, by defining the relative increase in the ion temperature as

$$\frac{\Delta T_i}{T_i} = \frac{\Delta W_w}{n_0 k_B T_i} \approx 2 \left(1 + \frac{\omega_{pe}^2}{\omega_{ge}^2}\right) \frac{W_{\text{eq}}}{n_0 k_B T_i} \tag{8.86}$$

or using the wave saturation level

$$\Delta T_i / T_i \approx 0.05 (v_{de}/v_{\text{th}i})^5 \ll 1 \tag{8.87}$$

which in the slow drift limit used here is small. Therefore the bulk plasma heating by the lower-hybrid drift instability is only weak.

However, the equivalent investigation for electrons shows that electron heating can be quite substantial. In response to the lower-hybrid drift instability the electrons gain energy as described by

$$\frac{n_0 m_e}{2} \frac{\partial}{\partial t} \int dv_\perp (v_\perp - v_{de})^2 f_{0e}(v_\perp, t) = \frac{\omega_{pe}^2}{\omega_{ge}^2} \frac{\partial W_E}{\partial t} \tag{8.88}$$

Thus the heating rate of the electrons is quite substantial if the plasma-to-cyclotron frequency ratio is large. The electrons gain heat at a very fast rate, faster than the wave fluctuations grow. But the heating of the bulk of the electrons still remains small. Energy is only fed to the few resonant electrons which are accelerated in the wave field.

Weak Whistler Turbulence

We now turn to electromagnetic waves below the electron-cyclotron frequency. For simplicity we consider only parallel propagation of the R- and L-modes (see Sec. 10.6 of our companion book). The linear distortion of the distribution function caused by these waves is given in Eq. (I.10.91) in terms of the initial non-equilibrium distribution, f_{0s}. For parallel propagation, performing the integration over the gyrating particle orbit, we get for the linear disturbance

$$\begin{aligned}\delta f_{s\mathrm{R,L}}(k,\mathbf{v},t) &= -\frac{iq_s}{2m_s}\frac{\delta E_{\mathrm{R,L}}(k,t)}{\omega(k,t)-kv_\parallel \pm \omega_{gs}} \\ &\left\{\left[1-\frac{kv_\parallel}{\omega(k,t)}\right]\frac{\partial}{\partial v_\perp}+\frac{kv_\perp}{\omega(k,t)}\frac{\partial}{\partial v_\parallel}\right\} f_{0s}(v_\perp,v_\parallel,t) \quad (8.89)\end{aligned}$$

where $k = k_\parallel$ is the parallel wavenumber, and we have indicated explicitly the dependence on wavenumber and time in the frequency, ω, and in the undisturbed background distribution, $f_{0s}(t)$. The time dependence in the frequency leads to the time evolution of the growth rate, $\gamma(k,t)$, which we again expect to vanish asymptotically in the long-time limit, when the quasilinear stable state is reached. This growth rate must be positive, $\gamma(k,t) > 0$, because otherwise the wave field dies out and there is no remarkable distortion of the distribution function. Growing whistlers and electromagnetic ion-cyclotron waves are obtained, for instance, for an excess temperature in the perpendicular direction, $A_s > 0$. But there are also other possibilities to drive these waves unstable, such as loss cones and beams.

The linear growth rate is determined from the linear dispersion relation, which for an unspecified distribution function is given as

$$\begin{aligned}D_{\mathrm{R,L}} &= 1-\frac{k^2c^2}{\omega^2(k,t)}+\sum_s\frac{\omega_{ps}^2}{2\omega^2(k,t)}\int d^3v\frac{v_\perp}{\omega(k,t)-kv_\parallel \mp \omega_{gs}} \\ &\cdot\left\{\left[1-\frac{kv_\parallel}{\omega(k,t)}\right]\frac{\partial}{\partial v_\perp}+\frac{kv_\perp}{\omega(k,t)}\frac{\partial}{\partial v_\parallel}\right\} f_{0s}(v_\perp,v_\parallel,t)=0 \quad (8.90)\end{aligned}$$

Finding the quasilinear equation of weak turbulence in even this simple case of parallel propagation requires quite some algebra. The steps of the calculation are as follows. Take the quasilinear equation (8.38) and write it for parallel propagation and a gyrotropic undisturbed distribution function, which does not depend on the gyration angle, ϕ, of the particles. Then replace the wave fields on the right-hand side in the ensemble-averaged part by the right- and left-polarized electromagnetic fields

$$\delta E_{\mathrm{R,L}}(k,t) = \delta E_x(k,t) \mp i\delta E_y(k,t) \tag{8.91}$$

thereby expressing the magnetic through the electric wave field using Maxwell's second equation. Now expand the wave field and the disturbance of the distribution on the right-hand side into Fourier series both in space and time, using different k and ω in the two Fourier integrals. At this point insert δf from Eq. (8.89) to express the right-hand side in terms of the electric wave field and the background distribution function only.

The next step is to interpret the average as integrals over the fast space-time variation of the wave field, keeping the background distribution and frequency constant and assuming that averages over linear quantities vanish in the random-phase approximation. Performing this integration over all space and time reduces the two Fourier integrals to just one, because the random-phase approximation produces δ-functions of the type $\delta(\omega - \omega_1 - \omega_2)\delta(k - k_1 - k_2)$. Remembering that the wave fields are real, $\delta E(k,\omega)\delta E(-k,-\omega) = |\delta E(k,\omega)|^2$, we define

$$\epsilon_0 \left\langle |\delta E(k,t)|^2 \right\rangle = 2W_E(k,t) \tag{8.92}$$

where the time dependence refers to the slow variation of the averages only, since fast time variations have been averaged out. Finally, we average over the gyration angle, ψ, from 0 to 2π using that due to gyrotropy of the initial background distribution

$$\frac{1}{2\pi}\int_0^{2\pi} d\psi \frac{\partial f_{0s}}{\partial \psi} = 0 \tag{8.93}$$

which shows that the final average distribution is also gyrotropic, despite the presence of the waves and their modifying effect. We then obtain

$$\begin{aligned}\frac{\partial f_{0s}}{\partial t} = \frac{i\omega_{ps}^2}{2m_s n_s}\sum_{\pm}\int dk W_E(k,t)\left\{\left[1 + \frac{kv_\parallel}{\omega(-k,t)}\right]\frac{1}{v_\perp}\frac{\partial}{\partial v_\perp}\right.\\ \left. - \frac{kv_\perp}{\omega(-k,t)}\frac{\partial}{\partial v_\parallel}\right\}\frac{1}{\omega(k,t) - kv_\parallel \pm \omega_{gs}}\left\{1 - \frac{kv_\parallel}{\omega(k,t)}\frac{\partial}{\partial v_\perp} + \frac{kv_\perp}{\omega(k,t)}\frac{\partial}{\partial v_\parallel}\right\} f_{0s}\end{aligned} \tag{8.94}$$

for the general parallel kinetic equation of weak electromagnetic plasma turbulence theory. The sum in this equation is over the two resonant contributions, corresponding to the two directions of propagation parallel and antiparallel to the magnetic field, $\pm k$.

The complicated expression on the right-hand side, in front of the slowly varying background distribution function, is the velocity space diffusion term. Obviously, in the magnetized plasma it depends on the two velocity derivatives parallel and perpendicular to the magnetic field. The important conclusion drawn from the above quasilinear weak turbulence kinetic equation (8.94) is that the background resonant distribution function experiences a two-dimensional velocity space diffusion in $v_{\parallel}$ and $v_{\perp}$ under the action of the parallel propagating electromagnetic R- and L-modes, while the evolution of the wave spectral density in this theory follows Eq. (8.50)

$$\partial W_E(k,t)/\partial t = 2\gamma(k,t)W_E(k,t) \tag{8.95}$$

We will now apply the present theory to whistlers excited by a temperature anisotropy, under the assumption that $|\omega(k) - \omega_{ge}| \gg kv_{\mathrm{the}}$ and $|\gamma(k)/\omega(k)| \ll 1$. The growth rate of the whistler, expressed through the initial undisturbed distribution is

$$\begin{aligned}\gamma(k,t) &= \frac{\pi}{2}\omega(k)\omega_{pe}^2 \int d^3v \frac{v_{\perp}\delta(\omega - kv_{\parallel} - \omega_{ge})}{\omega^2 + c^2k^2 + \omega^2\omega_{pe}^2/(\omega - \omega_{ge})^2} \\ &\cdot \left[(\omega - kv_{\parallel})\frac{\partial}{\partial v_{\perp}} + kv_{\perp}\frac{\partial}{\partial v_{\parallel}}\right] f_{0e}\end{aligned} \tag{8.96}$$

Taking into account only resonant interactions leads to the replacement of the denominator by the δ-function. The same replacement must be used also in the diffusion term of Eq. (8.94), which becomes

$$\begin{aligned}\frac{\partial f_{0e}(t)}{\partial t} &= \frac{\pi\omega_{pe}^2}{2m_e n_0} \int_{k\geq 0} dkW_E(k,t) \left[\left(1 - \frac{kv_{\parallel}}{\omega}\right)\frac{1}{v_{\perp}}\frac{\partial}{\partial v_{\perp}}v_{\perp} + \frac{kv_{\perp}}{\omega}\frac{\partial}{\partial v_{\parallel}}\right] \\ &\delta(kv_{\parallel} - \omega + \omega_{ge}) \left[\left(1 - \frac{kv_{\parallel}}{\omega}\right)\frac{\partial}{\partial v_{\perp}} + \frac{kv_{\perp}}{\omega}\frac{\partial}{\partial v_{\parallel}}\right] f_{0e}(t)\end{aligned} \tag{8.97}$$

where $\omega(k)$ is the whistler frequency, given as the solution of the real part of the dispersion relation in Eq. (5.10). The kinetic equation (8.97) describes the nonlinear evolution of the resonant part of the unstable electron distribution function. As for the Langmuir and lower-hybrid waves, we multiply Eq. (8.97) by the distribution function, f_{0e}, and integrate over the resonant part of the phase space to find

$$\begin{aligned}\frac{d}{dt}\int d^3v f_{0e}^2(t) &= -\frac{\pi\omega_{pe}^2}{m_e n_0}\int \frac{dk}{\omega(k)} W_E(k,t) \int d^3v\, \delta(kv_{\parallel} - \omega + \omega_{ge}) \\ &\cdot \left[(\omega - kv_{\parallel})\frac{\partial f_{0e}}{\partial v_{\perp}} + kv_{\perp}\frac{\partial f_{0e}}{\partial v_{\parallel}}\right]^2 \leq 0\end{aligned} \tag{8.98}$$

The right-hand side is always negative. Hence, the evolution of the resonant part of the distribution function tends towards an equilibrium, where the expression in the brackets vanishes for all resonant velocities

$$v_{\mathrm{res}} = [\omega(k) - \omega_{ge}]/k \tag{8.99}$$

In the long-time asymptotic limit this bracket must vanish, simply because the trivial possibility that the wave spectral density becomes zero makes sense only for damped waves, in which case the quasilinear theory is of no interest. One thus requires as the final state of the distribution function in the resonant interval of phase space

$$kv_\perp \partial f_{0e}(\infty)/\partial v_\parallel|_{v_\parallel = v_{res}} + (\omega - kv_\parallel)\partial f_{0e}(\infty)/\partial v_\perp|_{v_\parallel = v_{res}} = 0 \tag{8.100}$$

This condition is equivalent to the requirement that in the final state of quasilinear evolution of the whistler turbulence the linear growth rate has turned zero, $\gamma(\infty) = 0$.

In order to determine what this conclusion implies for the evolution of the distribution function in the resonant phase space interval, we observe that Eq. (8.100) has the characteristic differential equation

$$dv_\parallel / dv_\perp = v_\perp / [(\omega/k) - v_\parallel] \tag{8.101}$$

where $\omega(k)/k$ must be taken at the resonant velocity (8.99). Integration yields that

$$\frac{v_\perp^2 + v_\parallel^2}{2} - \int_0^{v_{res}} dv_\parallel \frac{\omega(v_\parallel)}{k} = \text{const} \tag{8.102}$$

is a constant of motion of the particles in the resonant region in the whistler wave field in the final state. Hence, the final resonant velocity distribution is given by

$$f_{0e}(v_\perp, v_\parallel, \infty) = f_{0e}\left[\frac{v_\parallel^2 + v_\perp^2}{2} - \int_0^{v_{res}} dv_\parallel \frac{\omega(v_\parallel)}{k}\right] \tag{8.103}$$

We know, however, that the resonant strip in $(v_\perp, v_\parallel)$-space is centered at the negative wave phase velocity $v_\parallel = -(\omega/k)$ but otherwise parallel to the $v_\perp$-axis (see the unstable right-hand part of Fig. 5.2). This strip has the coordinates

$$[\omega(k_1) - \omega_{ge}]/k_1 < v_\parallel < [\omega(k_2) - \omega_{ge}]/k_2 \tag{8.104}$$

and $0 < v_\perp < \infty$. The curves where the argument of the asymptotically stable resonant electron distribution function is constant are the sections of the circles in $(v_\perp, v_\parallel)$-space centered at $v_\parallel = -(\omega/k)$, $v_\perp = 0$. During quasilinear evolution the distribution evolves in phase space along these stable orbits.

8.4. Resonance Broadening

In the derivation of quasilinear wave-particle interaction we have continuously assumed that the orbits of the particles remain undisturbed by the wave field. This assumption holds for very weak wave fields. However, when the wave grows to large amplitudes, the orbits of the particles may become distorted. Particle trapping in the wave is one of the possible effects. In a broad random-phase wave spectrum this trapping is not very important, but distortion of the particle orbits will nevertheless occur. In such a case one cannot consider these orbits as unperturbed, but must include the average wave field in calculating the particle orbits. This is a very involved task. It is considerably simpler to assume that the distortion of the particle orbits produces a shift in the resonant wave frequencies and thus causes *resonance broadening* in a wide wave spectrum containing a large number of harmonics. Resonance broadening leads to leakage of particles out of the resonance and is thus it is another important mechanism of wave saturation and stabilization of instabilities.

Mechanism

The equation for the linear variation of the distribution function in random phase interaction between particles and waves can be written

$$\left[i(\mathbf{k}\cdot\mathbf{v}-\omega)-\frac{\partial}{\partial\mathbf{v}}\cdot\left(\mathsf{D}\frac{\partial}{\partial\mathbf{v}}\right)\right]\delta f_s(\mathbf{k}) = -\frac{q_s}{m_s}\delta\mathbf{E}(\mathbf{k})\cdot\frac{\partial f_{0s}}{\partial\mathbf{v}} \tag{8.105}$$

where, for purely electrostatic variations with $\delta\mathbf{E}||\mathbf{k}$, the diffusion coefficient, D, is

$$\mathsf{D} = -\mathrm{Re}\,\frac{iq_s^2}{m_s^2}\sum_{k'}\frac{\mathbf{k}'\mathbf{k}'}{k'^2}\frac{|\delta\mathbf{E}(\mathbf{k}')|^2}{(\mathbf{k}'+\mathbf{k})\cdot\mathbf{v}-(\omega'+\omega)} \tag{8.106}$$

That the equation for δf already contains a term which depends on the diffusion coefficient of the plasma in the wave field, arises from the interaction of the particles with their distorted orbits with the broadband wave field and is a deviation from linear theory.

When operating with this new expression for δf and building the modified quasilinear equations, one arrives at a quasilinear diffusion equation which contains the diffusion coefficient in a much more complicated version. To demonstrate this behavior let us, for the sake of simplicity, restrict ourselves to one dimension only. In addition we assume that the diffusion coefficient is constant. Then Eq. (8.105) simplifies to

$$i(kv-\omega)\delta f_s(k) - D\frac{\partial^2\delta f_s(k)}{\partial v^2} = -\frac{q_s}{m_s}\delta E(k)\frac{\partial f_{0s}}{\partial v} \tag{8.107}$$

Fourier-transforming the disturbed distribution function in velocity space

$$\delta f(k,v) = \frac{1}{2\pi}\int \delta f(\tau)\exp(iv\tau)d\tau \tag{8.108}$$

yields for the Fourier-transform, $\delta f(\tau) = \int \delta f(k,v)\exp(-iv\tau)dv$

$$\frac{d\delta f_s(\tau)}{d\tau} + \left(\frac{i\omega}{k} - \frac{\tau^2}{k}D\right)\delta f_s(\tau) = \frac{q_s}{m_s}\frac{\delta E(k)}{k}\int\frac{\partial f_{0s}}{\partial v}\exp(-iv\tau)dv \tag{8.109}$$

The solution of this equation is obtained by direct integration. The result is

$$\delta f_s(\tau) = \frac{q_s \delta E(k)}{m_s k} \exp\left[-\left(i\omega\tau/k - D\tau^3/3k\right)\right] \int_{-\infty}^{\tau} d\tau'$$
$$\left\{\exp\left[-\left(i\omega\tau'/k - D\tau'^3/3k\right)\right] \int dv' \frac{\partial f_{0s}}{\partial v} \exp(-iv'\tau')\right\} \quad (8.110)$$

This shows the exponential dependence of the disturbed distribution function on the diffusion coefficient, D, destroying the delta-function character of the resonance. We define the velocity half-width of the particle resonance as $\delta \approx k^{-1}(k^2 D/3)^{1/3}$, solve the Fourier integral over τ, and interpret the result as a resonant denominator

$$\delta f_s(k) \approx -\frac{iq_s}{m_s} \frac{\delta E(k)}{\omega - kv + ik\delta} \frac{\partial f_{0s}}{\partial v} \quad (8.111)$$

The distortion of the particle orbits introduces an additional term into the resonant denominator in δf_s, which causes the broadening of the resonance during the interaction between waves and particles. The resonance diffuses until the instability saturates.

Ion-Acoustic Mode

In particular, for resonant excitation of ion-acoustic waves by hot field-aligned electron currents and cold ions (see Sec. 4.2), the resonance broadening appear on the electrons only. In this case the imaginary part of the dielectric constant, including the resonance broadening $\delta = (k^2 D/3)^{1/3}$, is proportional to

$$\mathrm{Im}\,\epsilon \propto \int \frac{dv\,(\partial f_{0e}/\partial v)\,\delta}{[\omega - k(v_0 + u)]^2 + \delta^2} \quad (8.112)$$

where v_0 is the current speed, and u is the width of the resonance. For large resonance broadening, $\delta \gg kv_0$, the integral tends to zero as $1/\delta$. This implies a vanishing growth rate and thus saturation. The instability considered is resonant in the interval $c_{ia} < v_0 < v_{\mathrm{the}}$. Thus stabilization is achieved when the resonance broadens beyond this domain. This happens when the resonant width, $\Delta\omega$, approaches the value

$$\Delta\omega = kv_{\mathrm{the}} = (k^2 D/3)^{1/3} \quad (8.113)$$

Using the bandwidth, Δk, the diffusion coefficient can be expressed as

$$D \approx \left(\frac{m_i}{m_e}\right)^{1/2} \frac{\pi\omega_{pe}^2}{\Delta k v_{\mathrm{the}}} \frac{W_E}{n_0 k_B T_e} v_{\mathrm{the}}^2 \quad (8.114)$$

This enables us to determine the saturation level of the ion-acoustic instability, whenever the saturation is provided by simple resonance broadening in the turbulent wave field

$$\frac{W_{E,ia}}{n_0 k_B T_e} \approx \left(\frac{m_e}{m_i}\right)^{1/2} k\lambda_D \tag{8.115}$$

This saturation level is surprisingly low. It indicates that the distortion of the electron orbits in the wave field of the current-driven ion-acoustic wave is very strong. The particle orbits are readily deformed, resonance broadening sets on, and the instability saturates on the low level given by Eq. (8.115).

Ion-Cyclotron Resonance Broadening

Another important case is the excitation of electrostatic ion-cyclotron waves by field-aligned currents which has been discussed in Sec. 4.4. This case includes the effects of the magnetic field on the particle motion and is considerably more involved than resonance broadening in ion-acoustic waves. We write the dielectric response susceptibility of species s including the resonance broadening in the denominator

$$\chi_s(\mathbf{k},\omega) = \frac{1}{k^2\lambda_{Ds}^2}\left[1-(\omega-k_\parallel v_{ds})\sum_{l=-\infty}^{\infty}\int d^3v\, f_s(\mathbf{v})\frac{J_l^2(k_\perp v_\perp/\omega_{gs})}{\omega+i\Delta\omega_s(\mathbf{v})-k_\parallel v_\parallel - l\omega_{gs}}\right] \tag{8.116}$$

The resonance broadening in frequency symbolized by the imaginary frequency term in the denominator contains of course the full magnetized diffusion tensor. But for $k_\parallel/k \ll 1$ the resonance broadening is predominantly transverse to the magnetic field. Resonance broadening is now due to the ions because at these low frequencies the electrons behave adiabatically. Restricting to $D_\perp$ only, it can be estimated to

$$k_\perp^2 D_\perp \approx \left(\frac{k_\perp \delta E}{B_0}\right)^2 \frac{\Delta\omega_i G(k_\perp v_\perp/\omega_{gi})}{(\omega-k_\parallel v_\parallel-\omega_{gi})^2+\Delta\omega_i^2} \tag{8.117}$$

where we used only the first harmonic, l=1, neglected electrons, and introduced

$$G(\zeta_i) = \tfrac{1}{4}\left[J_0^2(\zeta_i)+2J_1^2(\zeta_i)+J_2^2(\zeta_i)\right] \tag{8.118}$$

Saturation occurs for $\Delta\omega_i = k_\perp^2 D_\perp$, a condition which yields for the ion-cyclotron mode the wave amplitude at saturation

$$\frac{e\delta E_{sat}}{k_\perp k_B T_e} \approx \frac{\omega_{gi}^2 \Lambda_1(k_\perp^2 v_{\mathrm{th}i}^2/\omega_{gi}^2)}{k_\perp^2 v_{\mathrm{th}i}^2[G(2^{1/2}k_\perp v_{\mathrm{th}i}/\omega_{gi})]^{1/2}} \tag{8.119}$$

Here we used the function $\Lambda_l(\zeta) = I_l(\zeta)\exp(-\zeta)$. This saturation amplitude is independent of the growth rate which at saturation is zero, and the above result is valid as long as the linear growth rate satisfies the condition $\gamma/(\omega - \omega_{gi}) \ll 1$. This saturation amplitude of current driven electrostatic ion cyclotron waves plays an important role in the calculation of anomalous collision frequencies in ion-cyclotron turbulence, for resonance broadening is the main saturation mechanism in this type of waves.

8.5. Pitch Angle Diffusion

The quasilinear theory developed in the last section can be applied to the particle dynamics in the magnetosphere. One must, however, distinguish between energetic particles and particles of lower energy. Energetic particles interact better with electromagnetic waves, simply because their energy may exceed the threshold resonance energy for this kind of interaction. The low energy component requires a different mechanism.

Whistler Turbulence

Assume that an energetic electron component is trapped in the dipolar geomagnetic field close to the Earth, performing bounce motions along the magnetic field between the mirror points of the particles, and at the same time gyrating about the geomagnetic field. Electrons having their mirror point in the deep ionosphere will be lost from the magnetic mirror due to collisions with the neutrals and the dense ionospheric plasma. Therefore the energetic particle distribution function of the trapped electrons will be a loss cone distribution. The trapped electrons will thus have some kind of temperature anisotropy, $A_{eh} > 0$. In addition, the radiation belts are at comparably low geomagnetic latitudes inside the dense plasmasphere with $n_0 \gg n_{eh}$.

Under these conditions, for sufficiently large trapped radiation belt electron fluxes, the hot electron component will excite whistler wave turbulence at frequencies below the threshold frequency, $\omega < \omega_c$, where

$$\omega_c/\omega_{ge} = A_{eh}/(A_{eh} + 1) \tag{8.120}$$

The growth rate of these waves is proportional to the ratio of trapped-to-plasmaspheric electrons, n_{eh}/n_0, and the whistlers propagate on the background plasmaspheric plasma. The whistler waves, being trapped in the geomagnetic field and bouncing in space between the lower-hybrid resonance points, readily reach large amplitudes, thereby entering the quasilinear regime. They start scattering the electrons and building up a platform in the resonant velocity space. Intuitively the only way to deplete the resonant electrons is to scatter them into the loss cone. This mechanism explains in a simple and satisfactory way the enhanced precipitation of electrons from the radiation belts in all cases when the flux of electrons in the radiation belts is enhanced, e.g., during substorms when transversely accelerated energetic electrons are injected into the radiation belts from the Earth's tail.

Consider the distribution function in the quasilinear limit given in Eq. (8.103). In this readily reached final stage the change in total energy of an electron, W_e, associated with the change in the parallel energy, $W_{e\|}$, is given by

$$\frac{dW_e}{dW_{e\|}} = \frac{\omega}{kv_\|} = \frac{\omega}{\omega - \omega_c} \tag{8.121}$$

As a consequence, one expects that the particle energy is nearly unchanged for frequencies sufficiently far below the critical frequency. At such frequencies, which are highly plausible because their resonant particle energy is lowest, the diffusion caused by the interaction of the electron distribution and the whistlers is a *pitch angle diffusion*, turning the particle orbits into the loss cone. The quasilinear diffusion equation for the electrons then becomes a pure pitch angle diffusion equation

$$\frac{\partial f_{0e}(v,\alpha,t)}{\partial t} = \frac{1}{\sin\alpha}\frac{\partial}{\partial\alpha}\left[D(v,\alpha,t)\sin\alpha\frac{\partial f_{0e}(v,\alpha,t)}{\partial\alpha}\right] \tag{8.122}$$

where $\alpha = \tan^{-1}(v_\|/v_\perp)$ is the pitch angle of a particle with velocity v and the diffusion coefficient for constant energy, rewritten in terms of the magnetic fluctuation field component instead of the electric field, is given by

$$D(v,\alpha,t) = \pi^2\omega_c^2\sum_k \frac{1}{|k_\||}\left|\frac{\delta B(k)}{B_0}\right|^2 \delta\left(v\cos\alpha - \frac{\omega-\omega_c}{k_\|}\right) \tag{8.123}$$

In order to describe the stationary state achieved in equilibrium between injection and precipitation, it is convenient to assume that the particles are injected in the nightside equatorial plane with pitch angles outside the equatorial loss cone angle, $\alpha > \alpha_\ell$. With an injection source, $S_e(v,\alpha)$, the stationary diffusion equation becomes

$$\frac{1}{\sin\alpha}\frac{\partial}{\partial\alpha}\left[D(v,\alpha)\sin\alpha\frac{\partial f_{0e}(v,\alpha)}{\partial\alpha}\right] = S_e(\alpha,v) \tag{8.124}$$

Within the loss cone at $\alpha < \alpha_\ell$, and for $\alpha_\ell \ll 1$, the diffusion equation must be supplemented by a loss term which can simply be modeled as $f(\alpha,v)/\tau_\ell(\alpha,v)$. Here τ_ℓ is the characteristic loss time of the precipitating particles which is determined by their bounce motion along the magnetic field and is for immediate loss equal to one quarter bounce period. This choice allows to write

$$\frac{1}{\alpha}\frac{\partial}{\partial\alpha}\left[\alpha D(\alpha,v)\frac{\partial f(\alpha,v)}{\partial\alpha}\right] - \frac{f(\alpha,v)}{\tau_\ell(\alpha,v)} = 0 \tag{8.125}$$

These two equations together with the appropriate boundary condition, that the distribution functions and their derivatives be continuous across the boundary of the loss cone at $\alpha = \alpha_\ell$, and an appropriate expression for the diffusion coefficient describe the pitch angle scattering in a spatially homogeneous radiation belt.

The choice made for the diffusion coefficient depends on the interaction mechanism between the whistlers and the particles. A simple choice is to assume that $D \approx D_0\alpha^q$ for small pitch angles inside the loss cone. Integrating Eq. (8.124) twice with respect to α and defining

$$\langle S_e(v)\rangle = \int_{\alpha_\ell}^{\pi/2} d\alpha' \sin\alpha' S_e(\alpha', v) \tag{8.126}$$

and assuming for simplicity that the injection is at $\alpha = \pi/2$ only, such that $S_e(v, \alpha') = S(v)\delta(\alpha' - \pi/2)$, the external solution is

$$f_e(\alpha, v) = \langle S_e(v)\rangle \left[\int_{\alpha_\ell}^{\alpha} \frac{d\alpha''}{\sin\alpha'' D(\alpha'')} + h(\alpha_\ell)\right] \tag{8.127}$$

The arbitrary function, $h(\alpha)$, has to be determined from the boundary conditions. The interior solution may be expressed in terms of modified Bessel functions. Demanding that $D_0\alpha_\ell^{q+1}\partial h/\partial\alpha|_{\alpha_\ell} = \langle S\rangle$, one finds

$$h(\alpha) = \frac{1}{D_0\alpha^{q/2}}\left(\frac{D_0\tau_\ell}{\alpha_\ell^2}\right)^{1/2} \frac{I_{q/(2-q)}\left[(1-q/2)^{-1}(\alpha^{2-q}/D_0\tau_\ell)^{1/2}\right]}{I_{q/(2-q)}\left[(1-q/2)^{-1}(\alpha_\ell^{2-q}/D_0\tau_\ell)^{1/2}\right]} \tag{8.128}$$

Two limiting cases are of interest. The first is *weak pitch angle diffusion*, when the particles drift into the loss cone at a much slower rate than they are lost, implying that the diffusion time is longer than one quarter bounce period. In this limit $D_0\tau_\ell \ll 1$ and

$$h_{wd}(\alpha) \approx \frac{1}{D_0\alpha^{q/2}}\left(\frac{D_0\tau_\ell}{\alpha_\ell^2}\right)^{1/2}\left(\frac{\alpha_\ell}{\alpha}\right)^{\frac{2-q}{4}} \exp\left[\left(\frac{\alpha^{2-q}}{D_0\tau_\ell}\right)^{1/2} - \left(\frac{\alpha_\ell^{2-q}}{D_0\tau_\ell}\right)^{1/2}\right] \tag{8.129}$$

One realizes that this solution contains one special limiting case, $q = 2$, when the solution changes character. For $q > 2$ the solution decays exponentially from the loss cone boundary down to $\alpha = 0$. Because there are only few particles inside the loss cone and few particles per unit time enter it, the particle precipitation is a slow drizzle of particles into the ionosphere. This is the usual weak diffusion precipitation case. Qualitatively this behavior is the same for all q.

The other case of interest is when the diffusion into the loss cone is so strong that the particles enter the loss cone in a time comparable to or shorter than the quarter bounce period. Then $D_0\tau_\ell \gg 1$ and the Bessel functions can then be expanded in the small-amplitude limit, yielding

$$h_{sd}(\alpha) \approx 2\tau_\ell/\alpha_\ell^2 \gg 1 \tag{8.130}$$

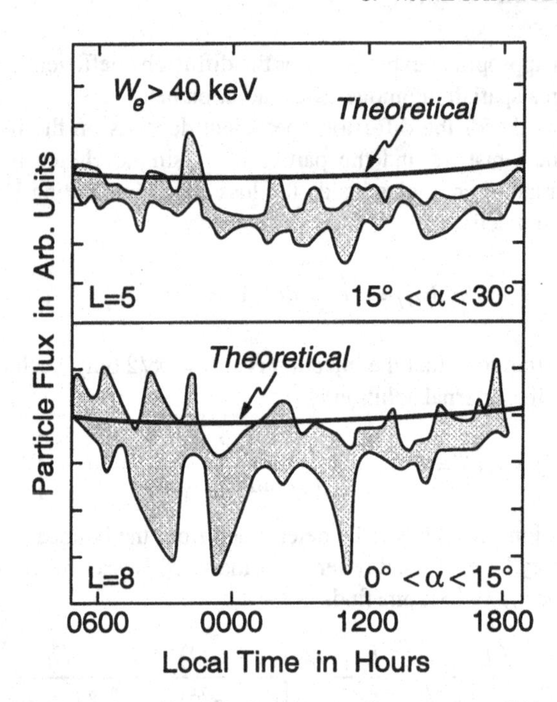

Fig. 8.6. Measured weakly diffusive electron fluxes in the radiation belts.

for all q. This expression does not depend on α inside the loss cone. Hence, the loss cone distribution is flat at $\alpha \leq \alpha_\ell$ for *strong pitch angle diffusion*, and the loss cone is filled with particles. Of course, the strength of the pitch angle diffusion depends on the intensity of the wave spectrum and thus on the number of particles in resonance with the waves, which implies that it depends on the injected flux of particles with energies above resonance energy. Predominantly, this should occur during injection of energetic particles due to enhanced convection from the tail during storms.

The lifetime of the injected trapped particles can be estimated formally from

$$\tau_L = \int_{\alpha_\ell}^{\pi/2} \sin\alpha d\alpha \int_{\alpha_\ell}^{\alpha} \frac{d\alpha'}{D(\alpha')\sin\alpha'} + (1 - \sin\alpha_\ell)h(\alpha_\ell) \tag{8.131}$$

In weak diffusion the last term can be neglected insofar as it is small and the lifetime is long. The life time in weak diffusion is some kind of weighted average of the inverse diffusion coefficient, $\tau_{L,wd} \approx 1/\langle D_0 \rangle$. Weak diffusion removes the particles in stationary state at the same rate they are injected.

The weakly diffusive equilibrium described by the weak-diffusion pitch angle distribution can be used to determine the maximum fluxes of energetic electrons trapped

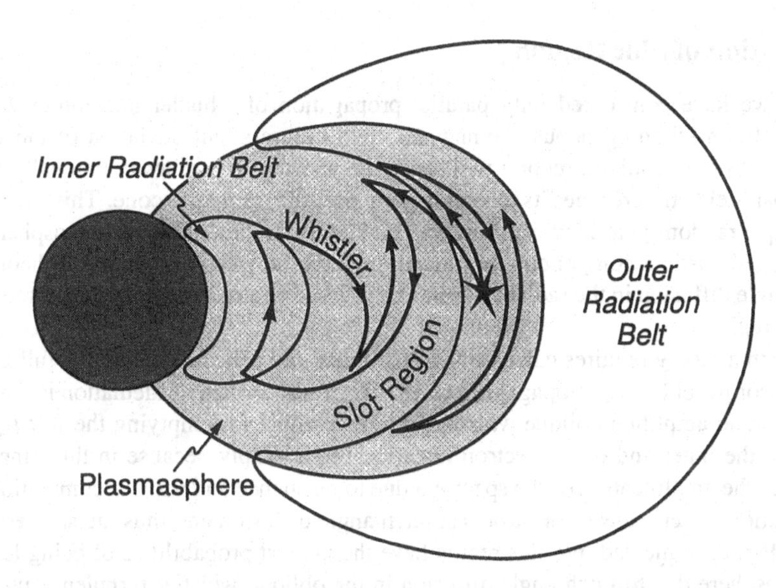

Fig. 8.7. Formation of slot region between inner and outer radiation belt.

in the Earth's radiation belt under the action of whistler turbulence wave-particle interaction. These omnidirectional fluxes as measured by early spacecraft are very close to the theoretical limits (see Fig. 8.6) and provide a nice confirmation of the relevance of quasilinear whistler turbulence for radiation belt dynamics.

When the diffusion coefficient increases towards strong diffusion, the second term in the above expression for the lifetime increases rapidly until the integral can be neglected. In this case the minimum lifetime becomes independent of D_0 and is given by

$$\tau_{L,sd}(\alpha, v) \approx h_{sd}(\alpha) = 2\tau_\ell(v)/\alpha_\ell^2 \tag{8.132}$$

Thus in strong diffusion the lifetime depends only on the quarter bounce time, τ_ℓ, and the width of the equatorial loss cone. In addition, it is a function of the energy through the energy dependence of the bounce period. Closer to the Earth the lifetime is shorter due to the shorter bounce period and the wider loss cone.

The present pitch angle diffusion theory is quite general and can be applied to electromagnetic ion-cyclotron waves in resonant interaction with energetic protons as well. It explains precipitation from the ion ring current into mid-latitudes and some part of auroral ion precipitation.

Formation of Slot Region

So far we have considered only parallel propagation of whistler and ion-cyclotron waves. In an inhomogeneous geomagnetic field even a slight deviation of the wave vector from the parallel direction will cause the whistler to turn away gradually from being parallel until it reaches its reflection point near the resonance cone. This spectrum of oblique random-phased whistler waves is naturally excited in the magnetosphere by the trapped particle component, and must be taken into account for a full theory of pitch angle diffusion in the radiation belts and of ion-cyclotron waves in the proton ring current region.

Such a theory requires calculating the different ray paths and using the full diffusion tensor for obliquely propagating waves. The result of such a calculation is that the reflected and amplified oblique whistlers are responsible for emptying the *slot region* between the inner and outer electron radiation belts, simply because in this range of L-values the amplification of the spectrum due to resonance and wave accumulation by propagation effects causes the strongest pitch angle diffusion and thus the shortest life times. Particles injected into this region have the highest probabilities of being lost to the ionosphere due to pitch angle diffusion in the oblique whistler turbulence present here.

This is schematically shown in Fig. 8.7. In this figure a whistler accidentally excited near the inner edge of the outer radiation belt region performs a complicated path in the magnetosphere. Thereby it crosses the slot region many times and interacts with the electron population therein. As result of this interaction the slot region becomes emptied of energetic radiation belt electrons and appears as a region where the trapped fluxes are very faint and the life times are short.

Electron-Cyclotron Turbulence

Whistler turbulence affects the energetic component of the magnetospheric particle distribution trapped in the radiation belts. On auroral field lines the energies of the particles are generally below the required resonance energies for excitation of whistler turbulence. On the other hand, strong electrostatic wave emission in the electron-cyclotron harmonic bands with amplitudes of up to several tens of mV/m are frequently observed during times when auroral electron fluxes dominate. We have shown in Sec. 4.5 that such waves can be excited by loss cone instabilities. The required resonant energies are lower than those for whistler mode noise, since it is easier to excite potential fluctuations than electromagnetic waves by oscillations in the particle densities. These electron-cyclotron waves extract the free energy from the loss cone distribution and tend to deplete the loss cone, leading to enhanced electron precipitation.

This mechanism can be described by quasilinear theory. Take the lowest electron-cyclotron harmonic, which is found near $\omega \approx 1.5\omega_{ge}$ in the oblique Bernstein mode.

The phase velocity of this mode is $\omega/k_\| \approx 3v_{\mathrm{th}\|e}$. The particles diffuse in phase space resonantly along orbits which are circles around the phase velocity of the wave

$$(v_\| - \omega/k_\|)^2 + v_\perp^2 = \mathrm{const} \tag{8.133}$$

and which for velocities larger than the phase velocity are surfaces of constant energy, such that energy diffusion is comparable to pitch angle diffusion. This requires to use the full quasilinear diffusion equation to describe the nonlinear interaction of electrostatic electron-cyclotron waves with the electron component. The diffusion equation written in terms of pitch angle and velocity takes the form

$$\begin{aligned}\frac{\partial f}{\partial t} &= \frac{1}{v\sin\alpha}\frac{\partial}{\partial\alpha}\left[\sin\alpha\left(D_{\alpha\alpha}\frac{1}{v}\frac{\partial f}{\partial\alpha} + D_{\alpha v}\frac{\partial f}{\partial v}\right)\right] \\ &\quad + \frac{1}{v^2}\frac{\partial}{\partial v}\left[v^2\left(D_{\alpha v}\frac{1}{v}\frac{\partial f}{\partial\alpha} + D_{vv}\frac{\partial f}{\partial v}\right)\right]\end{aligned} \tag{8.134}$$

where the components of the diffusion coefficient are defined as

$$\begin{aligned}D_{\alpha\alpha} &= \sum_l \int k_\perp dk_\perp \Psi_l(k_{\|\mathrm{res}}) \left[\frac{\pm l\omega_{ge}/\omega(k_{\|\mathrm{res}}) - \sin^2\alpha}{\sin\alpha\cos\alpha}\right]^2 \\ D_{\alpha v} &= \sum_l \int k_\perp dk_\perp \Psi_l(k_{\|\mathrm{res}}) \left[\frac{\pm l\omega_{ge}/\omega(k_{\|\mathrm{res}}) - \sin^2\alpha}{\sin\alpha\cos\alpha}\right] \\ D_{vv} &= \sum_l \int k_\perp dk_\perp \Psi_l(k_{\|\mathrm{res}})\end{aligned} \tag{8.135}$$

Here the resonant wavenumber is given by $k_{\|\mathrm{res}} v_\| = \omega \mp l\omega_{ge}$, and the wave spectral function, $\Psi_l(\mathbf{k})$, is defined as

$$\Psi_l(\mathbf{k}) = \frac{e^2}{4\pi m_e^2}\frac{\omega^2}{k^2v^2}\frac{J_l^2(k_\perp v_\perp/\omega_{ge})}{|v_\| - \partial\omega/\partial k_\||}|\delta\mathbf{E}(\mathbf{k})|^2 \tag{8.136}$$

where $J_l(z)$ is the Bessel function of order l. The diffusion coefficients given here have dimensions of v^2/t. Dividing them by v^2 gives diffusion coefficients which have the dimension of an inverse time. Thus $D_{\alpha\alpha}$ is the pitch angle diffusion coefficient, and D_{vv} is the velocity or energy diffusion coefficient, while the other coefficient is mixed in pitch angle and velocity.

It is convenient to assume a Gaussian spectrum of wavenumber spread $\Delta k_\|^2 = 0.25/r_{ge\|}$ as

$$|\delta E(k)|^2 = Ak_{\perp 0}^{-1}\delta(k_\perp - k_{\perp 0})\exp[-(|k_\|| - k_{\|0}/\Delta k_\|^2] \tag{8.137}$$

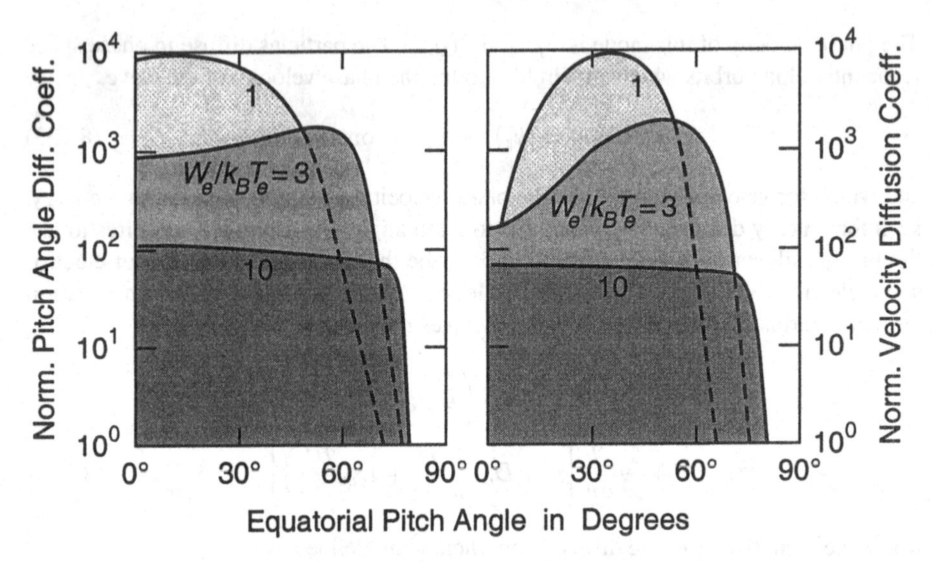

Fig. 8.8. Quasilinear normalized diffusion coefficients of electron-cyclotron waves.

where $\omega = 1.5\omega_{ge}$, $k_{\|0} = 0.5/r_{ge\|}$, and $k_{\perp 0} = 2/r_{ge\perp}$ are suggested by observations in the magnetosphere and yield an amplitude given by $A = 2\pi^{3/2}W_E/[\Delta k_{\|}\mathrm{erf}(1.6)]$ according to

$$\tfrac{1}{2}\epsilon_0|\delta\mathbf{E}(\mathbf{k})|^2 = A[\delta(k_\perp - k_{\perp 0})/k_{\perp 0}]\exp[-(|k_\|| - |k_{\|0}|)^2/(\Delta k_\|)^2] \qquad (8.138)$$

The zero-indexed wavenumbers are assumed to be given. For magnetospheric applications it is appropriate to integrate the diffusion coefficients along the bounce path of the particles. These average values are given by integrals of the type

$$\langle D_{\alpha\alpha}\rangle = \frac{1}{\tau_b(\alpha_{eq})}\int_0^{\lambda_m} g_{\alpha\alpha}(\lambda)D_{\alpha\alpha}(\lambda)\cos^7\lambda\, d\lambda \qquad (8.139)$$

Here $\lambda_m(v)$ is the mirror latitude of the particles of a certain energy, λ the geomagnetic latitude, α_{eq} the equatorial plane loss cone angle, τ_b the quarter bounce period, and the angular weights for each of the diffusion coefficients are

$$\begin{aligned} g_{\alpha\alpha} &= \cos\alpha/\cos^2\alpha_{eq} \\ g_{\alpha v} &= \sin\alpha/\sin\alpha_{eq}\cos\alpha_{eq} \\ g_{vv} &= \sin^2\alpha/\cos\alpha\sin^2\alpha_{eq} \end{aligned} \qquad (8.140)$$

For given particle energy and wave amplitude these equations allow to calculate the dependence of the diffusion coefficients on the pitch angle and for different resonances,

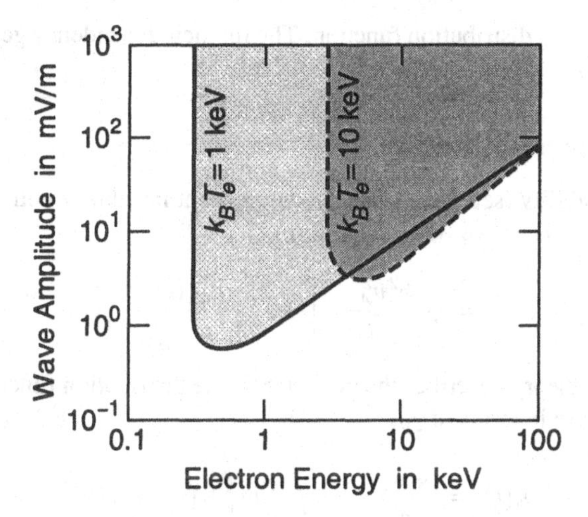

Fig. 8.9. Minimum wave amplitudes for strong diffusion in electron-cyclotron waves.

l, at different equatorial distances, L. Diffusion coefficients can be normalized to the particle thermal energies, W_e, and wave energies, W_E, as $\langle D\rangle k_BT_e/W_E$. Figure 8.8 shows these normalized coefficients at $L = 7$ for given energies. Largest diffusion is obtained for particles close the thermal energy. Moreover, for these energies the particles experience an energy diffusion which is of the same order as pitch angle diffusion at energies up to a few times the thermal particle energy.

As in whistler quasilinear theory, one can also consider the strong diffusion limit. This limit occurs for $\langle D_{\alpha\alpha}(\alpha_{eq})\rangle \geq \alpha_{eq}^2/\tau_b$ at the edge of the loss cone. The minimum electrostatic cyclotron wave amplitude to cause strong pitch angle diffusion can be calculated numerically for given diffusion coefficients by satisfying this condition. Figure 8.9 shows the result of such a calculation for $L = 7$. Thermal energies of $k_BT_e = 1\,\mathrm{keV}$ require electron-cyclotron wave fields of only about 1 mV/m to be in the strong pitch angle diffusion regime. Since such particles also undergo energy diffusion, they are accelerated at the same time. Hence, the particle component precipitated by electron-cyclotron waves will have an energetic tail like the measured auroral electron distributions. Hence, electron-cyclotron waves may contribute to auroral electron precipitation and to their energization and acceleration.

8.6. Weak Macro-Turbulence

The quasilinear theory developed in the previous sections can also be applied to macroinstabilities. This requires a different kind of technique, since the macro-variables depend only implicitly on the distribution function. The implicit dependence generates a quasi-linear time variation of the macroscopic variables.

Firehose Mode Stabilization

The firehose instability (see Sec. 3.4) is a non-resonant instability with the growth rate given in Eq. (3.78). Here we write in the form

$$\gamma_{\text{fh}}^2 = \frac{k^2 v_A^2}{1 + v_A^2/c^2} \left[\frac{p_\parallel(t) - p_\perp(t)}{B_0^2/\mu_0} - 1 \right] \tag{8.141}$$

Since quasilinear theory describes the evolution of the distribution function, we need to express the pressure in terms of f_{0s}

$$\begin{aligned} p_\parallel(t) &= \sum_s n_{0s} m_s \int d^3v v_\parallel^2 f_s(v_\perp, v_\parallel, t) \\ p_\perp(t) &= \tfrac{1}{2} \sum_s n_{0s} m_s \int d^3v v_\perp^2 f_s(v_\perp, v_\parallel, t) \end{aligned} \tag{8.142}$$

Instability results whenever

$$p_\parallel(t) > p_\perp(t) + B_0^2/\mu_0 \tag{8.143}$$

The reaction of the plasma to the growing wave and the distortion of the magnetic field will be such that the parallel pressure decreases and the perpendicular pressure increases until the growth rate asymptotically approaches zero for large times. To calculate this effect one must take into account the finite gyroradius corrections given in the growth rate Eq. (3.81). Firehose modes propagate parallel to the magnetic field. Hence, the evolution of the distribution functions is given by Eq. (8.94). Because the frequency of the mode is small and the wave is non-resonant, we expand the denominator for small frequencies. This expansion gives

$$\frac{\partial f_{0s}}{\partial t} = \frac{\omega_{ps}^2}{n_{0s} m_s c^2 \omega_{gs}^2} \left(v_\perp^2 \frac{\partial^2}{\partial v_\parallel^2} + \frac{v_\parallel^2}{v_\perp} \frac{\partial}{\partial v_\perp} v_\perp \frac{\partial}{\partial v_\perp} - 2v_\perp \frac{\partial}{\partial v_\parallel} v_\parallel \frac{\partial}{\partial v_\perp} \right) f_{0s} \int dk\, \gamma W_B \tag{8.144}$$

The spectral magnetic energy density introduced in this expression is related to the electric field density of the electromagnetic wave by

$$W_B(k,t) = c^2 k^2 W_E(k,t)/\omega^2 \tag{8.145}$$

and $\gamma(k,t) > 0$ is assumed, because only growing waves will cause quasilinear effects on the distribution function. To proceed, one must calculate the velocity moments of the perturbed equation. This yields the quasilinear evolution equations for the hydrodynamic pressure

$$\begin{aligned}\frac{dp_\perp(t)}{dt} &= p_\parallel(t)\int dk\frac{\gamma(k,t)W_B(k,t)}{B_0^2/2\mu_0}\\ \frac{dp_\parallel(t)}{dt} &= -2[2p_\parallel(t)-p_\perp(t)]\int dk\frac{\gamma(k,t)W_B(k,t)}{B_0^2/2\mu_0}\\ \frac{\partial W_B(k,t)}{\partial t} &= 2\gamma(k,t)W_B(k,t)\end{aligned} \tag{8.146}$$

Initially, as long as the instability works, one has $2p_\parallel > p_\perp$, and consequently

$$\begin{aligned} dp_\parallel/dt &\leq 0\\ dp_\perp/dt &\geq 0\end{aligned} \tag{8.147}$$

indicating that the parallel pressure decreases from the very beginning while the perpendicular pressure increases. The instability cools the plasma in the parallel and heats it in the perpendicular directions as one expects from simple logic. In the long-time limit the parallel pressure decreases as long until the growth rate vanishes, and one obtains in the final state some kind of equipartition law for the firehose mode

$$p_\parallel(\infty) = p_\perp(\infty) + B_0^2/\mu_0 \tag{8.148}$$

Because, in addition, the cut-off wavenumber is proportional to the growth rate (see Sec. 3.4), $k_0(\infty) \to 0$ together with the growth rate. The wavelength of the firehose mode thus increases with time in nonlinear interaction with the plasma.

From Eq. (8.146) we find that the evolution of the two pressure terms follows

$$\frac{dp_\perp}{dp_\parallel} = -\frac{p_\parallel/2}{2p_\parallel - p_\perp} \tag{8.149}$$

If the changes in the pressure close to the stable state at $t \to \infty$ are small, this can be expanded to obtain

$$\begin{aligned} p_\parallel(\infty) - p_{\parallel 0} &\approx -Z(0)2p_{\parallel 0}(2p_{\parallel 0} - p_{\perp 0})\\ p_\perp(\infty) - p_{\perp 0} &\approx Z(0)p_{\parallel 0}^2\end{aligned} \tag{8.150}$$

where

$$Z(0) = (p_{\parallel 0} - p_{\perp 0} - B_0^2/\mu_0)/p_{\parallel 0}(5p_{\parallel 0} - 2p_{\perp 0}) \tag{8.151}$$

The final spectral energy density in the magnetic field fluctuation follows from the energy conservation law

$$\tfrac{1}{2}p_{\parallel 0} + p_{\perp 0} + \int dk W_B(k,t) = \text{const} \tag{8.152}$$

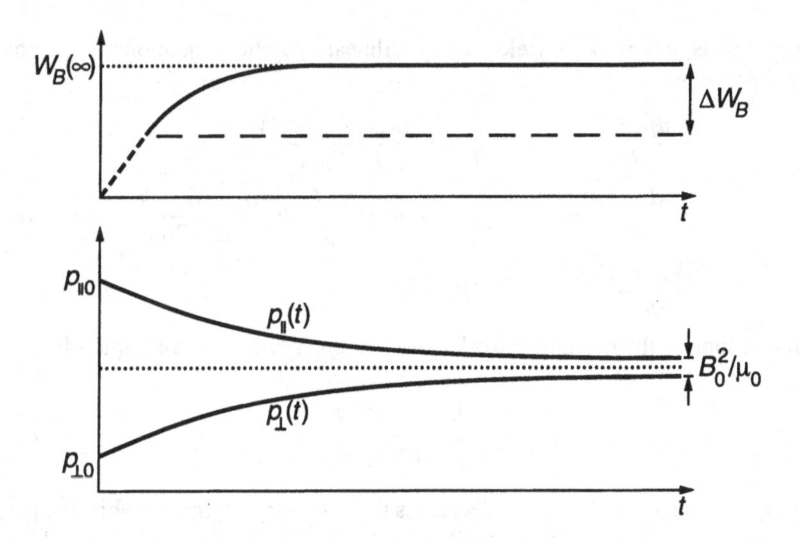

Fig. 8.10. Evolution of wave field and pressure in the firehose mode.

Using the above expressions for the changes in the pressure the gain in magnetic field energy can be estimated. It is given by the expression

$$\Delta W_B = \int dk[W_B(k,\infty) - W_B(k,0)] \approx p_{\parallel 0} Z(0)[p_{\parallel 0} - p_{\perp 0}] \tag{8.153}$$

Schematically, the evolution of the two pressure terms and the wave intensity is shown in Fig. 8.10. The decrease in the parallel pressure and the increase in the transverse pressure are not until equilibration. The difference left is just twice the original magnetic field pressure, while the wave intensity saturates at the level given by the above equation. The energy needed to keep this level is taken from the excess in the parallel pressure, but only part of this excess is devoted to raise the wave field intensity. The remaining part is attributed to heating of the plasma in the perpendicular direction. Clearly, this latter part is the irreversible part increases the entropy of the system.

Mirror Mode Stabilization

The other magnetohydrodynamic instability of interest is the mirror mode instability driven by transverse pressure excess (see Sec. 3.5). The quasilinear evolution equation for this perpendicular mode requires solution of the quasilinear equation including the perpendicular components of the diffusion tensor. When we denote its components by $D_{\perp\perp}$, $D_{\parallel\perp}$, $D_{\perp\parallel}$, and $D_{\parallel\parallel}$, and form the temperature moments of the distribution function, we find by integrating the full quasilinear equation the following evolution equations for the pressure

$$\begin{aligned} \frac{dp_{s\perp}}{dt} &= m_s \int d^3v f_s \left[\frac{1}{v_\perp}\frac{\partial}{\partial v_\perp}(v_\perp^2 D_{\perp\perp}) + v_\perp \frac{\partial D_{\perp\parallel}}{\partial v_\parallel} \right] \\ \frac{dp_{s\parallel}}{dt} &= 2m_s \int d^3v f_s \left[\frac{v_\parallel}{v_\perp}\frac{\partial}{\partial v_\perp}(v_\perp D_{\parallel\perp}) + \frac{\partial}{\partial v_\parallel}(v_\parallel D_{\parallel\parallel}) \right] \end{aligned} \tag{8.154}$$

where the diffusion coefficients can be written (indices a, b can be either of $\parallel$ or $\perp$)

$$D_{ab} = \frac{i\omega_{pi}^2}{n_0 m_i} \int d^3k \frac{J_1(\eta_i) W_E(\mathbf{k}, t)}{\omega(\mathbf{k}) - k_\parallel v_\parallel + i\gamma(\mathbf{k}, t)} A_a^* A_b \tag{8.155}$$

and $\eta_i = k_\perp v_\perp / \omega_{gi}$. The remaining abbreviations are

$$\begin{aligned} A_\perp &= (v_\parallel / v_\perp)(\omega_{gi}/\omega) \\ A_\parallel &= 1 - (\omega_{gi}/\omega) \end{aligned} \tag{8.156}$$

The electric and magnetic field intensities are related through

$$W_E(\mathbf{k}, t) = c^2 k^2 W_B(\mathbf{k}, t) / \gamma^2(\mathbf{k}, t) \tag{8.157}$$

with γ the mirror mode growth rate. Of course, the magnetic wave intensity evolves according to the usual wave evolution equation

$$\partial W_B(t)/\partial t = 2\gamma(t) W_B(t) \tag{8.158}$$

Contrary to the firehose mode, it is not possible to solve these combined equations analytically. The difficulty lies in the higher dimensionality of the problem.

Numerical solutions found that one needs a suprathermal ion component to force the mirror mode into instability, since the mirror mode is strongly damped by parallel Landau damping due to parallel ions, which eat up the energy pumped into the wave by the excess in perpendicular pressure. This Landau damping increases the parallel pressure, until the mirror mode saturates and its growth rate vanishes. The wavelengths of the mode which are damped away first are the short wavelengths, both in the parallel and in the perpendicular direction. Long parallel wavelengths grow longest, while perpendicular wavelengths have a wider growing spectrum, reaching from long to medium lengths, longer than the ion inertial length, $\lambda_\perp > c/\omega_{pi}$. In saturation the peak in the wave spectrum is found at $k_{\perp\mathrm{sat}} \approx 0.25\omega_{pi}/c$ and $k_{\parallel\mathrm{sat}} \approx 0.5\omega_{pi}/c$.

The evolution of suprathermal pressure and magnetic wave intensity is shown in Fig. 8.11. The mirror mode saturates rather fast within a few ion-cyclotron periods. The background thermal parallel pressure remains about unaffected. The suprathermal pressure responsible for the instability varies in the expected way, with the perpendicular pressure decreasing and the parallel pressure increasing due to Landau damping.

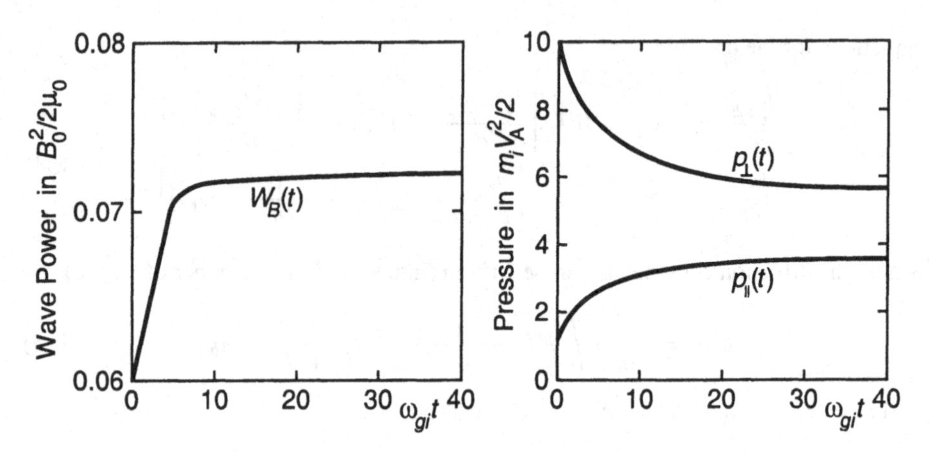

Fig. 8.11. Evolution of wave field and suprathermal pressure in the mirror instability.

These results are relevant for the mirror mode observations in the Earth's magnetosheath shown in Fig. 3.11, which showed that for $\beta_\perp \geq 1$ the mirror instability is marginally stable in the magnetosheath. To a certain extent this effect can be attributed to the quasilinear stabilization of the mirror instability described above. Concluding from the increasing discrepancy between $\beta_\perp$ and $\beta_\parallel$ when approaching the magnetopause, one may argue that the amplitude of the mirror mode waves should increase towards the magnetopause, but so far the observations give no clear indication of such an effect. An important reason for the failure of the interpretation in terms of quasilinear mirror mode stabilization is the fact that in the magnetosheath the electromagnetic ion-cyclotron instability competes with the mirror mode. But the mirror instability survives when a number of heavier ions are mixed into the plasma. The 3–5% Helium ions detected in the magnetosheath are sufficient to account for this effect.

Concluding Remarks

BGK modes are just one example of exact nonlinear waves arising in the interplay of wave evolution and particle trapping in single modes. These waves are of large amplitude and separate the initial particle distribution function into two parts, trapped and untrapped distributions which achieve a subtle equilibrium between the number of trapped particles and the wave amplitude. However, owing to the complexity of the kinetic description, the derivation of BGK modes has been successful in the one-dimensional homogeneous electrostatic case only. Below, when treating strong plasma turbulence, we will use a different approach to obtain exact stationary and non-stationary wave solutions in plasmas. Such an approach takes advantage of the simpler hydrodynamic description of the plasma.

Trapping of particles in waves may change the wave dynamics in many ways. The most obvious one is that trapping extracts part of the resonant particles from the distribution function. These particles loose their capability of further feeding instability. Hence trapping will cause stabilization of the instability. But this applies to instabilities of single modes. When the bandwidth of the mode becomes large, trapping becomes less important, and other effects set on which we are going to discuss in the next chapters. The main remaining effect of trapping is then the distortion of the particle orbit by the broadband spectrum. This causes resonance broadening in wave-particle interaction.

In this chapter we presented only the wave-particle aspect the weak turbulence theory, the zero-order response of the plasma to the presence of a random-phased broadband spectrum of unstable waves. In addition to quasilinear stabilization of the wave spectrum as result of this response, the deformation of the particle distribution function implies a number of effects like particle heating, resonant acceleration, energy diffusion, and pitch-angle diffusion, of which only the latter two have been discussed so far. Clearly, quasilinear weak wave-particle turbulence has a much wider application than mere pitch-angle diffusion. Examples not treated here are the diffusion of cosmic rays across astrophysical plasmas and acceleration of particles in their self-consistently excited low-frequency wave spectra in front of shock waves. In particular, in particle acceleration theory the quasilinear approach has proved to be very useful.

But quasilinear theory has also its drawbacks. For instance, in the context of gentle beam stabilization it causes severe problems when applying it to electron streams ejected from the solar corona which are the cause of solar type III radio bursts. Quasilinear theory would not permit these streams to propagate large distances in the corona, contrary to what is suggested by observation. Thus, quasilinear theory and weak wave-particle turbulence cannot be the last word spoken in nonlinear plasma theory. In the following chapters we explore other nonlinear effects which are ignored in quasilinear theory.

Further Reading

There is a small number of excellent books on nonlinear plasma theory, e.g., [1], [3], [7], and [9], which all were written in the sixties and early seventies. Meanwhile, nonlinear plasma physics has grown into a wide field, but no comprehensive modern text is available. The reason is that analytical theory has not developed much during the last twenty years, while the activities and interests have turned to numerical simulations. A contemporary summary of numerical plasma simulations is given in [8].

Particle trapping is described, at least in rudimentary form, in most books on nonlinear plasma physics as for instance [1], [7] and [2]. Single wave stabilization for

Bernstein modes and whistler waves by particle trapping may be found in the original literature. The former have been applied in space plasma physics context to radio emission from the Sun during solar type IV bursts. For references consult [4] and [6]. BGK modes are treated extensively by Davidson [1]. Quasilinear theory is the main subject of most books on nonlinear theory, e.g., [1] and [7]. The ion-cyclotron resonance broadening saturation theory has been taken from Dum and Dupree, *Phys. Fluids* **13** (1970) 2064. Pitch-angle diffusion, as one of the results of quasilinear theory, has wide application in space, and even in astrophysics, where it has been applied successfully to the diffusion theory of cosmic rays. The readers interested more deeply in space applications of pitch angle scattering are referred to [5].

[1] R. C. Davidson, *Methods in Nonlinear Plasma Theory* (Academic Press, New York, 1972).

[2] A. Hasegawa, *Plasma Instabilities and Nonlinear Effects* (Springer Verlag, Heidelberg, 1975).

[3] B. B. Kadomtsev, *Plasma Turbulence* (Academic Press, New York, 1965).

[4] D. J. McLean and N. R. Labrum (eds.), *Solar Radiophysics* (Cambridge University Press, Cambridge, 1985).

[5] L. R. Lyons and D. J. Williams, *Quantitative Aspects of Magnetospheric Physics* (D. Reidel Publ. Co., Dordrecht, 1984).

[6] D. B. Melrose, *Plasma Astrophysics* (Gordon and Breach, New York, 1980).

[7] R. Z. Sagdeev and A. A. Galeev, *Nonlinear Plasma Theory* (W. A. Benjamin, New York, 1969).

[8] T. Tajima, *Computational Plasma Physics* (Addison-Wesley Publ. Co., Redwood City, 1989).

[9] V. N. Tsytovich, *Nonlinear Effects in Plasmas* (Plenum Press, New York, 1970).

9. Weak Wave Turbulence

Quasilinear theory of wave-particle interaction is only one side of weak turbulence theory. It is that part dealing with the simplest random-phase reaction of the unstably excited wave field on the initial non-equilibrium distribution function. If only this kind of interaction would occur in a plasma, the plasma would approach thermal equilibrium in a straight way. In real plasmas this is by no means the case. In most cases, and long before quasilinear interaction manages to stabilize the plasma, other modes of interaction between the waves themselves come into play. Waves of certain frequencies and wavenumbers start colliding and exchange momentum and energy. These mutual interactions among the waves are described by the theory of *wave-wave interaction*, the other part of weak turbulence theory.

Wave-wave interaction theory or weak wave turbulence theory, as it is also called, ignores the particle response to the presence and mutual interaction of the waves. The plasma plays the role of an exciter and wave carrier but is otherwise a passive background, on top of which the various wave processes take place. In this chapter we treat the waves and their interaction in a manner as if the plasma would be absent, but the properties of the waves are still determined by the plasma dielectric properties. We do not account for wave dissipation in the plasma but for simplicity assume that any dissipation appears only among the waves themselves.

Dissipation is, however, a very important item insofar as it imposes a natural threshold on the interaction of the waves and on the nonlinear instabilities which set on when the waves may transform into each other. In most of space plasma physics these processes are still out of reach. The presentation given here ignores them because of this reason. But under certain conditions, such as heating of plasma in the ionosphere and the lower corona, and in dissipative wave transformation, neglecting wave dissipation is not justified.

The present chapter presents essentially three different approaches to wave turbulence in a plasma: the coherent approach, the incoherent or random-phase approach, and one particular form of incoherent wave interaction theory for the very-low frequency range of drift waves, which may be important for the evolution of low-frequency turbulence in space plasmas.

9.1. Coherent Wave Turbulence

Wave-wave interaction itself can be divided into two separate fields, *coherent interactions* and *incoherent interactions*. The theory of coherent interactions considers the interaction of single waves, while the theory of incoherent interactions considers the interaction among broad wave spectra and uses the random-phase approximation. The latter corresponds to a kind of quasilinear theory of wave interaction, where particles are considered as a background only providing the existence of plasma waves and allowing for their propagation and dispersion. Expressed in terms of time-scales, one can say that wave-wave interaction dominates over wave-particle interaction whenever the collision time between the waves is considerably shorter than the wave-particle quasilinear time-scale. Such conditions are frequently realized when the particle distribution function does not provide free energy and when the amplitude of at least one of the interacting waves is large. Formally, both wave-wave interaction processes can be described as collisions between pseudo-particles, with energy and momentum conservation in the pseudo-particle collision. The only additional condition is that in this process the dispersive properties of the participant waves come into play through their dispersion relations.

Coherency

To demonstrate what is meant by coherent wave-wave interaction it is sufficient to consider the simplest possible case of a collision between waves. This is the case when two waves experience a head-on collision and decay or merge into one third wave (Fig. 9.1), the case of coherent three-wave interaction. For comparison, wave scattering where two waves hit each other and escape as two other waves is already a four-wave process. Hence, in three-wave interaction the number of waves participating in the interaction is not conserved.

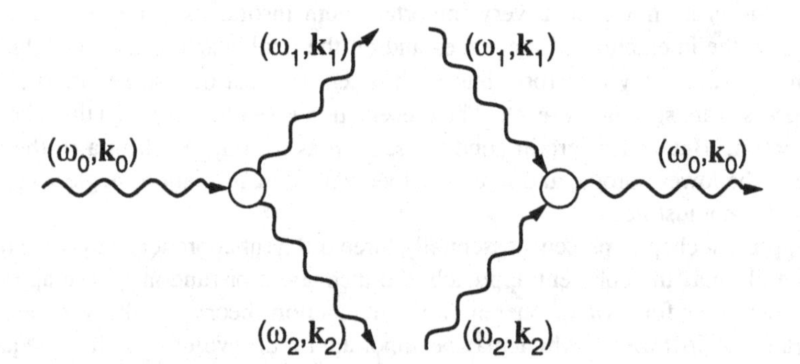

Fig. 9.1. Decay and merging in three-wave interaction.

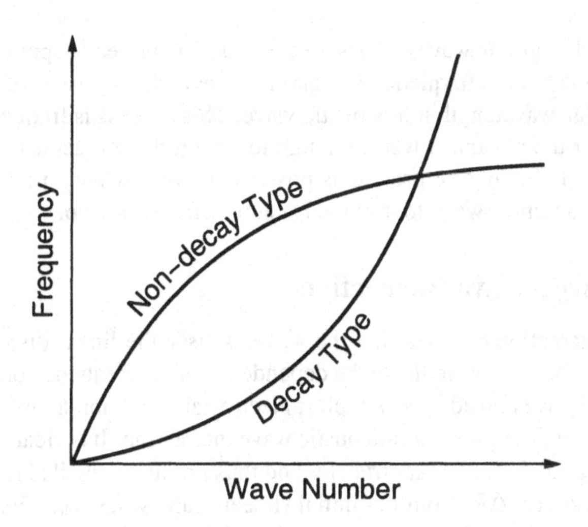

Fig. 9.2. Dispersion curves which may and may not yield wave-wave interaction.

Interaction of waves makes only sense if the waves are in resonance. This implies the following. Each wave is described by its frequency, ω, and its wavenumber, $\mathbf{k}$. The first corresponds to its energy, $\hbar\omega$, the second to its momentum, $\hbar\mathbf{k}$. Conservation of total energy and momentum during the collision of the three waves, considered as a collision between three quasi-particles, requires that

$$\begin{aligned} \omega - \omega' - \omega'' &= 0 \\ \mathbf{k} - \mathbf{k}' - \mathbf{k}'' &= 0 \end{aligned} \tag{9.1}$$

Here the three waves are distinguished by the primes. The two conditions (9.1) have the form of resonant denominators and are therefore called the resonance conditions or matching conditions, but physically spoken they are the two conservation laws of energy and momentum.

Of course, not all waves of arbitrary dispersion can participate in such an interaction, because satisfying the resonance conditions demands that $|\mathbf{k}' + \mathbf{k}''| = |\mathbf{k}| \leq |\mathbf{k}'| + |\mathbf{k}''|$. Hence, three waves belonging to the same dispersion branch can interact only if this branch is convex from below as shown in Fig. 9.2. Only in this case the triangle inequality can be satisfied. A high energy mode may decay spontaneously into two lower energy modes (left part of Fig. 9.1) of the same kind if the dispersion relation is convex from below, while it cannot decay if it is concave from below.

Decay is, of course, not restricted to the same mode. A given high-energy wave mode with convex dispersion can decay into another wave of its own mode and a low-frequency wave of a different mode. This would be the case for the convex dispersion relation of Langmuir waves. These waves can easily decay spontaneously into a longer-wavelength, lower-frequency Langmuir wave and, for instance, a much lower-frequency, similar-wavelength ion-acoustic wave. This process is frequently realized in nature, whenever the plasma is warm enough to permit the ion-acoustic wave to grow. On the other hand, the inverse process is possible as well, when an ion-acoustic wave collides with a Langmuir wave to mix the Langmuir frequency up.

Resonant Three-Wave Interaction

The resonant interaction requires that the waves satisfy the linear dispersion relation, $\epsilon(\mathbf{k}, \omega_{\mathbf{k}}) = 0$; for brevity we indicate the dependence of the frequency on the wavenumber by an index $\mathbf{k}$. We consider the simplest electrostatic and unmagnetized three-wave interaction case, i.e., Langmuir-ion-acoustic wave interaction. It is clear that the interaction causes energy and momentum transfer and thus produces small shifts in frequency, $\Delta\omega$, and wavenumber, $\Delta\mathbf{k}$, from the initial (linear) values, $\mathbf{k}_0, \omega_{\mathbf{k}_0}$. Expansion of the dispersion relation around the initial values gives for the variation of ϵ

$$\delta\epsilon(\mathbf{k}) = \mathrm{Re}\,\delta\epsilon(\mathbf{k}_0) + i\mathrm{Im}\,\delta\epsilon(\mathbf{k}_0) + \left.\frac{\partial\delta\epsilon}{\partial\omega}\right|_{\mathbf{k}_0} \Delta\omega + \left.\frac{\partial\delta\epsilon}{\partial\mathbf{k}}\right|_{\mathbf{k}_0} \cdot \Delta\mathbf{k} \tag{9.2}$$

For electrostatic waves it is most convenient to consider the variation of the electrostatic potential. The latter becomes a function of time, τ, related to the frequency shift, and space, $\boldsymbol{\xi}$, related to the wavenumber shift. With this in mind we can replace the shifts by the equivalent operators $\Delta\omega \to i\partial/\partial\tau$ and $\Delta\mathbf{k} \to -i\nabla_\xi$, where $\nabla_\xi = \partial/\partial\boldsymbol{\xi}$ is the spatial gradient operator acting on the displacement vector, $\boldsymbol{\xi}$. Moreover, using the group velocity

$$\mathbf{v}_{\mathrm{gr}0} = \partial\omega_{\mathbf{k}}/\partial\mathbf{k}|_{\mathbf{k}=\mathbf{k}_0} \tag{9.3}$$

and interpreting the variation of the dielectric function as an operator acting on the variation of the potential, one obtains the following equation for the potential amplitude

$$\delta\epsilon(\mathbf{k})\delta\phi(\mathbf{k}_0) = i\frac{\partial\epsilon}{\partial\omega_0}\left(\frac{\partial}{\partial\tau'} - \gamma_0\right)\delta\phi(\mathbf{k}_0) \tag{9.4}$$

where the following abbreviations have been used

$$\begin{aligned} \frac{\partial}{\partial\tau'} &= \frac{\partial}{\partial\tau} - \mathbf{v}_{\mathrm{gr}0}\cdot\nabla_\xi \\ \frac{\partial\epsilon}{\partial\omega_0} &= \left.\frac{\partial[\mathrm{Re}\,\delta\epsilon(\mathbf{k})]}{\partial\omega_{\mathbf{k}}}\right|_{\mathbf{k}=\mathbf{k}_0} \\ \gamma_0 &= -\frac{\mathrm{Im}\,\delta\epsilon(\mathbf{k})}{(\partial\epsilon/\partial\omega_0)} \end{aligned} \tag{9.5}$$

The quantity γ_0 is clearly the linear growth rate of the wave with frequency $\omega(\mathbf{k}_0)$. This interpretation is easy to understand. The dispersion relation of electrostatic waves results from the equation $\epsilon\delta\phi = 0$ for non-vanishing potential amplitude. Thus, when the linear dispersion relation is satisfied for the initial wave, the remaining relation is the product of the two distortions. This nonlinear term has been neglected in the development of the linear theory. The wave-wave interaction theory is thus, from the point of view of an expansion with respect to amplitudes, the first step of a true nonlinear theory. Since the distortion of the dielectric function depends on space and time, it becomes an operator acting on the distortion of the wave field. This is the content of the above nonlinear differential equation for $\delta\phi$.

Coupled Wave Equations

We have found a representation of the nonlinear operator with the unprimed wave as the modified wave. The wave system consists of three waves, however. Hence, the full dispersion relation depends on all tree waves, implying that we must vary it with respect to the three wave amplitudes. This yields up to the second-order variation

$$\delta\epsilon(\mathbf{k})\delta\phi(\mathbf{k}) + \sum_{\mathbf{k}=\mathbf{k}'+\mathbf{k}''} \delta\epsilon(\mathbf{k}',\mathbf{k}'')\delta\phi(\mathbf{k}')\delta\phi(\mathbf{k}'') = 0 \tag{9.6}$$

where we made use of the condition that in the coupling term only such couplings appear which satisfy the conservation laws. Moreover, taking only a sum over three waves in the second term restricts the consideration to three-wave interactions. Higher order interactions would provide addition sums. If we now substitute Eq. (9.5) into Eq. (9.6), we obtain the following coupled equations

$$\begin{aligned}
\left(\frac{\partial}{\partial\tau'} - \gamma_0\right)\frac{\partial\epsilon}{\partial\omega_0}\delta\phi(\mathbf{k}_0) &= i\delta\epsilon(\mathbf{k}_0,\mathbf{k}_1,\mathbf{k}_2)\delta\phi(\mathbf{k}_1)\delta\phi(\mathbf{k}_2) \\
\left(\frac{\partial}{\partial\tau'} - \gamma_1\right)\frac{\partial\epsilon}{\partial\omega_1}\delta\phi(\mathbf{k}_1) &= i\delta\epsilon(\mathbf{k}_1,-\mathbf{k}_2,\mathbf{k}_0)\delta\phi(-\mathbf{k}_2)\delta\phi(\mathbf{k}_0) \\
\left(\frac{\partial}{\partial\tau'} - \gamma_2\right)\frac{\partial\epsilon}{\partial\omega_2}\delta\phi(\mathbf{k}_2) &= i\delta\epsilon(\mathbf{k}_2,\mathbf{k}_0,-\mathbf{k}_1)\delta\phi(\mathbf{k}_0)\delta\phi(-\mathbf{k}_1)
\end{aligned} \tag{9.7}$$

This system results by taking each of the three waves as the initial one and letting it collide with the two others. Moreover, since the wave fields are real, $\delta\phi(-\mathbf{k}) = \delta\phi(\mathbf{k})^*$. Because each of the three waves can play the role of the initial wave, it is convenient to symmetrize these three evolution equations for the three wave amplitudes under the action of the two other waves. We define a normalized complex wave amplitude

$$A(\mathbf{k}_j;\tau,\xi) = \delta\phi(\mathbf{k}_j;\tau,\xi)\left|\frac{\partial(\omega\epsilon)}{\partial\omega(\mathbf{k}_j)}\frac{\epsilon_0 k_j^2}{2\omega(\mathbf{k}_j)}\right|^{1/2} \tag{9.8}$$

The derivative of ϵ in this expression is the expression which enters the fluctuation energy in one wave mode given in Eq. (2.19). The sign of the wave amplitude can be positive or negative and is determined as the sign of the derivative of the dielectric function

$$\mathrm{sgn}(\partial\epsilon/\partial\omega_j) = S(\mathbf{k}_j) \tag{9.9}$$

such that the spectral energy density is $W(\mathbf{k}_j)$, and the wave action can be written as

$$S(\mathbf{k}_j)|A(\mathbf{k}_j)|^2 = W(\mathbf{k}_j)/\omega(\mathbf{k}_j) = \hbar N(\mathbf{k}_j) \tag{9.10}$$

where $N(\mathbf{k}_j)$ is the number of wave quanta contained in the mode of wavenumber $\mathbf{k}_j$. Since the time derivative of the action is the energy, the new abbreviations can be used to write the above system of equations in its simplest symmetrized form

$$\begin{aligned} S(\mathbf{k}_0)\partial A(\mathbf{k}_0)/\partial\tau' &= -iVA(\mathbf{k}_1)A(\mathbf{k}_2) \\ S(\mathbf{k}_1)\partial A(\mathbf{k}_1)/\partial\tau' &= -iVA(\mathbf{k}_0)A(\mathbf{k}_2^*) \\ S(\mathbf{k}_2)\partial A(\mathbf{k}_2)/\partial\tau' &= -iVA(\mathbf{k}_1^*)A(\mathbf{k}_0) \end{aligned} \tag{9.11}$$

The common factor, $V = V(\mathbf{k}_0, \mathbf{k}_1, \mathbf{k}_2)$, on the right-hand sides of these equations is the three-wave coupling coefficient. The three wave modes have been numbered consecutively here. The coupling coefficient written satisfies the following symmetries

$$V(\mathbf{k}_1, -\mathbf{k}_2, \mathbf{k}_0) = V(\mathbf{k}_2, \mathbf{k}_0, -\mathbf{k}_1) = V(\mathbf{k}_0, \mathbf{k}_1, \mathbf{k}_2) \tag{9.12}$$

and is given by the rather complicated expression

$$V(\mathbf{k}_0, \mathbf{k}_1, \mathbf{k}_2) = \frac{\sqrt{2}k_0^2\epsilon(\mathbf{k}_0, \mathbf{k}_1, \mathbf{k}_2)}{\epsilon_0\left|[k_0^2\partial\epsilon/\partial\omega(\mathbf{k}_0)][k_1^2\partial\epsilon/\partial\omega(\mathbf{k}_1)][k_2^2\partial\epsilon/\partial\omega(\mathbf{k}_2)]\right|^{1/2}} \tag{9.13}$$

Wave Decay

These expressions close the system of equations for the amplitudes of the three waves. All three waves are coupled, but because we have neglected all other ingredients in the interaction apart from just these three waves, the total energy in this interaction and the total momentum is necessarily conserved. This conservation can be checked by expressing the complex amplitude by its modulus and a phase as

$$A = |A(\tau)|\exp[i\theta(\tau)] \tag{9.14}$$

Introducing this ansatz into the above system of equations, it can be reduced to the following form (for simplicity we use the subscripts 0, 1, 2, instead of the wavenumbers)

$$\begin{aligned} S_0\partial|A_0|/\partial\tau &= -V|A_1||A_2|\sin\theta \\ S_1\partial|A_1|/\partial\tau &= +V|A_0||A_2|\sin\theta \\ S_2\partial|A_2|/\partial\tau &= +V|A_0||A_1|\sin\theta \end{aligned} \tag{9.15}$$

and another equation for the time variation of the phase

$$\partial\theta/\partial\tau = V\cos(\theta_0 - \theta_1 - \theta_2)\partial[\ln(|A_0||A_1||A_2|)]/\partial\tau \tag{9.16}$$

These equations determine the constants of motion during the interaction of the three waves. The last equation immediately gives

$$|A_0||A_1||A_2|\cos\theta = \text{const} \tag{9.17}$$

The amplitude equations can, after multiplication with the appropriate amplitudes of the waves and addition as well as using the resonance conditions, be managed to be brought into the form of an energy density conservation law

$$W_0 + W_1 + W_2 = \text{const} \tag{9.18}$$

Also, integration of the amplitude equations (9.15) yields the conservation equations of the wave occupation numbers

$$\begin{aligned} N_0 + N_1 &= \text{const} \\ N_0 + N_2 &= \text{const} \\ N_1 - N_2 &= \text{const} \end{aligned} \tag{9.19}$$

These latter equations are equivalent to the creation and annihilation processes of occupations of states known from quantum field theory. Actually, taking one quantum away from the 0-mode, it appears in the 1- and 2-modes such that $\Delta N_0 = -1$ causes $\Delta N_1 = \Delta N_2 = 1$, a process which describes the decay of the 0-mode. Such decay processes may stabilize an instability of the 0-mode. To demonstrate how this happens, let us assume that $|A_0|$ is much larger initially than the amplitudes of the other two waves. In this case we can linearize the amplitude equations (9.15) in a decay process, in which the 0-mode can be considered to be a quasi-constant pump wave, the 1- and 2-modes both turn out to grow linearly in time according to

$$|A_{1,2}| \propto \exp[V|A_0|\tau] \tag{9.20}$$

Of course, this linear growth phase is valid only for the short initial time interval, as long as the 0-mode does not loose too much energy. The more exact solution to the above system of equation with initial condition $|A_0(0)|^2 = N_0(0) \gg N_1(0)$ and $|A_2(0)|^2 = N_2(0) = 0$ can be expressed in terms of elliptic functions. Clearly, the process involving only three waves is a purely reversible process. The two small amplitudes grow in time on the expense of the pump wave. When the energy of the pump wave is consumed, the daughter waves start pumping energy back into the mother wave, and so on, as long as no dissipation is involved. Figure 9.3 shows the strictly reversible evolution of the occupation numbers of the three waves.

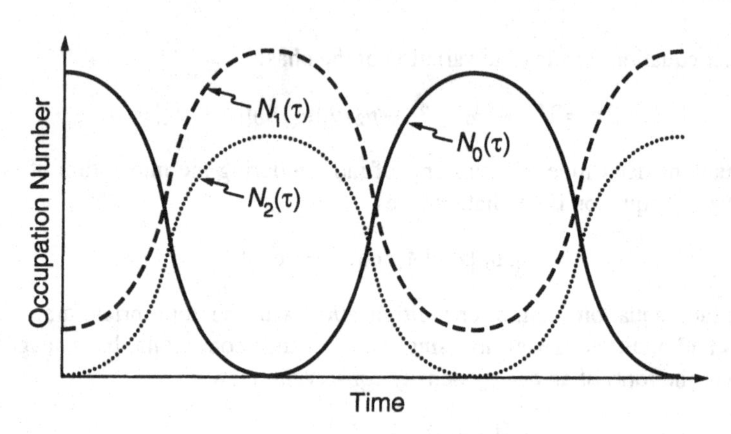

Fig. 9.3. Occupation number oscillation in coherent three-wave interaction.

Coherent processes depend heavily on the spectral width of the initial pump wave. If this width is larger than the inverse nonlinear oscillation time of the three-wave process, τ_{nl}, the pump wave cannot be considered as a single wave. The spectrum becomes broad, and the process looses its coherence. Interaction of this kind are the random-phase incoherent wave-wave processes considered in the next section. Hence, for a coherent three-wave interaction the bandwidth must be narrow

$$\Delta\omega \ll \tau_{nl}^{-1} \approx V|A_0| \tag{9.21}$$

If this condition is satisfied, the phase of the 0-mode changes slowly during the interaction process, and phase mixing can be neglected, just as required in a coherent process.

We should note, finally, that if a *negative energy mode* is involved in the three-wave interaction process, the character of the interaction changes completely, because extracting one quantum from the negative energy wave in order to add it to a positive energy wave, letting it grow, leads to growth of the negative energy wave as well. Such growth is faster than exponential and cannot be described by simple linear growth. The system explodes, and we encounter an irreversible *explosive instability*, which does not saturate in a three-wave process. This observation is not purely academic, because for instance, many of the drift modes in inhomogeneous plasmas turn out to be negative energy waves. When these waves are involved in three-wave interaction processes, they sometimes cause explosive instability, exciting other waves while growing themselves.

9.2. Incoherent Wave Turbulence

The previous section dealt with the coherent case, considering single waves. The situation becomes fundamentally different when many waves of different phases are involved in the interaction process. In such a case the conservation laws are not satisfied independently in each particular case of interaction, but in the average over all phases in the global interaction. This does not mean that they are not satisfied microscopically, it means only that there is no information available over the exact microscopic interaction to the outer world. The information is provided in such a way that the energy and momentum conservation laws are satisfied globally, while it is not important to know which waves exactly had interchanged their momenta and energies at a particular instant of time.

Random-Phase Wave-Wave Interaction

When Eq. (9.21) is strongly violated, the wave spectrum is broad enough to provide sufficient phase mixing, and the interaction between the waves can be described in a statistical version of weak turbulence theory, where the phases are considered to be randomly distributed in exactly the same manner as in weakly turbulent wave-particle interaction theory. In such a case the wave amplitude is represented as superposition of many waves, and the time averaged amplitude, $\langle A(\mathbf{k})\rangle = 0$, vanishes because the phase of the wave varies much faster than the amplitude factor

$$\langle A(\mathbf{k},\tau)\rangle = \lim_{T\to\infty}\frac{1}{T}\int_0^T A(\mathbf{k},\tau)\exp[i\theta(t)]dt = 0 \tag{9.22}$$

This integral is zero because the slowly variable amplitude factor with time scale τ can be extracted from the integral. The presence of many waves with different wave numbers in the random-phase approximation then implies that in the equations for the wave amplitudes (9.11) we sum over all possible combinations of wavenumbers satisfying the momentum conservation condition. But in addition we must permit for small frequency shifts, $\Delta\omega(\mathbf{k})$, which may arise in the interaction. These shifts correspond to energy uncertainties caused by the random phases. The interacting waves do not know precisely with which phase they have interacted. Of course, these mismatches in frequency are not free, but must satisfy the conservation laws as well.

The general theory for many-wave interaction is very complicated. We thus will discuss only the random-phase or incoherent weakly turbulent three-wave interaction process. Summing over all possible combinations of wavenumbers gives

$$S_0\frac{\partial A_0}{\partial t} \;=\; -i\sum_{\mathbf{k}_0=\Delta_0} V(\mathbf{k}_0,\mathbf{k}',\mathbf{k}'')A(\mathbf{k}')A(\mathbf{k}'')\exp(i\Delta\omega_0 t)$$

$$S_1 \frac{\partial A_1}{\partial t} = -i \sum_{\mathbf{k}_1=\Delta_1} V(\mathbf{k}_1, -\mathbf{k}'', \mathbf{k}) A(-\mathbf{k}'') A(\mathbf{k}) \exp(-i\Delta\omega_1 t) \tag{9.23}$$
$$S_2 \frac{\partial A_2}{\partial t} = -i \sum_{\mathbf{k}_2=\Delta_2} V(\mathbf{k}_2, \mathbf{k}, -\mathbf{k}') A(\mathbf{k}) A(-\mathbf{k}') \exp(-i\Delta\omega_2 t)$$

The definitions of the quantities are the same as in the foregoing section, and the mismatches and summation conditions are

$$\begin{aligned} \Delta_0 &= \mathbf{k}' + \mathbf{k}'' \\ \Delta_1 &= \mathbf{k} - \mathbf{k}'' \\ \Delta_2 &= \mathbf{k} - \mathbf{k}' \end{aligned} \tag{9.24}$$

and

$$\begin{aligned} \Delta\omega_0 &= \omega(\mathbf{k}_0) - \omega(\mathbf{k}') - \omega(\mathbf{k}'') \\ \Delta\omega_1 &= \omega(\mathbf{k}) - \omega(\mathbf{k}_1) - \omega(\mathbf{k}'') \\ \Delta\omega_2 &= \omega(\mathbf{k}) - \omega(\mathbf{k}_2) - \omega(\mathbf{k}') \end{aligned} \tag{9.25}$$

For a continuous broad spectrum one can replace the sums in these expressions by integrals over the primed wavenumbers and introduce a δ-function taking care of the resonances. Then the master equation for an amplitude is of the form

$$S_0 \frac{\partial A_0}{\partial t} = \int\int d^3k_1 d^3k_2 \delta(\mathbf{k}_0 - \mathbf{k}_1 - \mathbf{k}_2) V A_1 A_2 \exp[i\Delta\omega_0 t] \tag{9.26}$$

where $V = V(\mathbf{k}_0, \mathbf{k}_1, \mathbf{k}_2)$ is the three-wave coupling coefficient defined in Eq. (9.13). The equations for the two other wave amplitudes are obtained by permutation of the indices and observation of the symmetries in V.

Amplitude Equations

In the framework of a perturbation approach we assume that in any of the integrations or summations performed, the amplitudes under the integrals or sums can be considered to be slowly variable and can therefore be held constant in the time integration. Calculating the square of the amplitude by multiplication with the complex conjugate, one obtains for the energy density or occupation number of the 0-mode

$$\frac{\partial |A_0|^2}{\partial t} = -iS_0 \sum_{\mathbf{k}_0=\mathbf{k}_1+\mathbf{k}_2} \left[V \langle A_0^* A_1 A_2 \exp(i\Delta\omega_0 t)\rangle - V^* \langle A_0 A_1^* A_2^* \exp(-i\Delta\omega_0 t)\rangle\right] \tag{9.27}$$

and similar expressions for the energy densities of the other modes. In this formula there appear time averages over three amplitudes. The amplitudes to be used are calculated from the amplitude equations (9.23) in the manner indicated above by assuming that the product of amplitudes appearing in their integrands varies slowly and can be considered to be constant. The remaining integrals contain exponentials of the frequency mismatches and thus produce resonant denominators $\Delta\omega_{0,1,2} + i\delta$, where a small imaginary part is added to make the integrals converge. This imaginary part will later be set to zero. In the random-phase approximation one now assumes that any average of a product of amplitudes satisfies a condition like

$$\langle A_1 A_2 \rangle = |A_1|^2 \delta(\mathbf{k}_1 + \mathbf{k}_2) \tag{9.28}$$

where, since the wave amplitudes are real quantities, we also have $A(-\mathbf{k}) = A^*(\mathbf{k})$. All the coupling coefficients, V, become symmetric functions of the wavenumbers under this condition. Moreover, we replace the resonant denominators using Plemelj's formula (see Eq. (I.A.78) in the appendix of our companion book) by

$$(\Delta\omega \pm i\delta)^{-1} = P/\Delta\omega \mp i\pi\delta(\Delta\omega) \tag{9.29}$$

with P indicating the principal value, and the δ-function taking care of the exact resonance and energy conservation (note that the δ-function arises exclusively from the small imaginary part, $i\delta$, in the resonant denominator in the limit $\delta \to 0$). Carrying out all these calculations, Eq. (9.27) becomes

$$\frac{\partial |A_0|^2}{\partial t} = 4\pi S_0 \sum_{\Delta \mathbf{k}_0} |\delta(\Delta\omega_0) V|^2 \left[S_0 |A_1|^2 |A_2|^2 - S_1 |A_0|^2 |A_2|^2 - S_2 |A_0|^2 |A_1|^2 \right] \tag{9.30}$$

and similar equations for the energy densities of the remaining wave modes. For a broad spectrum it is convenient to replace the sum by an integral. In order to do this one defines, as in Eq. (9.26), continuous spectral energy densities, $|A_j|^2 \to \hbar N(\mathbf{k})_j d^3 k_j$, which are expressed in the number densities of the photons contained in the particular mode $j = 0, 1, 2$, as given in Eq. (9.10). Moreover, when transforming the sum into an integral one must take care of the restriction of the summation due to the requirement of the conservation of the wave momentum, which is expressed in the fact that the sums are taken only over the values $\Delta \mathbf{k}_0$, etc. Hence, in the integral appears another δ-function, i.e., $\delta(\mathbf{k}_0 - \mathbf{k}_1 - \mathbf{k}_2) = \delta(\Delta \mathbf{k}_0)$ and similar expressions for the other two modes

$$\begin{aligned} \Delta \mathbf{k}_0 &= \mathbf{k}_0 - \mathbf{k}_1 - \mathbf{k}_2 \\ \Delta \mathbf{k}_1 &= \mathbf{k}_1 - \mathbf{k}_2 - \mathbf{k}_0 \\ \Delta \mathbf{k}_2 &= \mathbf{k}_2 - \mathbf{k}_0 - \mathbf{k}_1 \end{aligned} \tag{9.31}$$

After all these preliminaries, we are now in the position to perform this transformation. As result the equation for the number density of wave quanta in the 0-mode is obtained

$$\frac{\partial N_0}{\partial t} = 4\pi S_0 \int\!\!\int d^3k_1 d^3k_2 \, |V|^2 \, [S_0 N_1 N_2 - S_1 N_0 N_2 - S_2 N_0 N_1] \, \delta(\Delta \mathbf{k}_0)\delta(\Delta\omega_0) \tag{9.32}$$

The corresponding equations for the other quantum occupation numbers are obtained by permutations (for $N_1 : 0 \to 1, 1 \to -2, 2 \to 0$, for $N_2 : 0 \to 2, 1 \to 0, 2 \to -1$) and observing the symmetry relations, $S_{-0} = -S_0$, $V(0,1,2) = V(1,-2,0) = V(2,0,-1)$, and $N(-\mathbf{k}) = N(\mathbf{k})$, with $S_{-0} = S(-\mathbf{k}_0)$, from Eq. (9.10)

$$\begin{aligned} \frac{\partial N_1}{\partial t} &= -4\pi S_1 \int\!\!\int d^3k_0 d^3k_2 |V|^2 \, [S_0 N_1 N_2 - S_1 N_0 N_2 - S_2 N_0 N_1] \, \delta(\Delta \mathbf{k}_0)\delta(\Delta\omega_0) \\ \frac{\partial N_2}{\partial t} &= -4\pi S_2 \int\!\!\int d^3k_0 d^3k_1 |V|^2 \, [S_0 N_1 N_2 - S_1 N_0 N_2 - S_2 N_0 N_1] \, \delta(\Delta \mathbf{k}_0)\delta(\Delta\omega_0) \end{aligned} \tag{9.33}$$

The integrands of these three last equations (9.32) and (9.33) are identical. Only signs and integration variables have changed. Taken together these are the *wave kinetic equations* for incoherent three-wave interaction.

The wave kinetic equations resemble the coherent wave equations (9.11). There are two important differences between the coherent and the incoherent descriptions. The first concerns the coupling time scale of evolution of the wave energies or amplitudes. In the coherent case the time scale is inversely proportional to the amplitude of the wave while in the incoherent random-phase case it is inversely proportional to the energy of the wave or, equivalently, the number density of wave quanta. This implies that the coherent time scale of the evolution is much shorter than the incoherent time scale. Coherent systems evolve much faster than incoherent systems. In addition, the coherent evolution was strictly reversible, while the incoherent interaction of the waves is *irreversible*. In the random-phase interaction the waves forget what their initial phase has been, and the systems can never return to this initial state. This can be demonstrated by defining an entropy function for the random-phase interaction as

$$\mathcal{S}(t) = \sum_{j=0,1,2} \int d^3k \ln N_j(\mathbf{k}_j) \tag{9.34}$$

Differentiating with respect to time and inserting the wave kinetic equations into this form yields an ever-positive right-hand side such that $d\mathcal{S}/dt \geq 0$, which implies that the entropy will always grow, and the system is irreversible. The condition for a stationary final state is found to be given by

$$\frac{S_0}{N_0} - \frac{S_1}{N_1} - \frac{S_2}{N_2} = 0 \tag{9.35}$$

which must be satisfied for each particular set of wavenumbers and frequencies satisfying the energy and momentum conservation laws or the resonance conditions. This stationary state implies some kind of equipartition between the quanta in the wave modes.

9.3. Weak Drift Wave Turbulence

The general theory of wave turbulence developed in the previous sections can be applied to all cases of wave interaction. Here we consider the particular case of drift waves in the plane perpendicular to the magnetic field, where a special two-dimensional kinetic wave equation can be derived. It describes drift wave turbulence and is important in space plasma applications where long wavelength modes are relevant in the particle and energy diffusion processes.

Drift Wave Equations

The drift wave equation applies to the very low frequency range below the ion-cyclotron frequency, $\omega < \omega_{gi}$, which is of interest for large-scale plasma turbulence. The long-wavelength range is defined as the domain of wavenumbers $kc_{ia}/\omega_{gi} \ll 1$. The wave spectrum at shorter wavelengths, where $kc_{ia}/\omega_{gi} \approx 1$, is caused by scattering of the long-wavelength mode. For the shorter wavelength part of the spectrum it is necessary to include the inhomogeneity of the plasma. What we are interested in is the evolution equation for this spectrum.

Turning to a fluid description, the following model equations are used as starting point. Electrons are considered in the magnetized drift approximation with negligible inertia parallel to the magnetic field

$$\frac{\partial \delta n}{\partial t} + \nabla_\perp \cdot (n_0 \mathbf{v}_d) = 0 \tag{9.36}$$

where n_0 is the unperturbed density, $\mathbf{v}_d = \mathbf{v}_E + \mathbf{v}_P$ is the sum of the electric field and polarization drifts defined in Eqs. (1.10) and (1.11), and $\nabla_\perp$ is the perpendicular gradient operator. The electric field and thus the polarization drift is expressed through the electric potential, ϕ, but includes the full nonlinearity caused by the transverse convective derivative

$$\begin{aligned} \mathbf{v}_E &= -\frac{\nabla_\perp \phi \times \mathbf{B}_0}{B_0^2} \\ \mathbf{v}_P &= -\frac{1}{\omega_{gi} B_0}\left[\frac{\partial}{\partial t}\nabla_\perp + (\mathbf{v}_E \cdot \nabla_\perp)\nabla_\perp\right]\phi \end{aligned} \tag{9.37}$$

Parallel to the magnetic field the electrons obey Boltzmann's equation

$$\frac{\delta n}{n_0} = \frac{e\phi}{k_B T_e} \tag{9.38}$$

The nonlinearity in the polarization drift of the ions is the driving force for the nonlinear coupling.

Hasegawa-Mima Equation

From this set of equations one can derive a nonlinear equation. Using the identity

$$\nabla \cdot [(\hat{\mathbf{e}}_{\parallel} \times \nabla\phi) \cdot \nabla]\nabla\phi = [(\hat{\mathbf{e}}_{\parallel} \times \nabla\phi) \cdot \nabla]\nabla^2\phi \tag{9.39}$$

and normalizing $n, t, \mathbf{x}, \phi$ to $n_0, 1/\omega_{gi}, c_{ia}/\omega_{gi}, k_B T_e/e$, respectively, the above system of equations can be reduced to the following nonlinear equation for ϕ

$$\frac{\partial}{\partial t}(\nabla^2\phi - \phi) = -[(\hat{\mathbf{e}}_{\parallel} \times \nabla\phi) \cdot \nabla]\nabla^2\phi \tag{9.40}$$

This is the *Hasegawa-Mima equation.* It describes the temporal and spatial evolution of the potential of nonlinear drift waves in the very low-frequency approximation. The Hasegawa-Mima equation is very similar to an equation describing incompressible hydrodynamic turbulence, but Eq. (9.40) describes turbulence of drift waves in compressible plasma physics, with the compression given by the field-aligned electron motion.

Constants of Motion

The Hasegawa-Mima equation (9.40) has two constants of motion which indicate that there are two different inertial ranges. These constants of motion are obtained by multiplying Eq. (9.40) by ϕ and integration over the entire volume. Then the nonlinear term becomes

$$\int d^3x\,\phi[(\hat{\mathbf{e}}_{\parallel} \times \nabla\phi) \cdot \nabla]\nabla^2\phi = \int d^3x\,\nabla \cdot [\phi\nabla^2\phi(\hat{\mathbf{e}}_{\parallel} \times \nabla\phi)] \tag{9.41}$$

This expression can be transformed into a surface integral. Thus one finds that

$$\frac{\partial W_1}{\partial t} = \frac{1}{2}\frac{\partial}{\partial t}\int d^3x\,[\phi^2 + (\nabla\phi)^2] = -\int \mathbf{j}_1 \cdot d\mathbf{A} \tag{9.42}$$

where the current density is given by

$$\mathbf{j}_1 = -\phi\nabla\frac{\partial\phi}{\partial t} + \frac{1}{2}(\nabla^2\phi)^2\hat{\mathbf{e}}_{\parallel} \times \nabla\phi \tag{9.43}$$

In non-normalized variables $n = n_0 e\phi/k_B T_e$ and $v_E^2 = (\nabla\phi)^2/B_0^2$, we find for the first conserved quantity, which turns out to be the total energy

$$W_1 = (n^2 k_B T_e/n_0 + m_i n_0 v_E^2)/2 \tag{9.44}$$

The second conserved quantity is obtained by multiplying Eq. (9.40) by $\nabla^2\phi$

$$\frac{\partial W_2}{\partial t} = \frac{1}{2}\frac{\partial}{\partial t}\int d^3x\,[(\nabla\phi)^2 + (\nabla^2\phi)^2] = -\int \mathbf{j}_2 \cdot d\mathbf{A} \tag{9.45}$$

where the second current density is given by

$$\mathbf{j}_2 = \nabla\phi\frac{\partial\phi}{\partial t} + \frac{1}{2}(\nabla^2\phi)^2\hat{\mathbf{e}}_\parallel \times \nabla\phi \tag{9.46}$$

It can also be shown that the second conserved quantity, $W_2 = W_{kin} + \Omega^2$, is the sum of the kinetic energy and the squared *vorticity*

$$\Omega^2 = (\nabla \times \mathbf{v}_E)^2 = (\nabla^2\phi)^2(k_B T_e/eB_0)^2 \tag{9.47}$$

Adding up both constants, $W_3 = W_1 + W_2$, gives as an alternative conserved quantity the expression

$$W_3 = \left[\frac{n}{n_0} - (\nabla \times \mathbf{v}_E)^2\frac{e^2 B_0^2}{k_B^2 T_e^2}\right]^2 \tag{9.48}$$

This is the squared *enstrophy* in a compressible two-dimensional medium as can be seen from the following derivation. Write the ion equation of motion

$$m_i\frac{d\mathbf{v}_i}{dt} = e(\mathbf{E} + \mathbf{v}_i \times \mathbf{B}) - \frac{1}{n}\nabla p \tag{9.49}$$

and consider an electromagnetic perturbation. Let us define the vorticity vector as $\mathbf{\Omega} = \nabla \times \mathbf{v}_i + (e/m_i)\mathbf{B}$. Taking the curl of the above equation, using Faraday's law, $\partial\mathbf{E}/\partial t = -\nabla \times \mathbf{B}$, and the identity $\mathbf{v}\cdot\nabla\mathbf{v} = \frac{1}{2}\nabla v^2 - \mathbf{v} \times \nabla \times \mathbf{v}$, one gets

$$\frac{\partial\mathbf{\Omega}}{\partial t} - \nabla \times \mathbf{v}_i \times \mathbf{\Omega} = \frac{1}{m_i n^2}\nabla n \times \nabla p \tag{9.50}$$

The divergence of the vorticity vanishes, $\nabla\cdot\mathbf{\Omega} = 0$. Therefore we have for the curl of the cross product of velocity and vorticity, $\nabla \times \mathbf{v} \times \mathbf{\Omega} = -\mathbf{\Omega}\nabla\cdot\mathbf{v} + \mathbf{\Omega}\cdot\nabla\mathbf{v} - \mathbf{v}\cdot\nabla\mathbf{\Omega}$. The second term in this expression vanishes because there is no variation of the velocity in the direction of the vorticity, and the above equation for the vorticity transforms into

$$\frac{d\mathbf{\Omega}}{dt} + \mathbf{\Omega}\nabla\cdot\mathbf{v}_i = \frac{1}{m_i n^2}\nabla n \times \nabla p \tag{9.51}$$

For a cold ion fluid the ion pressure vanishes. Using the continuity equation to express the divergence of the ion velocity $\nabla \times \mathbf{v}_i = -d(\ln n)/dt$, one finds

$$\frac{d}{dt}\left(\frac{\Omega}{n}\right)^2 = 0 \tag{9.52}$$

an equation which expresses conservation of enstrophy in a compressible fluid. Now using Boltzmann's distribution for the density n this expression reduces to the conservation of the quantity W_3, showing that the compressible enstrophy is conserved by the Hasegawa-Mima equation.

Kinetic Drift Wave Equation

The first step in the derivation of the nonlinear kinetic wave equation is standard and follows the procedure discussed in the previous sections. One expands the electric wave potential function into a spatial Fourier series

$$\phi(\mathbf{x}, t) = \tfrac{1}{2}\sum_{\mathbf{k}_\perp} [\phi(\mathbf{k}_\perp, t)\exp(i\mathbf{k}_\perp \cdot \mathbf{x}) + \text{c.c.}] \tag{9.53}$$

Let us normalize time to the inverse ion gyroperiod, $1/\omega_{gi}$, and lengths to c_{ia}/ω_{gi}. This implies that the frequencies are normalized to the ion-cyclotron frequency, $\omega/\omega_{gi} \to \omega$. The Fourier ansatz transforms the above set of nonlinear equations into the following evolution equation for the spatial Fourier amplitudes

$$\frac{\partial\phi(\mathbf{k}, t)}{\partial t} + i\omega_d(\mathbf{k})\phi(\mathbf{k}, t) = \tfrac{1}{2}\sum_{\Delta\mathbf{k}=0} V(\mathbf{k}, \mathbf{k}', \mathbf{k}'')\phi(\mathbf{k}', t)\phi(\mathbf{k}'', t) \tag{9.54}$$

where $\Delta\mathbf{k} = \mathbf{k} - \mathbf{k}' - \mathbf{k}''$ and where we explicitly indicated the $\mathbf{k}$- and t-dependences, but dropped the $\perp$-sign on $\mathbf{k}$. The interaction matrix is given by

$$V(\mathbf{k}, \mathbf{k}', \mathbf{k}'') = \frac{k''^2 - k'^2}{1 + k^2}\hat{\mathbf{e}}_\parallel \cdot (\mathbf{k}' \times \mathbf{k}'') \tag{9.55}$$

The new frequency, ω_d, is the drift wave frequency normalized to the ion-cyclotron frequency, ω_{gi}, and given by

$$\omega_d(\mathbf{k}) = -\frac{k_B T_e}{\omega_{gi} e B_0}\frac{k_\perp L_n}{1 + k^2} \tag{9.56}$$

with $L_n = 1/\nabla_\perp(\ln n_0)$ the gradient scale length.

Because of the assumption of very small parallel wavenumbers, $k_\parallel \approx 0$, Eq. (9.54) describes two-dimensional electrostatic low-frequency drift wave turbulence in magnetized plasmas. The drift frequency appearing in this equation plays somewhat the role of a viscosity. For very small k the right-hand side can be dropped and the problem becomes a linear one, with the dominating linear term yielding $\omega = \omega_d$ and no wave-wave interaction present. For very large normalized $k \gg 1$ the drift wave frequency vanishes, and the linear term must be replaced by some kind of viscosity or ion Landau damping. Here the wave energy is dissipated. This is logical because one expects that damping occurs at very short wavelengths. The turbulent state belongs to wavenumbers close to $k \approx 1$, when all the couplings are included.

One can solve the above equation using the methods of weak turbulence by assuming that a large-amplitude wave exists initially at very long wavelengths. The first step is to integrate Eq. (9.54) with respect to time to obtain

$$\phi(\mathbf{k}, t) = \tfrac{1}{2} \sum_{\Delta\mathbf{k}=0} V(\mathbf{k}, \mathbf{k}', \mathbf{k}'') \int_0^t dt' \, \phi(\mathbf{k}', t)\phi(\mathbf{k}'', t) \exp[-i\omega_d(\mathbf{k})(t - t')] \quad (9.57)$$

In order to find an equation for the root-mean square amplitude, we follow the usual procedure by multiplying Eq. (9.54) by the conjugate complex functions, $\phi^*(\mathbf{k}, t)$, and adding the complex conjugate product. Then one obtains the symmetric equation

$$\frac{\partial}{\partial t}|\phi(\mathbf{k}, t)|^2 = \frac{1}{2} \sum_{\Delta\mathbf{k}=0} V(\mathbf{k}, \mathbf{k}', \mathbf{k}'')[\phi(\mathbf{k}', t)\phi(\mathbf{k}'', t)\phi^*(\mathbf{k}, t) + \text{c.c.}] \quad (9.58)$$

This last equation determines the evolution of the wave intensity. One now substitutes the integrated version of the wave potential in Eq. (9.57) for each of the amplitudes, ϕ, into Eq. (9.58). Subsequently one takes the ensemble average of the resulting expression. Because we assume a broad spectrum, we can apply the random-phase approximation of the incoherent wave-wave interaction. This approximation produces a two-point correlation function on the right-hand side which can be written as

$$\langle \phi(\mathbf{k}, t)\phi(\mathbf{k}', t)\rangle = \delta(\mathbf{k} - \mathbf{k}')|\phi(\mathbf{k}, t)|^2 \exp[-(i\omega + \gamma)(t - t')] \quad (9.59)$$

The damping rate, γ, of the two-point correlation function has been introduced explicitly. As we know from incoherent interaction theory, it is necessary to introduce this as a decoherence parameter in order to obtain non-diverging results. If we now follow the indicated path, substituting Eq. (9.57) into Eq. (9.58), and using Eq. (9.59), we obtain the wave kinetic equation of weak low-frequency drift wave turbulence

$$\frac{\partial}{\partial t}|\phi(\mathbf{k}, t)|^2 = \frac{1}{2} \sum_{\mathbf{k}'} V(\mathbf{k}, \mathbf{k}', \mathbf{k} - \mathbf{k}') \left[\frac{V(\mathbf{k}, \mathbf{k}', \mathbf{k} - \mathbf{k}')}{\gamma(\mathbf{k}') + \gamma(\mathbf{k} - \mathbf{k}')} |\phi(\mathbf{k}')|^2 |\phi(\mathbf{k} - \mathbf{k}')|^2 \right.$$
$$\left. \frac{V(\mathbf{k}, \mathbf{k}, -\mathbf{k}')}{\gamma(\mathbf{k}) + \gamma(\mathbf{k}')} |\phi(\mathbf{k})|^2 |\phi(\mathbf{k}')|^2 + \frac{V(\mathbf{k}, \mathbf{k}, \mathbf{k}' - \mathbf{k})}{\gamma(\mathbf{k}) + \gamma(\mathbf{k}' - \mathbf{k})} |\phi(\mathbf{k})|^2 |\phi(\mathbf{k}' - \mathbf{k})|^2 \right] \quad (9.60)$$

This is the *drift wave kinetic equation* for the wave-coupling in two-dimensional drift wave turbulence in a magnetized plasma, with the turbulence being in the plane perpendicular to the magnetic field. It is the k-space version of the Hasegawa-Mima equation (9.40). Interpreting the three terms on the right-hand side shows that the first term is the mode coupling term, while the two other terms describe the self-interaction of the wave, which without the introduction of the imaginary damping part would lead to divergence.

Spectral Evolution

Let us assume that a large-amplitude drift wave of initial potential amplitude $\phi(\mathbf{k}_0)$ and $|\phi(\mathbf{k}_0)|^2 \gg |\phi(\mathbf{k})|^2$ has been excited by some instability mechanism. One particular example of such drift waves would be lower-hybrid drift modes. This wave is of long wavelength, $|\mathbf{k}_0| \ll 1$, corresponding to injection of wave energy at small wavenumbers. Moreover, we assume that the damping of the initial wave is much smaller than the damping of the shorter wavelengths waves, $|\gamma(\mathbf{k}_0)| \ll |\gamma(\mathbf{k})|$. Under these conditions one can linearize the above wave kinetic equation (9.60) with respect to the weak short-wave intensity in order to determine the spectral width of the shorter waves. To this end we need the intensity of the shorter waves. This is obtained exactly along the same lines as in the previous section. We multiply Eq. (9.54) by its conjugate complex and use Eq. (9.57) to obtain an equation for the correlation of the small amplitude short-wave spectrum

$$\left[\frac{\partial}{\partial t} + i\omega_d(\mathbf{k}) - \frac{|\phi(\mathbf{k}_0)|^2}{4\gamma(\mathbf{k})} V(\mathbf{k}, \mathbf{k}_0, \mathbf{k} - \mathbf{k}_0) V(\mathbf{k}, \mathbf{k}, -\mathbf{k}_0)\right] \langle \phi(\mathbf{k}, t)\phi^*(\mathbf{k}, t)\rangle = 0 \tag{9.61}$$

Using Eq. (9.59) yields the following equality

$$2\gamma^2(\mathbf{k}) = \frac{1}{2}\frac{(\mathbf{k} \times \mathbf{k}_0)^2}{1 + k^2}\frac{(k^2 - 2\mathbf{k}\cdot\mathbf{k}_0)(k^2 - k_0^2)}{1 - (\mathbf{k} - \mathbf{k}_0)^2}|\phi(\mathbf{k}_0)|^2 \tag{9.62}$$

from where one obtains for the damping rate

$$\gamma(\mathbf{k}) = \frac{1}{2^{3/2}}\frac{k_0 k^3}{1 + k^2}\left[1 - \frac{k_0^2}{k^2}\frac{(k^2 - k_1^2)(k^2 - k_2^2)}{1 + k^2}\right]|\phi(\mathbf{k}_0)| \tag{9.63}$$

where $k_{1,2}^2 = -1 \pm 1/\sqrt{2}$ and we have assumed that $|\phi(\mathbf{k})|^2$ is isotropic, such that it is allowed to average over the angle between $\mathbf{k}$ and $\mathbf{k}_0$. Only the lowest-order term in k_0^2/k^2 is retained. As one observes, for small wavenumbers the spectral width behaves like $\gamma \propto k^3$. On the other hand $\omega \propto k$. Hence, the width may become smaller than the frequency, in contradiction to our assumption of small frequency. This sets a limitation on the theory.

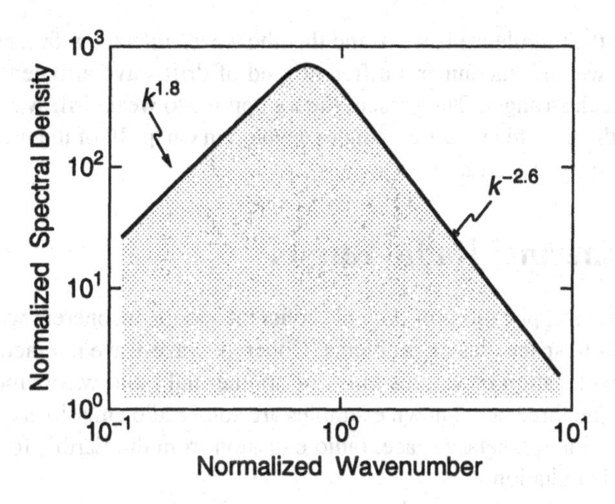

Fig. 9.4. Drift wave turbulent spectrum.

In the range of validity of the theory we are interested in the shape of the stationary spectrum, $|\phi(\mathbf{k})|^2$. We set the time derivative on the left-hand side of Eq. (9.60) to zero and equate the damping rate, $\gamma(\mathbf{k})$, from Eq. (9.63) to the growth rate provided by the mode-coupling term in Eq. (9.60). We then expand the spectral density, $|\phi(\mathbf{k}-\mathbf{k}_0)|^2$, around the small spectral density, $|\phi(\mathbf{k})|^2$, in powers of $\mathbf{k}_0$. This yields

$$|\phi(\mathbf{k}-\mathbf{k}_0)|^2 \approx |\phi(\mathbf{k})|^2 - \mathbf{k}_0 \cdot \nabla_{\mathbf{k}} |\phi(\mathbf{k})|^2 + \tfrac{1}{2}(\mathbf{k}_0 \cdot \nabla_{\mathbf{k}})^2 |\phi(\mathbf{k})|^2 \tag{9.64}$$

The operator $\nabla_{\mathbf{k}} = \partial/\partial\mathbf{k}$ is the k-space gradient operator. Using the explicit form of the coupling constant, V, it can be shown that the leading term of the mode coupling cancels with the corresponding term of the damping. Thus balancing the two terms of the order of k_0^2, one obtains a particular ordinary differential equation for the isotropic spectral density

$$\left(\frac{d^2}{dk^2} + \frac{2}{k}\frac{1+3k^2}{1+k^2}\frac{d}{dk} + \frac{15k^4+18k^2-5}{k^2(1+k^2)}\right)|\phi(k)|^2 = 0 \tag{9.65}$$

This equation describes the stationary spectrum of the spectral density of drift wave turbulence and yields for the short-wavelength spectrum, $k < \min[10, (T_e/T_i)^{1/2}]$

$$|\phi(k)|^2 \approx k^{1.8}/(1+k^2)^{2.2} \tag{9.66}$$

This spectrum has a well-pronounced peak at normalized wavenumber $k_{\max}^2 \approx 0.7$ and decays as $k^{-2.6}$. Figure 9.4 shows a sketch of the stationary drift wave turbulent spectrum. One should, however, keep in mind that this spectrum has been obtained under the assumption that the turbulence is weak and that the waves interact with random phases. At a later stage we will encounter a different kind of drift wave turbulence where the interaction is much stronger. The present theory applies to weak drift wave turbulence only, while the theory which we are going to develop in Chap. 10 of the present volume applies to strong drift wave turbulence.

9.4. Nonthermal Radio Bursts

There are numerous applications of both the coherent and the incoherent wave-wave interaction theories to space plasma problems. Coherent wave-wave interaction has been favored in the explanation of various kinds of nonthermal radio wave emissions from space plasmas. The three best-known examples are solar radio emissions originating in the corona and in interplanetary space, radio emission from the Earth's foreshock, and auroral kilometric radiation.

The solar coronal examples include escaping radio radiation mostly in the X-mode during solar type III events when electron beams travel across the low density corona, during solar type II events, when matter ejected from the solar atmosphere drives a fast shock wave in front of the ejecta, during solar type IV bursts when particles trapped in magnetic mirror configurations generated electrostatic electron-cyclotron harmonics, and during solar type I bursts, when local reconnection events occur. In all these cases different types of plasma waves are generated which by interaction with Langmuir waves excited by fast electrons produce an escaping radio wave at a frequency higher than the local plasma frequency. These examples are ideal examples of three-wave interactions in which the two electrostatic wave frequencies mix up into a high-frequency electromagnetic wave.

Solar Type III Bursts

During a type III solar radio burst, when the merging waves are two oppositely directed Langmuir waves (ℓ), excited by a gentle electron beam in the solar corona, which undergo a head-on collision, such a process can be symbolically described as

$$\ell_1 + \ell_2 \rightarrow t_0 \tag{9.67}$$

Here t_0 is the transverse electromagnetic radio wave. Clearly, because the wavelength of the transverse wave is much longer than the wavelengths of the two Langmuir waves, the two wave vectors $\mathbf{k}_1$ and $\mathbf{k}_2$ must be about of equal length but opposite direction. Since for Langmuir waves the dispersion curve is convex from below, this process is possible. One could object that such a process is strictly reversible and thus would not lead to an escaping wave. However, since the solar corona is an open system, once the transverse wave comes into life during the interaction, it will, because of its high

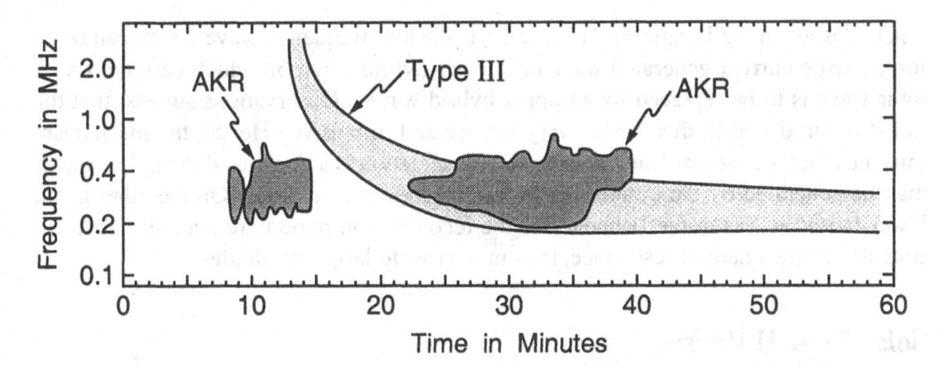

Fig. 9.5. Spectrogram of interplanetary type III burst and auroral kilometric radiation.

phase and group velocities close to the velocity of light, c, immediately escape from the generation region. In this case there will be no chance for the wave to re-decay into the Langmuir waves by which it had been created. The coherent interaction works here as a generator of radiation very similar to a coherent laser.

The same mechanism applies to the generation of type III burst radiation in interplanetary space from travelling 10 keV electron beams ejected from the sun. An example is shown in Fig. 9.5 where an electron beam encounters the Earth's orbit and emits local radiation at twice the local plasma frequency by the above mechanism. Close to the Sun the density is high and the radiation has high frequencies. Close to the Earth the density is low and nearly constant in the solar wind. Therefore the frequency at later times, when the beam is closer to the Earth, is low and becomes nearly constant. From the slope of the emission the velocity of the beam travelling across the solar wind can be determined. It corresponds to a beam energy of about 10 keV. Moreover, the density profile in the solar wind can be deduced, showing that the solar wind density decays with distance from the sun as $n_{sw}(r) \propto r^{-2}$.

The auroral kilometric radiation shown in Fig. 9.5 is generated close to the Earth in the auroral zone and escaping into the solar wind. This radiation is not related to the type III emission. Most of it is generated by a cyclotron maser mechanism in the auroral magnetosphere, as described in Sec. 5.5.

Solar Type I Bursts

During type I solar radio bursts the radiation properties are different from type III burst. Then presumably low-frequency electrostatic (or electromagnetic) waves generated in the reconnection region may merge with Langmuir waves. The corresponding symbolic equation reads

$$\ell_1 + s_2 \rightarrow t_0 \tag{9.68}$$

where ℓ is again the Langmuir wave, and s is the low-frequency wave which can be an ion-acoustic current generated wave or a lower-hybrid wave (in which case the Langmuir wave is to be replaced by an upper-hybrid wave). Observations suggest that the radiation produced in this case is very intense and impulsive. Hence, the mechanism must be strictly coherent. Ion-acoustic wave interaction is a good candidate. The waves may be generated by the current in the reconnection region itself. On the other hand, lower-hybrid waves at the boundary of the reconnection region are another candidate because, being a natural resonance, they may grow to large amplitudes.

Solar Type II Bursts

Solar type II burst radiation is caused by travelling shock waves. Here the radiation is produced by the reflected electrons in the foreshock region which excite Langmuir waves. But shock radiation has also another so far unexplained strong component which is attributed to the shock front itself and may arise from lower-hybrid wave interactions with upper-hybrid waves. Part of the emission might be random-phased.

A nice example of an observation of this kind of wave-wave interaction, locally producing an electromagnetic wave in the foreshock region at twice the electron plasma frequency, is included in Fig. 4.2. The upper frequency band above the locally observed plasma frequency shows the weak radiation. The much higher intensity of the Langmuir wave emission indicates that the process does not have a very high efficiency. In fact, referring to the equations for the three-wave interaction, we see that the efficiency is proportional to the square of the wave amplitude in the coherent case, leading to a much lower wave amplitude in the upmixed electromagnetic wave than in the electrostatic wave.

Solar Type IV Bursts

Solar type IV emission is predominantly generated by a cyclotron maser mechanism, whenever the density is low enough or the particle energies are relativistic, but random-phased interaction between electron-cyclotron modes (oblique Bernstein modes) with upper-hybrid modes, i.e., interaction among different harmonics of these waves, may also play some role. Saturation of the electrostatic waves may be due to resonance broadening.

Concluding Remarks

Though space plasma physics is inherently nonlinear and though most of the processes in the solar atmosphere, solar corona, solar wind, and magnetosphere are nonlinear processes, only a few practical applications of nonlinear plasma physics and weak turbulence theory exist. The reasons are manyfold.

First, observations are often not sophisticated enough to resolve the higher-order nonlinear effects, and linear or macroscopic theory does in many cases explain the observations well enough. Second, the methodological difficulties in the nonlinear analytical treatment of nonlinear effects have, with the exception of the simplest quasilinear ones, prevented people from wide application of these methods.

Third and most important, in most cases space plasmas reach a final long-term state before they are observed by instrumentation. As a result, the real situations encountered by spacecraft are final states of evolved nonlinear equilibria, which to a certain extent have entirely forgotten their initial state. It is therefore nearly impossible to extract information about this initial state from observations and to properly reconstruct the whole nonlinear chain of interaction which led to the formation of this state. Only when observing bursty and explosive effects, and when asking for anomalous transport coefficients and the origin of nonlinear equilibria leading to stable non Maxwellian distribution functions, nonlinear theory comes into play.

Last, researchers are now treating nonlinear phenomena by numerical simulations. This new kind of approach to nonlinear plasma physics encompasses the difficulties encountered when trying to solve the complicated nonlinear equations analytically. It is therefore not only very attractive but also successful, if its results are properly interpreted.

Because of the scarceness of applications, we also do not extend our presentation of weak turbulence to the next step in perturbation theory by including higher-order wave-particle interactions. Usually these theories are collected under the name of *induced scattering* or *nonlinear wave-particle interaction* to distinguish them from quasilinear theory.

Inclusion of dissipative effects into wave turbulence theory would extend the theory into another important direction, which we have ignored here. Dissipation sets a threshold on wave-wave instabilities, both in the coherent and in the incoherent cases. A wave which would become unstable in the dissipationless interaction among waves must, in addition, overcome the damping provided by the dissipative processes in a plasma. Hence, dissipationa introduces a threshold into the interaction, just as in linear instability theory.

Further Reading

The theory of coherent wave-wave interactions is described in many of the books cited below. An extensive presentation is given in [9], but here we follow the shorter presentation in [3]. The general theory of incoherent plasma wave turbulence is contained in [1]. A simplified shorter version is given in [8]. For a lucid derivation of the wave kinetic equations and discussion of their properties see [1]. Drift wave effects are contained in [4]. The Hasegawa-Mima drift theory is taken from the original papers by Hasegawa and Mima in *Phys. Rev. Lett.* **39** (1977) 205, and in *Phys. Fluids* **21** (1978) 87. Dissipative effects in wave-wave turbulence in application to ionospheric physics are described in [2]. The general theory is given in [5]. Finally, the wave-wave theory and observations of solar radio bursts can be found in [6] and [7].

[1] R. C. Davidson, *Methods in Nonlinear Plasma Theory* (Academic Press, New York, 1972).

[2] A. V. Gurevich, *Nonlinear Phenomena in the Ionosphere* (Springer Verlag, Heidelberg, 1978).

[3] A. Hasegawa, *Plasma Instabilities and Nonlinear Effects* (Springer Verlag, Heidelberg, 1975).

[4] B. B. Kadomtsev, *Plasma Turbulence* (Academic Press, New York, 1965).

[5] P. K. Kaw et al., in *Advances in Plasma Physics*, eds. A. Simon and W. B. Thompson (Wiley Interscience, New York, 1976).

[6] D. J. McLean and N. R. Labrum (eds.), *Solar Radiophysics* (Cambridge University Press, Cambridge, 1985).

[7] D. B. Melrose, *Plasma Astrophysics* (Gordon and Breach, New York, 1980).

[8] R. Z. Sagdeev and A. A. Galeev, *Nonlinear Plasma Theory* (W. A. Benjamin, New York, 1969).

[9] J. Weiland and H. Wilhelmsson, *Coherent Nonlinear Interaction of Waves in Plasmas* (Pergamon Press, Oxford, 1977).

10. Nonlinear Waves

When perturbation theory cannot be applied to describe the evolution of the wave spectrum and its interaction with the particle component, one is forced to look for methods which avoid the conventional WKB perturbation technique. In such cases it is difficult to find a small parameter which can serve as an expansion parameter. The interaction is then subject to long-range forces. Methods to treat conditions of this kind are sparse. They have been developed in the past few decades and have provided insight into some aspects of the physics, which is closely related to so-called *strong plasma turbulence*. As in the previous chapter, the reader should consult the original literature for particular applications. Here we give a brief overview of the basic ideas with only a few illustrative applications.

One may naively believe that strong plasma turbulence sets in simply when the wave amplitudes have grown to high values. But this condition was also at the fundament of weak turbulence theory and is thus not the decisive one to distinguish between weak and strong turbulence. Rather it is decisive how the turbulence presents itself to the observer. If it can be considered as a mixture of a large number of waves, with the probability of collisions between the waves decreasing with the number of waves involved in a collision, or as an ensemble of waves and particles, with the probability of the interaction falling with the number of participants, the turbulence is weak.

On the other hand, this condition ceases to be applicable in many cases. Consider, for instance, a single wave which grows in the absence of any other waves. Growth of the wave must be limited by some process which becomes important at large amplitudes. At very large wave amplitudes an expansion of the disturbance with respect to the ratio of the wave amplitude to the background equilibrium values fails, and the resulting nonlinear evolution of the wave will not be describable by weak turbulence theory. Cases like this one are typical for strong plasma turbulence. To treat them one seeks for non-perturbative methods.

Because it is very difficult to treat strong turbulence on the basis of kinetic theory, one usually returns to a fluid description of the plasma. The dominant nonlinearity is then the nonlinearity in the convective derivative term. But even this comparably simple and for more than two centuries well-known nonlinearity introduces a large number of new and unexpected effects. It is, in principle, responsible for the whole field of hydrodynamic turbulence in fluid dynamics. In plasma physics new effects arise due to the relatively short-range Coulomb interactions. Sometimes this makes investigation of strong turbulence easier than in hydrodynamics.

In the present chapter we consider several basic nonlinear equations which lead to the development of strong turbulence. The basic ingredient is that strong turbulence consists of a large number of localized wave solutions of these equations, so-called *solitons* or sometimes also *cavitons*. We prefer to call them solitons and to reserve the name cavitons for the structures produced in strong plasma turbulent interaction (see Chap. 11). The number of equations leading to such solutions is surprisingly small, and we discuss only those which are of relevance for space plasma physics.

10.1. Single Nonlinear Waves

When perturbation theory cannot be applied to describe the interactions in a plasma, one speaks of strong plasma turbulence. This happens when the wave amplitudes have grown to such high values that an expansion of the disturbance with respect to the small parameter, i.e., the ratio of the wave amplitude to the background equilibrium values, fails. Hence, the primary concern of strong plasma turbulence is the treatment of the evolution of single large-amplitude nonlinear waves. Surprisingly, such an approach readily leads to the evolution of turbulence.

Wave Steepening

We have already been familiar with one type of exact nonlinear waves, the BGK modes treated in Sec. 8.2. These were based on kinetic theory. A considerably more simple approach is to treat nonlinear waves in terms of the hydrodynamic picture. Consider, for example, the following one-dimensional nonlinear equation for the velocity amplitude of a wave in a medium supported by some force, F

$$\frac{\partial v}{\partial t} + v\frac{\partial v}{\partial x} = F \tag{10.1}$$

The left-hand side of this equation is nothing else but the convective derivative in a fluid-like medium and thus of fairly general validity. The nonlinear term on the left-hand side, when expanded into a Fourier series, contains a large number of wave-wave interaction terms. Therefore this term can be considered as a coupling term between many waves of different wavelengths.

Superposition and coupling of waves of different wavelength implies deformation of the wave profile. This is easily understood. Assume that an initial wave, $\delta v(x,t) = \delta v \cos[k(x - v_0 t)]$, is injected into the plasma, with v_0 being the convection speed.

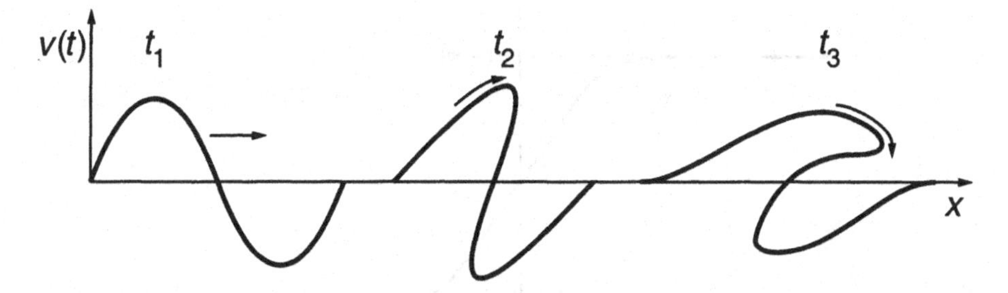

Fig. 10.1. Steepening and breaking of a wave in a dissipationless plasma.

Together with the disturbance, the velocity amplitude will be

$$v(x,t) = v_0 + \delta v \cos[k(x - v_0 t)] \quad (10.2)$$

The cosine function represents an oscillation between positive and negative values. In the wave maxima on the top of the crest the total velocity will be larger than the streaming velocity, v_0, while in the wave minima, the valleys of the disturbance, the velocity is reduced and thus is smaller. The wave minima stay behind the flow while the maxima run away. Necessarily, the minima must be overtaken by the maxima, leading to wave steepening, and ultimately to wave breaking, if there is no dissipative effect which counteracts the breaking. Figure 10.1 shows schematically how this breaking evolves in time in a dissipationless fluid.

That the nonlinear term produces higher harmonics and thus shorter wavelengths can be seen analytically. The product $v\partial v/\partial x$, with $v \propto \cos(kx)$, generates the term $-k\cos(kx)\sin(kx) = -k\sin(2kx)/2$, which is a wave of half the original wavelength. When further steepening occurs, these higher harmonics produce even shorter wavelengths, until the wave near its crest consists of a large number of nonlinearly superimposed waves of ever shorter wavelengths. This superposition generates a strongly curved wave front which ultimately must break.

Burgers Equation

When the right-hand side of Eq. (10.1) is non-zero, wave steepening and breaking may balance each other. Assume, for instance, that the dominant term on the right-hand side is proportional to the second derivative of the velocity amplitude with respect to x. In this case Eq. (10.1) can be written as

$$\frac{\partial v}{\partial t} + v\frac{\partial v}{\partial x} = \alpha\frac{\partial^2 v}{\partial x^2} \quad (10.3)$$

This is *Burgers equation* with $\alpha > 0$. From the structure of Eq. (10.3) one recognizes that it is a nonlinear diffusion equation, with α playing the role of the diffusion co-

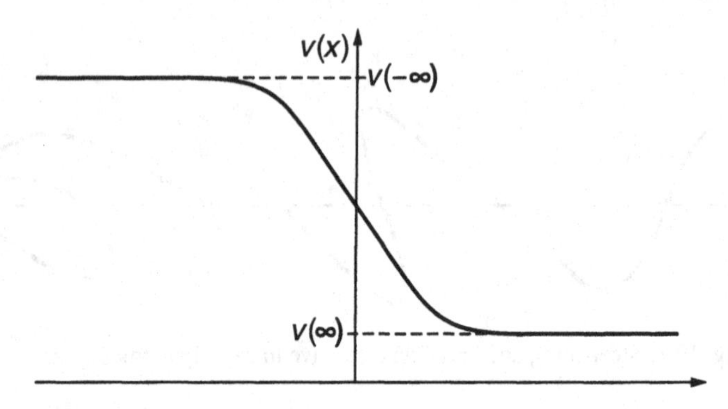

Fig. 10.2. Shock ramp solution of Burgers dissipative equation.

efficient and the convective term providing the nonlinearity. However, in comparison with a linear diffusion equation lacking the nonlinear term the character of the solutions change totally. The reason is that the diffusive second order derivative may compensate for the nonlinearity. For increasing nonlinear steepening the second derivative starts contributing more and more until it becomes comparable to the nonlinear term and the steepening ceases. In this state the solution is stationary with vanishing time derivative.

The stationary solution can be found when transforming to a coordinate system moving with the wave. Let the velocity of the wave be u and introduce the new coordinate, $y = x - ut$, Burgers equation is written

$$\alpha \frac{\partial^2 v}{\partial y^2} = (v - u)\frac{\partial v}{\partial y} \tag{10.4}$$

We are only interested in localized solutions which are regular at infinity. The solution obtained is

$$v - u = -u \tanh[u(x - ut)/2\alpha] \tag{10.5}$$

The form of this solution is a *shock ramp*. Figure 10.2 shows the form of the shock ramp as solution of Burgers equation. The height of the ramp is u, the thickness of the ramp is α/u, and the shock propagates with velocity u along x. Hence, Burgers equation yields as stationary solutions propagating shock waves where the dissipation contained in the diffusion coefficient becomes effective in the steep gradient inside the ramp where it balances the nonlinearity.

Korteweg-de Vries Equation

Balancing the nonlinearity with the help of dissipation is only one possibility. When there is no dissipation in the medium, the right-hand side of Eq. (10.1) may contain a third-order derivative with respect to x. Third-order spatial derivatives describe dispersion. The equation obtained is the *Korteweg-de Vries equation*

$$\frac{\partial v}{\partial t} + v\frac{\partial v}{\partial x} + \beta\frac{\partial^3 v}{\partial x^3} = 0 \tag{10.6}$$

The simplest way to understand the origin of nonlinear equations of the type of the Burgers and the Korteweg-de Vries equations is to consider the general one-dimensional dispersion relation, $\omega(k) = 0$, where the frequency is an arbitrary function of the wavenumber. With a Taylor expansion of the frequency with respect to k we get

$$\omega(k) = \omega(0) + v_{gr}k + \frac{1}{2}\frac{\partial^2\omega}{\partial k^2}k^2 + \frac{1}{6}\frac{\partial^3\omega}{\partial k^3}k^3 + \ldots \tag{10.7}$$

If we now interpret frequency and wavenumber as one-dimensional operators

$$\begin{aligned} \omega &= i\partial/\partial t \\ k &= -i\partial/\partial x \end{aligned} \tag{10.8}$$

acting on the wave function, $\psi(x,t)$, it is easy to see that the terms up to the third-order derivative just reconstitute an equation which contains both the Burgers and Korteweg-de Vries equations. In such an approach one can ignore the linear term in the above expansion of the frequency, because this term describes only a Doppler shift. Hence, all discussion which follows refers to a frame of reference which moves at the constant group velocity, v_{gr}.

The coefficients in front of the operators are derivatives of the frequency with respect to wavenumber. It is the frequency which contains all information about the nonlinear evolution of the system. This can be understood by analogy to quantum mechanics when one recognizes that the frequency is the energy and, hence, plays the role of the Hamiltonian of the system in which all the information is stored. Burgers equation is obtained when cutting the expansion after the second derivative term, while the Korteweg-de Vries equation follows when dropping the second derivative but keeping the third-order derivative of $\omega(k)$.

Similar to Burgers equation, the Korteweg-de Vries equation also allows for stationary localized solutions. But because dispersion causes only reversible effects, such a solution will be restricted to a finite spatial interval. Assume that it exists and moves at speed u. Again introducing the coordinate $y = x - ut$, measured from the center of the localized solution, the Korteweg-de Vries equation (10.6) can be transformed into the third-order ordinary differential equation

$$(v-u)\frac{\partial v}{\partial y} + \beta\frac{\partial^3 v}{\partial y^3} = 0 \tag{10.9}$$

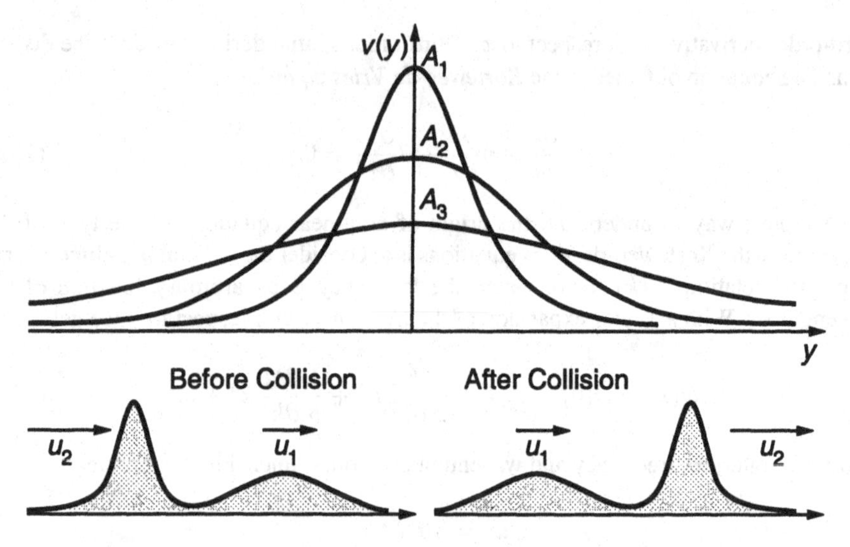

Fig. 10.3. Soliton solutions of the stationary Korteweg-de Vries equation.

Solution of this stationary equation requires the prescription of three boundary conditions. Accounting for reversibility and localization we choose $v = \partial v/\partial y = 0$ at $y \to \pm\infty$. It can then be shown by substitution that the following function solves the stationary Korteweg-de Vries equation

$$v_{\rm sol}(x - ut) = 3u \,\mathrm{sech}^2 \left[(u/\beta)^{1/2} (x - ut)/2\right] \tag{10.10}$$

This function describes a stationary bell-shaped curve propagating at velocity u along x without any change of its form. The width of this pulse is given by $\Delta = 2\sqrt{\beta/u}$. For the same β, high-speed solutions have a narrower width than low-speed solutions. There is a distinct relation between the amplitude, A, of the pulse and its width

$$\Delta = 2(3\beta/A)^{1/2} \tag{10.11}$$

Figure 10.3 sketches how such pulses looks like. Since the pulses do not not change their profile during propagation, do not slow down, and are stable against disturbances, they have earned the name *solitary waves* or (topological) *solitons*.

It can be shown that a whole chain of different solitons with different u_j solves the stationary Korteweg-de Vries equation. All these solitons move at their constant speeds along x without changing their shapes. Because the narrower and faster solitons overtake the slower and wider solitons, one would expect that an interaction should occur. But calculations and numerical experiments show that collisions between two or more solitons have practically now effect on the solitons. After collision they reappear and separate at their initial speeds as sketched in the lower part of Fig. 10.3.

This behavior is characteristic for topological solitons and stationary structures. In systems described by equations of the same family as the Korteweg-de Vries equation, characterized by dispersion effects only, topological solitons constitute a natural fundamental eigenmode system of waves. On the other hand, when a wave evolves into a chain of solitons of different amplitudes, widths and speeds, the plasma assumes a grainy quasi-irregular structure, which cannot be described anymore by weak turbulence theory. This is the origin of the notion that the soliton state of a plasma corresponds to the state of *strong plasma turbulence.* In this state the plasma is filled with localized waves and assumes the character of a gas of particles with the solitons constituting the particles of different speeds, momenta and energies. One possibility to describe such a strongly turbulent state would thus be a kinetic quasi-particle description with the particles being solitons.

Stationary Solution of the Korteweg-de Vries Equation

We have guessed the solution of the stationary Korteweg-de Vries equation. Let us now demonstrate, how this solution is found. Equation (10.9) can be directly integrated once. Applying the boundary conditions at infinity, the integration constant turns out to be zero yielding the result

$$\beta\frac{\partial^2 v}{\partial y^2} = v(u - \tfrac{1}{2}v) \tag{10.12}$$

Inspection of this equation shows immediately that it is of the type of the equation of motion of a pseudo-particle with the left-hand side the acceleration, β the mass, and the right-hand side the force. Such an equation can be solved by multiplying it with the pseudo-velocity, $\partial v/\partial y$. The right-hand side can be represented as the derivative of a *pseudo-potential* or *Sagdeev potential*, $S(v)$. Integrating once more and again applying the boundary conditions at infinity gives the pseudo-energy conservation law

$$\frac{\beta}{2}\left(\frac{\partial v}{\partial y}\right)^2 = \frac{v^2}{2}\left(u - \frac{1}{3}v\right) = -S(v) \tag{10.13}$$

This equation allows for a first conclusion. Because the left-hand side is a positive quantity, solutions exist only under the condition that the pseudo-potential is negative

$$S(v) = \frac{v^2}{2}\left(\frac{1}{3}v - u\right) < 0 \tag{10.14}$$

Stationary topological soliton solutions of the Korteweg-de Vries equation exist only in the region of velocity space where this condition is satisfied. This implies that

$$v < 3u \tag{10.15}$$

which suggests that all Korteweg-de Vries solitons must have velocities smaller than the given characteristic speed $3u$. We will later demonstrate, for the particular example of ion-acoustic waves, what the meaning of this conclusion is. However, quite generally we have found a method by which a class of qualitative solutions can be obtained from equations similar to the Korteweg-de Vries equation by looking for the pseudo-potential and determining the regions where it becomes negative. Any initial condition leading to a physical solution must fall into those domains. Under these conditions Eq. (10.13) can be solved by quadrature to find $\partial v/\partial y = \sqrt{-2S(v)/\beta}$ and solving the integral

$$y - y_0 = \int_0^v \frac{dv}{[-2S(v)/\beta]^{1/2}} \tag{10.16}$$

which yields the above solution (10.10) of the Korteweg-de Vries equation.

10.2. Nonlinear Wave Evolution

Stationary solutions are stationary only in the wave frame of reference. The nonlinear Burgers and Korteweg-de Vries equations do allow not only for time evolution of these single stationary pulses, but they also contain information if such a stationary state is accessible or not from a given initial conditions. In order to decide what happens to a given initial condition, it is necessary to solve the time-dependent equations.

Time-Dependent Solution of Burgers Equation

So far we considered the stationary equations with traveling wave solutions. One of the solutions was the traveling shock wave, showing that in the fluid picture a shock ramp can be a quite natural stable solution in the presence of nonlinear wave steepening and dissipation. Such a wave necessarily travels across the plasma. On the other hand, this solution identifies a shock as a nonlinear steepened wave which is subject to dissipative equilibrium. The other type of solution was a traveling solitary structure which can be considered as an eigenstate of a dispersive medium.

Such nonlinear structures must evolve from some initial state. One must thus solve the time-dependent problem to justify the accessibility of the final stationary nonlinear state. To solve the Burgers equation is comparably easy. It can be done by transforming to a new variable ϕ related to v through

$$v(y,t) = -2\alpha[\partial \ln \phi(y,t)/\partial y] \tag{10.17}$$

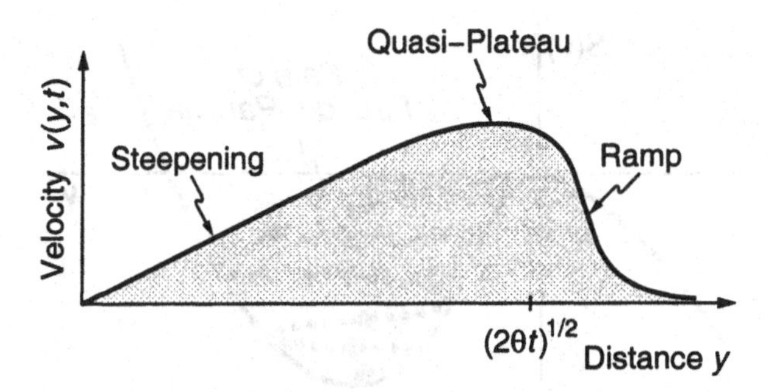

Fig. 10.4. Time-asymptotic solution of Burgers equation.

Burgers equation then becomes a diffusion equation for the new function ϕ

$$\frac{\partial \phi}{\partial t} = \alpha \frac{\partial^2 \phi}{\partial y^2} \tag{10.18}$$

and the solution can be immediately written down as

$$\phi(y,t) = \frac{1}{(4\pi\alpha t)^{1/2}} \int_{-\infty}^{\infty} d\eta \exp\left[-\frac{(y-\eta)^2}{4\alpha t} - \frac{1}{2\alpha} \int_0^{\eta} v_0(\tau) d\tau \right] \tag{10.19}$$

The initial disturbance v_0 must satisfy the condition of convergence $\int_0^y dy'\, v_0(y') \leq \mathrm{const} \cdot y$ for $y \to \infty$. Because of this condition one has

$$\int_{-\infty}^{\infty} dy'\, v_0(y') = \Theta < \infty \tag{10.20}$$

and the asymptotic solution for large times becomes

$$v(y, t \to \infty) \approx -2\alpha \frac{d}{dy} \ln G \left[\frac{y}{(4\alpha t)^{1/2}} \right] \tag{10.21}$$

where the function G is given by

$$G(x) = \frac{1}{\sqrt{\pi}} \left[\mathrm{e}^{-\Theta/4\alpha} \int_{-\infty}^{x} d\eta\, \mathrm{e}^{-\eta^2} + \mathrm{e}^{+\Theta/4\alpha} \int_{x}^{\infty} d\eta\, \mathrm{e}^{-\eta^2} \right] \tag{10.22}$$

This solution gives the asymptotic profile of the shock ramp as shown in Fig. 10.4. The quasi-plateau behind the shock ramp at the maximum of the shock is the plateau which in the stationary solution was assumed to exist at $y = -\infty$. The non-stationary solution shows that the time evolution of the wave asymptotically leads to the qualitatively predicted wave steepening and dissipative formation of a shock ramp.

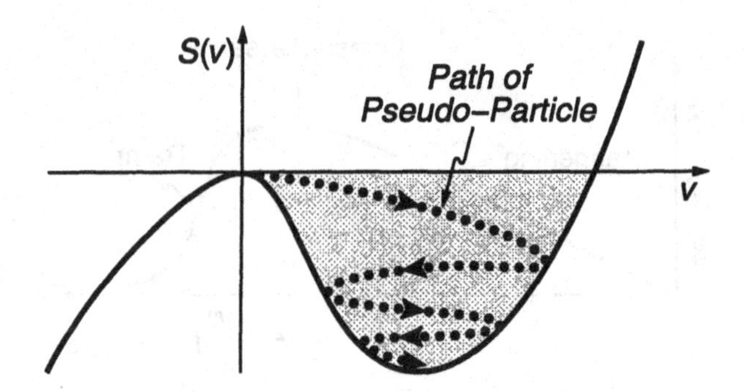

Fig. 10.5. Pseudo-potential for the Korteweg-de Vries-Burgers equation.

Laminar Shock Waves

Burgers equation led to a theory of shock waves as stationary solutions, caused by the balance of nonlinear and dissipative effects. Having solved the Korteweg-de Vries equation one may argue that another combined balance can be reached if both, dissipation and dispersion together, brake the nonlinear steepening of a wave. In this case one expects that the shock ramp will become modified by something like solitary wave-like structures. This is indeed the case when one looks for solutions of the Burgers-Korteweg-de Vries equation transformed to a system stationary in the shock frame

$$\beta\frac{\partial^3 v}{\partial y^3} - \alpha\frac{\partial^2 v}{\partial y^2} + \frac{\partial v}{\partial y}(v-u) = 0 \tag{10.23}$$

Assuming as for the Burgers equation that at $y \rightarrow \infty$, the boundary conditions are $v = \partial v/\partial y = \partial^2 v/\partial y^2 = 0$ and a first integration yields

$$\beta\frac{\partial^2 v}{\partial y^2} - \alpha\frac{\partial v}{\partial y} + \tfrac{1}{2}v^2 - uv = 0 \tag{10.24}$$

from which we derive the Sagdeev potential in the form

$$S(v) = \tfrac{1}{6}v^3 - \tfrac{1}{2}uv^2 \tag{10.25}$$

The first-order derivative in the above equation corresponds to a friction force with friction coefficient α, and time represented by $\tau = -y = -(x-ut)$. If the pseudo-particle was at the origin at $\tau = 0$, it will find itself at the ground of the pseudo-potential, $S(v)$, at $\tau = \infty$ or $y = -\infty$. The corresponding value of the velocity at the minimum of S is $v = 2u$. Figure 10.5 shows the form of the pseudo-potential and the oscillatory path of the pseudo-particle down from the origin to the bottom of the potential.

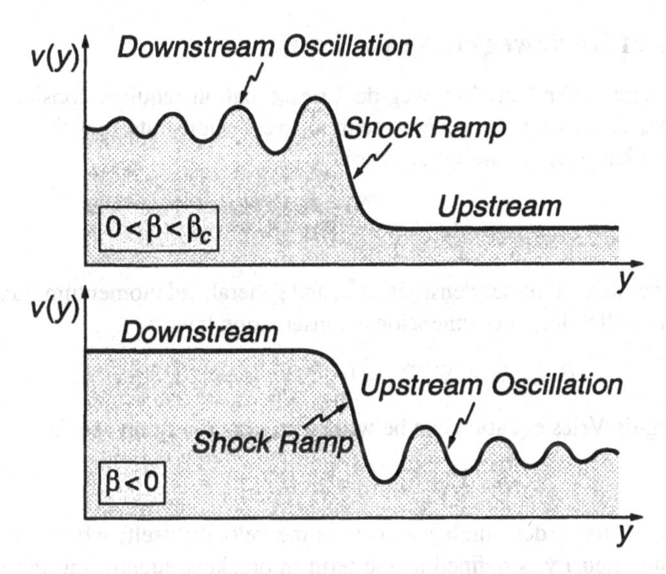

Fig. 10.6. Laminar Burgers-Korteweg-de Vries shock profiles.

This downwalk of the pseudo-particle to the bottom of the pseudo-potential is the irreversible dissipative process leading to the shock wave structure, which is provoked by the Burgers contribution to the Burgers-Korteweg-de Vries equation. That the pseudo-particle does not simply fall down to the ground, as in the pure Burgers case, but walks down in several steps indicates that the shock profile oscillates in space. The type of oscillation depends on the sign of the parameter β. In particular, for positive $\beta > 0$ the oscillatory part of the profile is behind the shock ramp in the downstream compressed region (Fig. 10.6, top panel). Above a certain critical value of β the pseudo-particle walks straight to the bottom of the potential, and the shock becomes the Burgers shock solution. For negative $\beta < 0$ the oscillatory part of the profile is found upstream of the shock ramp and is caused by disturbances which move against the incoming flow (Fig. 10.6, bottom panel).

The shock wave described by the solution of Eq. (10.24) has a jump in velocity of Δv, and its speed in the moving frame is given by $u = \Delta v/2$. Transforming to the system in which the medium is at rest, this speed becomes $\Delta v/2 + v_0$. Therefore, the Mach number of the shock flow is

$$M = 1 + \Delta v/2v_0 \tag{10.26}$$

and is independent of the dispersion parameter β. The oscillations behind or in front of the shock ramp are soliton-like and can be represented by the soliton solutions of the Korteweg-de Vries equation.

Invariants of Korteweg-de Vries' Equation

Solving the time-dependent Korteweg-de Vries equation requires considerably more effort. Before discussing the general method, we demonstrate that the Korteweg-de Vries equation has a large number of invariants

$$I_n = \int_{-\infty}^{\infty} \mathcal{N}_n(x,t)\,dx \tag{10.27}$$

where the generalized number densities, $\mathcal{N}_n$, and generalized momentum flux densities, $\mathcal{P}_n$, satisfy the following one-dimensional conservation laws

$$\partial\mathcal{N}_n/\partial t + \nabla_x \mathcal{P}_n = 0 \quad \text{for} \quad n = 1, 2, 3\ldots \tag{10.28}$$

The Korteweg-de Vries equation can be written as a conservation law for $n = 1$

$$\partial v/\partial t + \nabla_x \left(\tfrac{1}{2}v^2 + \beta\nabla_x^2 v\right) = 0 \tag{10.29}$$

The generalized first-order number density is the velocity itself, while the first-order momentum flux density is defined as the term in brackets, identifying the Korteweg-de Vries equation as the conservation law of the generalized momentum $\int v(x,t)\,dx$. The higher-order conservation laws can be obtained by successive multiplication of the Korteweg-de Vries equation by $v, v^2, \ldots$ The next conservation law is

$$\frac{1}{2}\frac{\partial v^2}{\partial t} = -\nabla_x \left[\frac{v^3}{3} + \beta\left(v\nabla_x^2 v - \frac{1}{2}\nabla_x^2 v\right)\right] \tag{10.30}$$

This equation corresponds to an energy conservation law. The other higher-order conservation laws have no direct physical interpretation. The point is that the Korteweg-de Vries equation functions similar to a kinetic equation. It is the basic equation of a hierarchy of conservation laws, which can be derived as moment equations in a way similar to the derivation of moment equations for the hydrodynamic variables from the Liouville equation.

It can be shown that the nth order 'density' invariants can be represented as polynomials in v, $\nabla_x v$ when expanding with respect to the small dispersion parameter, β

$$\mathcal{N}_n = v^n/n - \tfrac{1}{2}(n-1)(n-2)\beta v^{n-3}(\nabla_x v)^2 + O(\beta^2) \tag{10.31}$$

For the Korteweg-de Vries soliton (10.10) the solution for v is given, and the invariants can be calculated explicitly

$$I_{n,\mathrm{sol}} = \int_{-\infty}^{\infty} dx\,\mathcal{N}_n[v_{\mathrm{sol}}] = (12\beta)^{1/2}(3u)^{(n-1/2)}2^n[(n-1)!]^2/(2n-1)! \tag{10.32}$$

These invariants play a central role in the time-dependent solution of the Korteweg-de Vries equation.

10.3. Inverse-Scattering Method

In this section we present one very general method of solving time-dependent nonlinear equations, which applies to a whole class of equations. We use this method for solving the Korteweg-de Vries equation.

Inverse-Scattering Solution of Korteweg-de Vries Equation

It is quite simple to solve the linearized Korteweg-de Vries equation and to guess its stationary nonlinear condition, but it is much more involved to find analytical time-dependent solutions of the nonlinear equation. The idea of the method is as follows. In quantum mechanics the Schrödinger equation describes the energy states of a system exposed to an external potential. Knowing the full spectrum of these states it is possible to perfectly reconstruct this external potential by solving a particular GLM-integral equation given by Gelfand, Levitan and Marchenko. Assuming that this potential is the unknown solution of the Korteweg-de Vries equation and using the well-known methods of solving the Schrödinger equation and the GLM-equation, together with the knowledge of the initial condition and the form of the above invariants, the solution of the Korteweg-de Vries equation can be constructed. This way of solving nonlinear equations is called *inverse scattering method*, because the inverse problem of the Schrödinger equation arises in the investigation of light-scattering problems in astrophysics.

In order to sketch how this method works, consider Schrödinger's equation for the wave function, $\psi(x,t)$, in the particular form

$$\nabla_x^2\psi + \tfrac{1}{6}\left[\lambda(t) - v(x,t)\right]\psi = 0 \tag{10.33}$$

The factor 1/6 has been extracted for convenience. The potential $v(x,t)$ is chosen as the unknown function in the time-dependent KdV-equation. The Schrödinger equation is assumed to be stationary and depends on time only via the time dependence of v and the corresponding time dependence of the energy eigenvalues, λ. Thus the time dependence is only parametric and determined by the conditions on v

$$\frac{\partial v}{\partial t} + v\nabla_x v + \nabla_x^3 v = 0 \tag{10.34}$$

which is the Korteweg-de Vries equation with $\beta = 1$, and $v(x,0) = 0$ at $x \to \pm\infty$. Note that the solutions of the Korteweg-de Vries equation are reversible. The choice of β corresponds to the following rescaling: $v \to \beta^{1/5}v, t \to \beta^{1/5}t, x \to \beta^{2/5}x$. Inserting v from (10.33) into (10.34) gives

$$\nabla_x\left\{(\psi\nabla_x - \nabla_x\psi)\left[\frac{\partial\psi}{\partial t} - \frac{1}{2}(v+\lambda)\nabla_x\psi + \nabla_x^3\psi\right]\right\} = -6\psi^2\frac{\partial\lambda}{\partial t} \tag{10.35}$$

The left-hand side of this expression is a divergence which, when integrating over all space and taking into account that the solution of the Korteweg-de Vries equation and thus also that of the Schrödinger equation is bounded, must vanish. Hence, the eigenvalues of the discrete spectrum of the Schrödinger equation must be time-independent, whenever v is a solution of the Korteweg-de Vries equation, $\partial\lambda/\partial t = 0$. Consequently, one finds by integrating the last equation twice that the wave function satisfies the condition

$$\frac{\partial\psi}{\partial t} - \frac{1}{2}(v+\lambda)\nabla_x\psi + \nabla_x^3\psi = \psi\left(A + B\int dx\,\psi^2\right) \tag{10.36}$$

The coefficients A, B vanish for the discrete spectrum, since in the limit $x \to \pm\infty$ the left-hand side must vanish and because of the normalization condition of the wave function. In the case of a continuous spectrum one must, however, take into account that a wave can fall in from infinity. In this case $A, B \neq 0$.

In the discrete part of the spectrum ψ is represented as the expansion with respect to eigenfunctions, ψ_n, of the Schrödinger equation in the Hilbert space representation. These eigenfunctions must be regular for $x \to \pm\infty$, such that

$$\psi_n(x,t) \approx c_{n\pm}(t)\exp(\mp k_n x) \quad \text{for} \quad x \to \pm\infty \tag{10.37}$$

where the spatial decay decrement is $k_n^2 = -\lambda_n/6$. In order to find the coefficients, one inserts this asymptotic expression into the above condition on ψ and finds

$$c_{n\pm}(t) = c_{n\pm}(0)\exp(\pm 4k_n^3 t) \tag{10.38}$$

Similarly, the regular eigenfunctions of the continuous spectrum may be chosen as

$$\psi_{cont} \approx c_\pm(k,t)\exp(\pm ikx) + \delta_\mp \exp(-ikx) \quad \text{for} \quad x \to \pm\infty \tag{10.39}$$

Here $\delta_+ = 1$ and $\delta_- = 0$. Again inserting into Eq. (10.36) allows to determine the coefficients of the continuous spectrum as $B = 0$, $A = 4ik^3$, and one finds

$$\begin{aligned} c_+(k,t) &= c_+(k,0)\exp(4ik^3 t) \\ c_-(k,t) &= c_-(k,0) \end{aligned} \tag{10.40}$$

Clearly, c_+ is the reflection coefficient, while c_- is the transmission coefficient of the wave incident from $x = +\infty$, and we must have in addition

$$|c_-|^2 + |c_+|^2 = 1 \tag{10.41}$$

The former expressions show that the time dependence of the coefficients c_n and c is entirely determined by the spectrum. One has to find only the initial coefficients at $t = 0$. This can be done by solving the time independent Schrödinger equation using

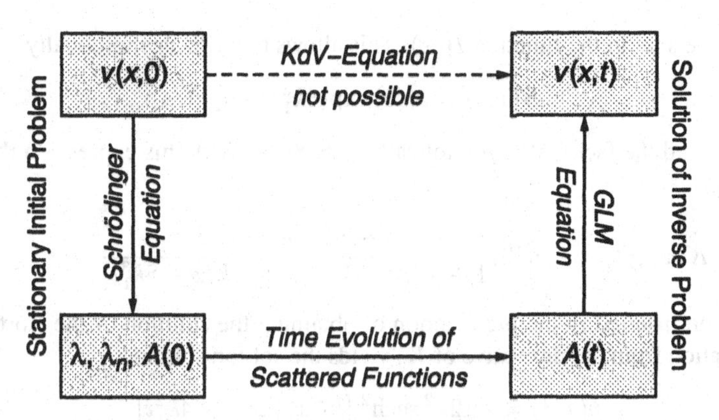

Fig. 10.7. Inverse-scattering solution of the Korteweg-de Vries equation.

a potential $v(x,0)$, which corresponds to the known initial condition of the solution of the Korteweg-de Vries equation. But the inverse scattering theory for the Schrödinger equation provides a direct solution for the potential v as solution of the Korteweg-de Vries equation when solving the GLM-equation

$$K(x,y) = -B(x+y) - \int_x^\infty K(x,z)B(y+z)\,dz \tag{10.42}$$

This solution is simply represented as the differential of the solution $K(x,x)$ of this integral equation taken at $z = x$

$$v(x,t) = -12[dK(x,x;t)/dx] \tag{10.43}$$

The kernel of the GLM-equation is hereby given as a function which is entirely expressed through the initial conditions and the spectrum of the Schrödinger equation

$$B(x) = \frac{1}{2\pi}\int c_+(k,t)\exp(ikx)\,dk + \sum_n c_{n+}^2(t)\exp(-k_n x) \tag{10.44}$$

and one has $k = (\lambda/6)^{1/2}$ and $k_n = (-\lambda_n/6)^{1/2}$.

The inverse-scattering procedure yields the solution of the time-dependent Korteweg-de Vries equation, at least by solving the GLM-equation numerically for given initial condition $v(x,0)$. Figure 10.7 shows schematically how the inverse scattering method works starting from an initial disturbance, bypassing the impossible direct way via solution of the stationary Schrödinger equation, boosting the scattered amplitudes in time and solving the GLM-equation.

Let us apply this method to soliton solutions. In this case the continuous spectrum contribution can be neglected. Otherwise we restrict ourselves to only the leading term,

$n = 1$, in the sum in the function $B(x)$. This allows to write asymptotically

$$B(x,t) = c_1^2(0)\exp(8k_1^3 t - k_1 x) \tag{10.45}$$

where we used the fact that asymptotically $x \approx 8k_1^2 t$. With this expression the GLM-equation gives

$$K(x,y;t) = -c_1^2(0)\frac{\exp\left[-k_1(x+y) + 8k_1^3 t\right]}{1 + [c_1^2(0)/2k_1]\exp\left[-2k_1 x + 8k_1^3 t\right]} \tag{10.46}$$

which, when inserted in the prescription of obtaining the solution of the Korteweg-de Vries equation from the derivative of K, yields the soliton solution

$$v(x,t) = -12k_1^2\,\mathrm{sech}^2\left[k_1(x - x_0) - 4k_1^2 t\right] \tag{10.47}$$

with $c_1^2(0) = 2k_1\exp(2k_1 x_0)$. The amplitude of the soliton is now $12k_1^2$, and its velocity one third of this, which coincides with the solution found earlier.

Self-Similar Solution

Let us return to the Korteweg-de Vries equation (10.34) and write the initial condition $v(x,0) = u\phi(x/\ell)$ such that the function ϕ describes the self-similar shape of the solution, $\xi = x/\ell$ is a dimensionless coordinate, and ℓ is the characteristic width of the localized solution which for the soliton solution v_{sol} is $\ell = (4\beta/u)^{1/2}$. It is then possible to rewrite the Schrödinger equation

$$\psi''(\xi) + \tfrac{1}{6}\sigma^2[W_n + \phi(\xi)]\psi(\xi) = 0 \tag{10.48}$$

where the double prime denotes the second spatial derivative, and the quantity $\sigma^2 = u\ell^2/\beta$ serves as a parameter. The amplitudes, c_n, of the soliton solutions of the Korteweg-de Vries equation are related to the energy eigenvalues of the Schrödinger equation

$$c_n = -2uW_n \tag{10.49}$$

The structure of the Schrödinger equation then shows that for $\phi(\xi) < 0$ no discrete spectrum and therefore no solitons exist. Initial disturbances of this kind do not evolve into localized solutions. The discrete spectrum requires that

$$\Phi = \int \phi(\xi)\,d\xi > 0 \tag{10.50}$$

and the number of solitons which evolve depends on the value of the similarity parameter, σ, such that for small σ only one eigenvalue and thus one soliton exists. This value is found by a lengthy perturbation calculation

$$W_1 = -(\sigma^2/24)\Phi^2 \quad \text{for} \quad \sigma^2 \ll 12 = \sigma_{\mathrm{sol}}^2 \tag{10.51}$$

and the amplitude of this solitons can be represented through

$$c_1 = 2u\Phi \tag{10.52}$$

The quantity σ_{sol} is the similarity parameter for the stationary soliton solution. This implies that under non-stationary conditions an initial disturbance will decay into many solitons of small amplitude. From this consideration it becomes clear that for large values of σ the number of solitons which evolve from a given initial disturbance becomes large by itself. The nature of the Korteweg-de Vries equation is thus to split a given large initial disturbance into a large number of small-amplitude nonlinear structures which, after formation, behave as independent non-interacting entities, while giving the plasma a grainy texture which we call strong turbulence. The large-amplitude single soliton found under stationary conditions is thus an idealized solution, which probably does not exist when evolving from an initial condition.

Soliton Distribution

In the case of large initial disturbances it is possible to calculate the number of solitons which evolve out of the disturbance. This number can be estimated using the infinite set of invariants of the Korteweg-de Vries equation. Because the amplitudes of the evolving solitons are determined by these invariants and are not equally large, one can find a distribution function of soliton amplitudes or, because the amplitudes are unambiguously related to the soliton speeds, a distribution function of the solitons with respect to their velocities. The number dN of solitons with amplitudes in the interval $[u, u + du]$ can be defined as

$$dN = F(u)\, du \tag{10.53}$$

On the other hand, the density of energy levels of the Schrödinger equation in the discrete spectrum is given as

$$\frac{dN}{dW} = \frac{\sigma}{\sqrt{24}\pi} \int \frac{d\xi}{[\phi(\xi) - W]^{1/2}} \tag{10.54}$$

in the region of positive initial disturbance, $\phi > 0$. These two expressions enable one to find the distribution function $F(u)$ of the soliton amplitudes

$$\boxed{F(u) = \frac{\sigma}{4\pi(3u_0)^{1/2}} \int d\xi \, [2u_0\phi(\xi) - 3u]^{-1/2}} \tag{10.55}$$

The integration is to be taken over the positive part of the argument of the square root, $2u_0\phi > 3u$. The lengthy derivation is not given here. Integrating from zero to infinity one can find the number of solitons

$$N = \int F(u) du = \frac{\sigma}{\sqrt{6}\pi} \int d\xi \, \phi^{1/2}(\xi) \tag{10.56}$$

These equations can serve to estimate the graininess of a Korteweg-de Vries turbulent plasma. They also play a role in more general theories of strong plasma turbulence caused by drift waves and electron acoustic waves.

10.4. Acoustic Solitons

To obtain nonlinear equations governing the evolution of plasma waves, it is necessary to reduce the full set of complicated nonlinear equations of plasma physics which cannot be solved by analytical methods to one of the nonlinear equations derived in the previous sections. This can in general not be done, but there are a number of cases when it is possible to apply some reduction methods which lead to the Korteweg-de Vries or Burgers equations. In this section we show how this reduction method works for a ion-acoustic, electron-acoustic, and kinetic Alfvén waves.

Reductive Perturbation Method

Both the Burgers and Korteweg-de Vries equations are general equations which do not apply exclusively to plasmas. As a first example of solitons in a plasma we consider nonlinear ion-acoustic waves in the two-fluid approximation. This means that we are not interested in the instability causing these waves, but assume that an ion-acoustic wave exists in the plasma and that this wave undergoes nonlinear evolution. What then is the dynamics of this evolution and which of the nonlinear equations governs it? Hence, we are interested in reducing the full nonlinear set of two-fluid equations to one master equation, which describes nonlinear ion-acoustic waves. The general reduction scheme of finding such a reduced equation is the reductive method of stretched coordinates.

Consider the invariance of the Korteweg-de Vries equation with respect to the following coordinate transformations

$$t \rightarrow \epsilon t \qquad x \rightarrow \epsilon^{1/3} x \qquad v \rightarrow \epsilon^{-2/3} v \tag{10.57}$$

All its solutions have the same shape if transformed accordingly and are called *self-similar*. They can be written

$$v(x,t) = \beta(3\beta t)^{-2/3} \psi[x/(3\beta t)^{1/3}] \tag{10.58}$$

A transformation of this kind is called a *self-similarity transformation*. A self-similarity transformation can also be found for the Burgers equation. It can be applied to reduce a system of partial differential equations of the kind for the column vector, U

$$\frac{\partial \mathrm{U}}{\partial t} + \mathsf{A} \cdot \nabla_x \mathrm{U} + \sum_{\mu=1}^{q} \prod_{\nu=1}^{p} \left(H_\mu^\nu \frac{\partial \mathrm{U}}{\partial t} + K_\mu^\nu \nabla_x \mathrm{U} \right) = 0 \tag{10.59}$$

with H^{ν}_{μ}, K^{ν}_{μ} matrices which themselves depend on $\mathbf{U}$, to a simpler nonlinear equation. One introduces a small parameter $\epsilon \ll 1$ and expands

$$\mathbf{U} = \mathbf{U}_0 + \epsilon \mathbf{U}_1 + \dots \tag{10.60}$$

while at the same time introducing the stretched coordinates

$$\xi = \epsilon^a (x - ut) \qquad \tau = \epsilon^{a+1} t \tag{10.61}$$

where $a = (p-1)^{-1}$. If we now require that $\mathbf{U}_1 \to 0$ for $x \to \infty$, write $\mathbf{U}_1 = \mathbf{R}\phi$, and chose the equation which determines the column eigenvector, $\mathbf{R}$, from the matrix equation

$$(\mathbf{A} - u\mathbf{I}) \cdot \mathbf{R} = 0 \tag{10.62}$$

we can reduce the above very complicated system asymptotically to the simpler self-similar equation for ϕ

$$\frac{\partial \phi}{\partial \tau} + \alpha \phi \nabla_{\xi} \phi + \beta \nabla^{p}_{\xi} \phi = 0 \tag{10.63}$$

This equation is a nonlinear dispersive equation with a dispersion of order p. Its coefficients can be expressed through the velocity u and $\mathbf{R}$. β is the solution of the dispersion relation of the linear system

$$\omega = uk + i^{p-1} \beta k^p + \dots \tag{10.64}$$

and $\alpha = \nabla_u \cdot (u_0 \mathbf{R}_0)$. Here we used that the wavenumber $k = \mathrm{O}(\epsilon^p)$ is of order p.

Ion-Acoustic Solitons

The first example of solitons relevant for space plasmas are acoustic solitons. We consider a one-dimensional system in which ion-acoustic waves have been excited by, for example, a field-aligned current instability. In the one-dimensional field-aligned case the physics is independent of the magnetic field. Normalizing all quantities, $n_j \to n_j/n_0$, $v_j \to v_j/c_{ia}$, $E \to eE/(m_e k_B T_e)^{1/2} \omega_{pe}$ (or $\phi \to e\phi/k_B T_e$), $x \to x/\lambda_D$, and $t \to \omega_{pi} t$, and assuming Boltzmann-distributed electrons, the equations describing the evolution of the plasma are

$$\begin{aligned} \frac{\partial n_i}{\partial t} + \frac{\partial (n_i v_i)}{\partial x} &= 0 \\ \frac{\partial v_i}{\partial t} + v_i \frac{\partial v_i}{\partial x} &= -\frac{\partial \phi}{\partial x} \\ \frac{\partial^2 \phi}{\partial x^2} &= -n_i + \exp(\phi) \end{aligned} \tag{10.65}$$

When linearizing these equations one obtains the dispersion relation

$$\omega^2 = k^2/(1+k^2) \tag{10.66}$$

which in the small wavenumber limit, $k^2 \ll 1$, can be expanded and reduced to

$$\omega = k(1-k^2/2) \tag{10.67}$$

One immediately recognizes that the dispersion coefficient will have the value $|\beta| = 1/2$. We now assume initial quasineutrality, $n_i = n_e$, and rewrite the equation of motion as

$$\frac{\partial v_i}{\partial t} + v_i \frac{\partial v_i}{\partial x} = -\frac{1}{n_i}\frac{\partial n_i}{\partial x} \tag{10.68}$$

This equation is identical with the momentum conservation equation of a gas with sound velocity $c_{ia} = 1$, in which the electrons exert a pressure on the plasma through the electric field, $E = -\nabla_x \phi$. Therefore the nonlinear wave is an ion-acoustic wave. But when the wave steepens, the left-hand side of the Poisson equation, $\nabla_x^2 \phi$, grows due to inertial effects of the steepening gradients, and quasineutrality becomes obsolete. This causes the dispersive effects in the dispersion relation, which ultimately lead to stabilization of the nonlinear steepening. One therefore expects that ion-acoustic waves may lead to the formation of topological solitons.

Let us look for a solution of the above nonlinear equations, which depends on x, t in the form $\xi = x - ut$, and let us further assume homogeneous boundary conditions for $|\xi| \to \infty$, i.e., $n_i \to 1, v_i = \phi = \nabla_\xi \phi \to 0$. Then we get

$$\tfrac{1}{2}(\nabla_\xi \phi)^2 = \exp(\phi) + u(u^2 - 2\phi)^{1/2} - (u^2+1) \tag{10.69}$$

Setting the left-hand side of this expression to zero and assuming that the potential has an extremum, $\phi_{\max}$, at the place where its derivative vanishes, we obtain for the speed of the structure

$$u^2 = \frac{[\exp\phi_{\max} - 1]^2}{2[\exp\phi_{\max} - 1 - \phi_{\max}]} \tag{10.70}$$

Further assuming that $\phi_{\max} \ll 1$, and $u - 1 = \delta u \ll 1$, one gets

$$(\nabla_\xi \phi)^2 = \tfrac{2}{3}\phi^2(3\delta u - \phi) = -S(\phi) \tag{10.71}$$

where the right-hand side is the pseudo-potential of the ion-acoustic wave. We immediately identify this equation as the first integral of the stationary Korteweg-de Vries equation. The solution is the Korteweg-de Vries soliton

$$\phi = 3\delta u\, \mathrm{sech}^2 \left[\left(\frac{\delta u}{2} \right)^{1/2} (x - ut) \right] \tag{10.72}$$

which is valid for small-amplitude slow ion-acoustic solitons. The maximum amplitude of the soliton is given by the condition that the pseudo-potential must be negative for solitons to occur. One finds that $\phi_{\max} = u^2/2$. Inserting for the speed into Eq. (10.70) yields $\phi_{\max} \approx 1.3$ and therefore as the maximum speed of the soliton $u_{\max} \approx 1.6$. Since this speed is normalized to the ion-acoustic velocity, we find that ion-acoustic solitons exist for Mach numbers

$$M_{ia,\mathrm{sol}} < 1.6 \tag{10.73}$$

The stationary solution obtained is, of course, not the full story because the system must evolve toward the state where the soliton has formed. We now apply the reductive perturbation method to the initial system of equations describing the nonlinear evolution of ion-acoustic waves in order to find the basic nonlinear equation. Eliminating the ion density and electric field from the equations, the system is reduced to the following two equations

$$\frac{\partial n_e}{\partial t} + \nabla_x(n_e v_i) - \nabla_x\left[\left(\frac{\partial}{\partial t} + v_i\nabla_x\right)\nabla_x \ln n_e\right] = 0 \tag{10.74}$$

$$\frac{\partial v_i}{\partial t} + v_i\nabla_x v_i + \nabla_x \ln n_e = 0 \tag{10.75}$$

From this equation we read that $p = 3, q = 1, a = 1/2$, and $\beta = 1/2$. It is then possible to find expressions for the vectors and matrices. To zero order in ϵ the two components of the column vector $\mathbf{U}_0$ are $(1, 0)$. The eigenvalues of the matrix $\mathbf{A}_0$ are then $u_0 = \pm 1$. The column vector $\mathbf{R}$ has the components $(1, 1)$, and $\nabla_u \cdot (u_0\mathbf{R}_0) = (0, 1)$, which yields $\alpha = 1$. The stretched coordinates are

$$\begin{aligned} \xi &= \epsilon^{1/2}(x - u_0 t) \\ \tau &= \epsilon^{3/2} t \end{aligned} \tag{10.76}$$

and the nonlinear equation obtained for the first-order in ϵ is a version of the Korteweg-de Vries equation written in terms of the electron density

$$\frac{\partial n_{e1}}{\partial \tau} + n_{e1}\frac{\partial n_{e1}}{\partial \xi} + \frac{1}{2}\frac{\partial^3 n_{e1}}{\partial \xi^3} = 0 \tag{10.77}$$

But because of the Boltzmann relation, $n_{e1} = \exp\phi_1 - 1 \approx \phi_1$, the same equation holds for the electrostatic potential of the nonlinear wave. Thus its stationary solution is the soliton found above.

The Korteweg-de Vries equation (10.77), derived for the ion-acoustic wave evolution, implies that ion-acoustic waves will under most conditions by their inner dynamics, which determines their dispersive properties, evolve into a chain of solitons of

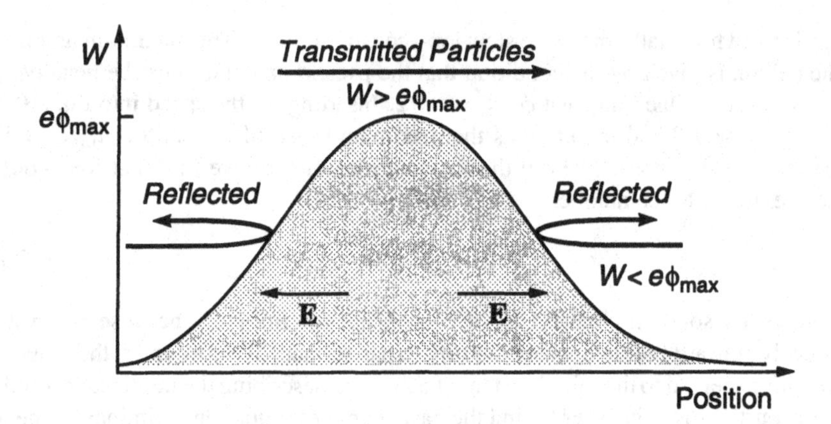

Fig. 10.8. Symmetric reflection and transmission of particles in soliton.

small amplitudes and different speeds. The number of solitons and the turbulent structure of the plasma are determined by the initial condition. Small-amplitude waves will decay into one or a few solitons, but when the mechanism of instability continuously produces growing waves, the plasma will end up with a large number of ion-acoustic solitons propagating across it. It will become striated perpendicular to the magnetic field, and these density and reversible potential striations will propagate at velocities well below the ion-acoustic speed. They will overtake each other, but will not interact significantly.

Conditions of this kind are believed to exist in places where the ion-acoustic wave instability is driven by field-aligned currents oder heat-fluxes. In the solar wind acoustic solitons may be responsible for the medium level of density fluctuations observed and may play a role in the generation of the radiation from foreshock electrons. In the magnetosphere they are involved in the formation of striations in the auroral and equatorial upper ionospheres.

Microscopic Double Layers

Another important observation is that the solution of the Korteweg-de Vries equation for the potential does not necessarily require the potentials on both sides to be the same. This gives the possibility for the description of microscopic double layers in terms of asymmetric solitons. The two particular families of solitons for which double layers seem to be possible are the ion-acoustic soliton and a pure electron fluid soliton, the *electron-acoustic soliton* arising from the electron-acoustic instability. In the latter the cold electrons play the role of the ions.

In a fluid picture description of the double layer the formation of the potential

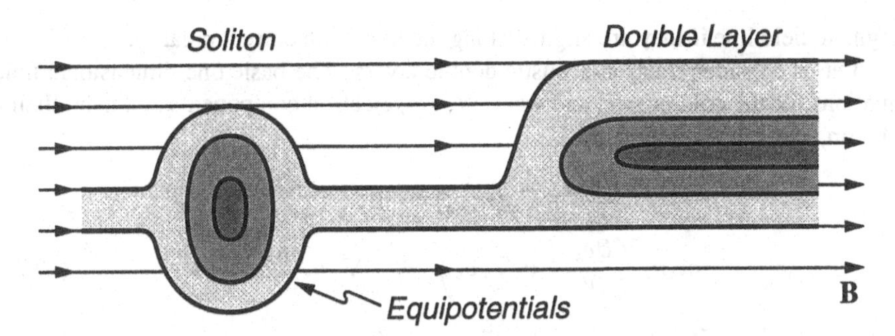

Fig. 10.9. Equipotentials in the combination of a soliton and a double layer.

difference corresponds to the possibility that the low energy component may become asymmetrically reflected from the potential well of the soliton with the asymmetry arising from asymmetries in the bulk velocities of the soliton or the particle fluids. The condition for this reflection is that the thermal energy of the reflected particles is less than the potential energy of the soliton. But in the magnetosphere, where energies are not below 1 eV, this requires large potential drops, and fluid-like double layers will barely evolve. Nevertheless, if thermal energies are low, the exclusion of the ions or the cold electron component in the case of the electron-acoustic solitons yields violation of the charge neutrality condition across the soliton. Clearly this happens only if the soliton moves because it is the motion which introduces the asymmetry. Thus fast solitons will easier be subject to double layer formation.

Figure 10.8 shows the reflection mechanism of particles of the right sign by the symmetric potential hump inside the soliton. Only particles with energies $< e\phi_{\max}$ are reflected, while higher energy particles are transmitted through the potential hump. Figure 10.9 shows the qualitative shape of equipotential contours in presence of a soliton and a microscopic double layer. The plasma is assumed magnetized with the equipotentials at infinity parallel to the field lines. The deviation of the equipotentials from the field inside the soliton and double layer implies convective motions of cold particles with gyroradii much less than the transverse extents of these structures while the fast particles are not affected.

The conditions change when kinetic effects are included. Then the low energy particles in the distribution function may become reflected from the soliton potential and asymmetry will arise in a more natural way. Moreover, the particles moving nearly in resonance with the soliton have very low energy and may experience one-sided reflection easiest. In summary, microscopic solitons of the kind of ion- and electron-acoustic solitons can in principle evolve into weak microscopic double layers containing potential drops, which in the magnetosphere should amount to a fraction of an electron volt. But such microscopic double layers may add up to large potential drops along a magnetic field line if they are aligned along the field with correct polarity.

Let us consider electron-acoustic double layers. The basic one-dimensional fluid equations for the cold, $s = c$, and warm, $s = h$, electron components and neutralizing cold ion, $s = i$, background are

$$\begin{aligned} \frac{\partial n_s}{\partial t} + \nabla_x(n_s v_s) &= 0 \\ m_s n_s \left(\frac{\partial v_s}{\partial t} + v_s \nabla_x v_s \right) &= -\nabla_x p_s - n_s \nabla_x \phi \\ \frac{\partial p_s}{\partial t} + v_s \nabla_x p_s + p_s \nabla_x v_s &= 0 \end{aligned} \tag{10.78}$$

which must be completed by the Poisson equation

$$\nabla_x^2 \phi = n_{0h} \exp \phi - \sum_{s=i,c} n_s \tag{10.79}$$

The quantities have been normalized for simplicity here. For instance, the normalization of the potential is $e\phi/k_B T_h \rightarrow \phi$ and so on. In addition one implies the following boundary conditions

$$\left. \begin{array}{lll} \phi \rightarrow 0 & \nabla_x \phi \rightarrow 0 & \nabla_x^2 \phi \rightarrow 0 \\ n_s \rightarrow n_{0s} & p_s \rightarrow p_{0s} & v_s \rightarrow 0 \end{array} \right\} \text{for} \quad x \rightarrow \infty \tag{10.80}$$

In order to derive the nonlinear equation describing the double layer formation we introduce the following stretching transformation

$$\begin{aligned} \xi &= \epsilon(x - ut) \\ \tau &= \epsilon^3 t \end{aligned} \tag{10.81}$$

which differs from the one used before. We now expand all fluid quantities and the potential in powers of ϵ and follow the procedure described in the previous section. We then obtain the following nonlinear equation for the first order expansion term of the potential

$$\frac{\partial \phi_1}{\partial \tau} + \frac{a_1}{2} \nabla_\xi \phi_1^2 + a_2 \nabla_\xi \phi_1^3 + a_3 \nabla_\xi^3 \phi_1 = 0 \tag{10.82}$$

This is the *modified Korteweg-de Vries equation.* It contains a third order nonlinearity and is valid as long as $|a_1/a_3| = O(\epsilon)$, and the coefficients are

$$\begin{aligned} a_1 &= -\frac{(n_{0c}/n_{0h})^2 + 3(n_{0c}/n_{0h} + 4T_c)}{2(n_{0c}/n_{0h})(n_{0c}/n_{0h} + 3T_c)^{1/2}} \\ a_2 &= -\frac{(n_{0c}/n_{0h})^4 - 15(n_{0c}/n_{0h})^2 - 180T_c(n_{0c}/n_{0h}) - 432T_c^2}{12(n_{0c}/n_{0h})^3(n_{0c}/n_{0h} + 3T_c)^{1/2}} \\ a_3 &= (n_{0c}/n_{0h})^2/2n_{0c}(n_{0c}/n_{0h} + 3T_c)^{1/2} \end{aligned} \tag{10.83}$$

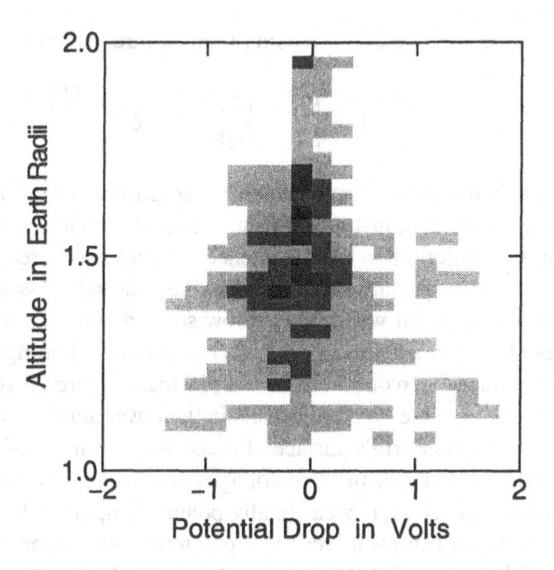

Fig. 10.10. Measured auroral double layer potentials.

The modified Korteweg-de Vries equation is in the form that another pseudo-potential can be defined. Multiplying by ϕ_1 and integrating once one arrives at the simplified form, using the new coordinate $\zeta = \xi - U\tau$ which is a stationary coordinate traveling with the double layer across the plasma

$$\frac{1}{2}\left(\frac{d\phi_1}{d\zeta}\right)^2 = \frac{U}{2a_3}\phi_1^2 - \frac{a_1}{6a_3}\phi_1^3 - \frac{q_2}{4a_3}\phi_1^4 = -S(\phi_1) \tag{10.84}$$

where double layer solutions require that $S(0) = S(\phi_{\max}) = 0$ and the derivatives of S satisfy the condition

$$\nabla_{\phi_1} S|_{\phi_{\max}} = \nabla_{\phi_1} S|_0 = 0 \tag{10.85}$$

in addition to the condition that the second derivatives of S at $\phi_{\max}$ and $\phi_1 = 0$ should be negative. The potential $\phi_{\max}$ is just the maximum achievable double layer potential.

One can satisfy the condition if $\phi_{\max} = -a_1/3a_2$ and the double layer speed is $U = -a_1^2/18a_2$. Then the modified Korteweg-de Vries equation has the stationary solution

$$\phi = -\frac{1}{2}\chi\left\{1 - \tanh\left[\left(-\frac{\chi 62a_2}{8a_3}\right)^{1/2}\left[x - (u - \tfrac{1}{2}\chi^2 a_2)t\right]\right]\right\} \tag{10.86}$$

where $\chi = \epsilon a_1/3a_2$ and the solution has been written in the original coordinates. In order that a solution exists the arguments of the roots must be positive definite. This leads to a condition on the normalized cold plasma temperature as function of the density ratio

$$T_c = -\frac{45\delta}{216}\left\{1 \pm \left[1 - \frac{108}{2025}(15 - \delta^2)\right]^{1/2}\right\} \tag{10.87}$$

where $\delta = n_{0c}/n_{0h}$ is the density ratio which for solutions to exist needs to be $\delta > \sqrt{15}$. Hence, as the theory demonstrates, electron-acoustic double layers are theoretically possible, but the restriction on their existence is rather severe requiring large cold electron densities in which case any instability may become suppressed.

This is an interesting result which may throw some doubt onto the real existence of stationary double layers in agreement with the experimental finding that only bursty and very low voltages have been observed in real plasmas. Figure 10.10 gives a feeling for the realistic values of double layer potentials in the lower auroral magnetosphere as function of height above the Earth's surface. Precise measurement of parallel electric potentials is very difficult to perform. The voltages are differences between measured and background potentials. Though occasionally potentials up to –2 V are detected, the distribution of the voltages centers at very low potential values near zero. These data indicate that most of the real double layers are very weak and microscopic double layers which may have arisen from particle reflection at solitary structures.

10.5. Alfvén Solitons

Solitons exist also for low-frequency Alfvén waves, both for the kinetic and shear kinetic modes. We will not go through the whole process of reductive perturbation, but look for solutions which are stationary in the frame of the traveling structure.

Kinetic Alfvén Solitons

As we know, there are two kinetic Alfvén modes, one for plasma beta $\beta > m_e/m_i$, the other under $\beta < m_e/m_i$ conditions. Remembering the two different dispersion relations

$$\omega = \begin{cases} k_\parallel v_A(1 + k_\perp^2 \tilde{r}_{gi}^2)^{1/2} & \text{for} \quad \beta > m_e/m_i \\ k_\parallel v_A(1 + k_\perp^2 c^2/\omega_{pe}^2)^{-1/2} & \text{for} \quad \beta < m_e/m_i \end{cases} \tag{10.88}$$

we first find that kinetic Alfvén solitons, if they exist, are intrinsically two-dimensional. The second observation is that solitons resulting from the first of these dispersion relation, which describes the proper kinetic mode, have dispersive properties different from those of the ion-acoustic and electron-acoustic modes discussed so far. The shear kinetic mode described by the second of the dispersion relations, on the other hand, has dispersive properties similar to the ion-acoustic wave. These differences will necessarily affect the properties of the solitons resulting from the underlying dynamics.

Let us consider the kinetic mode first, because it promises to confront us with some new effects. Referring to the fluid picture, we use the representation with two electric potentials (cf. Sec. 10.6 of our companion book, *Basic Space Plasma Physics*), $E_\perp = -\nabla_\perp\phi_\perp, E_\parallel = -\nabla_\parallel\phi_\parallel$. Assuming quasineutrality, $n_i = n_e$, and Boltzmann-distributed electrons

$$n_e = n_0 \exp(e\phi_\parallel/k_BT_e) \tag{10.89}$$

the basic equations describing the evolution of nonlinear kinetic Alfvén waves are

$$\begin{aligned}\frac{\partial B_\perp}{\partial t} &= \nabla_\perp\nabla_\parallel(\phi_\perp - \phi_\parallel)\\ \nabla_\perp^2\nabla_\parallel^2(\phi_\perp - \phi_\parallel) &= \mu_0\nabla_\parallel\frac{\partial j_\parallel}{\partial t}\\ \frac{\partial n_i}{\partial t} &= \frac{1}{\omega_{gi}B_0}\nabla_\perp\left(n_i\nabla_\perp\frac{\partial\phi_\perp}{\partial t}\right)\end{aligned} \tag{10.90}$$

The field-aligned current is entirely carried by the hot electron component. Hence

$$\nabla_\parallel j_\parallel = e(\partial n_e/\partial t) \tag{10.91}$$

In dimensionless variables, with $\xi \to x\omega_{gi}/c_{ia}$, $\zeta \to z\omega_{pi}/c$, $\tau \to \omega_{gi}t$, $n \to n/n_0$, $e\phi_\perp/k_BT_e \to \tilde{\phi}_\perp$, $e\phi_\parallel/k_BT_e \to \tilde{\phi}_\parallel$, this system of equations can be made stationary when transforming to the following co-moving coordinate

$$\eta = \kappa_\perp\xi + \kappa_\parallel\zeta - \tau \tag{10.92}$$

This coordinate is introduced in order to transform to a frame which moves together with the localized kinetic Alfvénic structure. However, in the present case it is important to note that this motion is neither parallel nor perpendicular to the magnetic field, but is a two-dimensional motion in a direction oblique to the external field. It turns out that the system of fundamental equations allows for such a transformation, reducing the physics to a one-dimensional problem.

Being interested in localized solutions, the boundary conditions at infinity are chosen such that the derivatives of the density parallel and perpendicular to the magnetic field vanish at infinity, $\nabla_\perp n = 0$ at $\xi \to \pm\infty$, and $\nabla_\parallel n = 0$ at $\zeta \to \pm\infty$. Hence, expressed in terms of the potentials, this implies that the potentials and their first, second, and third derivatives vanish at infinity. Under these conditions it is easy to see that the system of equations becomes stationary and does not explicitly depend on time. One may combine it and integrate it one to find the following equation

$$\kappa_\perp^2\kappa_\parallel^2 n\frac{d^2 \ln n}{d\eta^2} = (1-n)(n-\kappa_\parallel^2) \tag{10.93}$$

It is clear that when linearizing this equation and tries to find the linear dispersion relation, one arrives at the normalized dispersion relation for the kinetic Alfvén wave

$$\kappa_\parallel^2(1+\kappa^2\kappa_\perp^2)=1 \tag{10.94}$$

where $\kappa^2=\kappa_\parallel^2+\kappa_\perp^2$ is the oblique normalized wave vector. If we express the normalized quantities through their dimensional equivalents, we recover the dispersion relation

$$\omega^2=k_\parallel^2 v_A^2(1+k_\perp^2\tilde{r}_{gi}^2) \tag{10.95}$$

with $\tilde{r}_{gi}$ the modified ion gyroradius, containing the ratio of electron-to-ion temperature, as shown in Eq. (I.10.179) of the companion volume.

From Eq. (10.93) it is possible to obtain a first integral by integration with respect to η. This integral provides us with a pseudo-potential

$$\left(\frac{dn}{d\eta}\right)^2=-S(n,\kappa_\parallel,\kappa_\perp) \tag{10.96}$$

given by the following expression

$$S(n,\kappa_\parallel,\kappa_\perp)=-\frac{2n}{\kappa_\perp^2\kappa_\parallel^2}\left[(1-n)(n+\kappa_\parallel^2)+(1+\kappa_\parallel^2)n\ \ln\ n\right] \tag{10.97}$$

Since S must be negative for real solutions to exist, the expression in the brackets on the right-hand side of this equation is positive.

The pseudo-potential determines the regions of localized solutions in the normalized density plane. Because these solutions are both stationary and localized, they correspond to solitons. Note that the condition for existence of kinetic Alfvén solitons does not depend on the perpendicular wavenumber. This suggests that it is the parallel electric field in the wave, which is responsible for the formation of topological solitons. Dispersion of the wave parallel to the magnetic field causes the balance of the nonlinearity in the basic equations.

Figure 10.11 shows where the regions of existence of the solitons are located. The existence of solitons is heavily modulated by the parallel wavenumber dependence. As one finds, solitons may exist both for normalized densities $n\geq 1$ and $n\leq 1$ with $n=1$ included. In the region $n>1$ the solitons have a well-defined maximum amplitude given by the point $n=n_{\max}$, where the function representing the bracket in the pseudo-potential crosses the real axis to negative values. The corresponding wavenumbers in this region are all $\kappa_\parallel^2>1$, which means that these solitons propagate at a speed less than the Alfvén velocity, v_A, parallel to the magnetic field and constitute density compressions. Hence, such solitons are sub-Alfvénic.

There exists also a region of density depressions $n<1$, which propagate at sub-Alfvénic speeds as well. The minimum density of these solitons would be at $n=0$

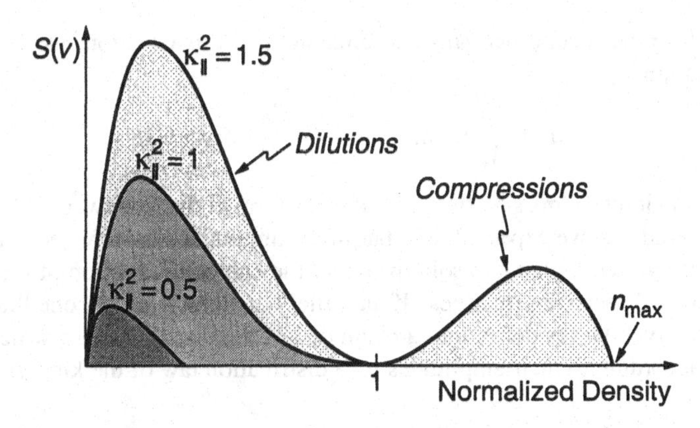

Fig. 10.11. Pseudo-potential and regions of solitons for the kinetic Alfvén wave.

and corresponds to total evacuation of the plasma from the region of the soliton. Such solitons may not exist. At least the method of their description applied here breaks down in this extreme case. Only the most dilute of these solitons would propagate at super-Alfvénic speeds. If they can exist at all, the form of the pseudo-potential predicts that they must have a minimum amplitude in the dilution of the density defined by the crossing point of the pseudo-potential with the n-axis at $n < 1$. Thus there cannot be a smooth transition in the stationary case from an undisturbed state to a dilute state. The transition must be catastrophic. But from a stationary theory it cannot be decided if such solitons can evolve or not.

The shape of the soliton solutions can be found from integration of Eq. (10.96). This calculation must be performed numerically, but for small-amplitude compressive sub-Alfvénic solitons, $n - 1 = \delta \ll 1$, an analytic solution is obtained when the pseudo-potential is expanded with resect to δn. This solution is given by

$$\delta n = \frac{3}{2} M \operatorname{sech}^2 \left(\frac{M^{1/2} \eta}{2|\kappa_\| \kappa_\perp|} \right) \tag{10.98}$$

which is identical to the stationary Korteweg-de Vries soliton solution obtained earlier. Hence, kinetic Alfvén solitons as stationary solutions when propagating at low Mach number, $0 < M = \kappa_\|^2 - 1 \ll 1$, and sub-Alfvénic speeds follow from the stationary Korteweg-de Vries equation.

This stationary equation is found by Taylor-expanding the pseudo-potential (10.97) up to third order in $\delta n = n - 1$. Due to the boundary conditions, $S(1) = S'(1) = 0$, we find

$$\left(\frac{d\delta n}{d\eta} \right)^2 = \frac{2}{\kappa_\|^2 \kappa_\perp^2} \left(M\, \delta n^2 - \frac{1}{3} \delta n^3 \right) \tag{10.99}$$

If we now take the second derivative of both sides of this equation, we arrive at the desired equation

$$\delta n' + \frac{2}{M}\delta n\,\delta n' - \delta n' - \frac{\kappa_\parallel^2 \kappa_\perp^2}{M}\delta n''' = 0 \tag{10.100}$$

Supplementing it by the missing time derivative, a form of the Korteweg-de Vries equation is recovered. As we expected, low but finite amplitude sub-Alfvénic compressive kinetic Alfvén waves evolve into solitons under the competitive action of their nonlinearity and their dispersive properties. If an initial condition is given, one thus expects that kinetic Alfvén waves decay into a chain of low amplitude solitons which may be distributed according to their amplitudes by a distribution law of the kind given in Eq. (10.55).

Because there is no threshold for this kind of evolution of kinetic Alfvén waves, they will always tend to evolve into solitons. Soliton formation will be restricted only by other factors as the dimensions of the system, inhomogeneities, wave transformation and wave coupling. The number of solitons which can be formed depends not only on the initial disturbance but also on the available length of the field lines. In the magnetosphere this length is restricted by two conditions, the value of the plasma beta and the distance between the point where $\beta > m_e/m_i$ and the plasma sheet. Along this length a magnetic pulse injected from the plasma sheet by, say, reconnection can break off into a number of kinetic Alfvén solitons, which travel as low amplitude pulses into the inner magnetosphere up to the point where $\beta < m_e/m_i$, where they possibly transform into shear kinetic Alfvén waves. The larger amplitude solitons will propagate at higher speed and may overcome the lower amplitude pulses. How the transformation works is unknown. The simplest way is that at the altitude of transition from $\beta > m_e/m_i$ to $\beta < m_e/m_i$ the incoming kinetic Alfvén solitons serve as initial conditions for the evolution of shear kinetic Alfvén solitons.

Shear Kinetic Alfvén Solitons

Shear kinetic Alfvén waves can evolve into solitons in a way entirely analog to kinetic Alfvén waves. The difference in the dispersive properties of the two modes suggests that shear kinetic waves behave similar to ion-acoustic solitons with a negative dispersion factor.

It is simple matter to derive the pseudo-potential for these waves and to discuss the conditions of existence of such solitons. The dynamics of low-amplitude solitons is also governed by a Korteweg-de Vries equation, but the solitons have slightly different properties. The basic equations for this case are the same as for the former case, with the exception that one replaces the Boltzmann law for the electrons with the full compressive electrons dynamics, as before in dimensionless variables

$$\begin{aligned} \frac{\partial n_e}{\partial \tau} + \nabla_\zeta (n_e v_e) &= 0 \\ \frac{\partial v_e}{\partial \tau} + \tfrac{1}{2}\nabla_\zeta v_e^2 &= \tfrac{1}{2}\alpha \nabla_\zeta [\phi_\parallel - \ln n_e] \end{aligned} \tag{10.101}$$

The coefficient α is defined as $\alpha = (m_i/m_e)\beta_e$, where β_e is the electron plasma beta. We change to the co-moving coordinates, $\eta = \kappa_\parallel \zeta + \kappa_\perp \xi - \tau$, and use homogeneous boundary conditions at infinity. The basic equations then reduce to

$$\begin{aligned} (n - \kappa_\parallel n v_e)' &= 0 \\ \left(v_e - \tfrac{1}{2}\kappa_\parallel v_e^2\right)' &= -\tfrac{1}{2}\alpha\kappa_\parallel (\phi_\parallel - \ln n)' \\ \left(n - \kappa_\perp^2 n \phi_\perp''\right)' &= 0 \\ \kappa_\perp^2 \kappa_\parallel (\phi_\perp - \phi_\parallel)'''' &= (n v_e)'' \end{aligned} \tag{10.102}$$

The primes indicate the number of total derivatives with respect to the co-moving coordinate η.

The last equation can be integrated twice, the other equations once in η from $-\infty$ to η, and the variables may be expressed by the quasineutral normalized density, n

$$\begin{aligned} v_e &= (n-1)/n\kappa_\parallel \\ \phi_\parallel &= \ln n + [(1-n^2)/\alpha\kappa_\parallel^2 n^2] \\ \phi_\perp'' &= (\kappa_\parallel/\kappa_\perp) v_e \\ \phi_\parallel'' &= -(1-n)(n-\kappa_\parallel^2)/n\kappa_\perp^2\kappa_\parallel^2 \end{aligned} \tag{10.103}$$

Since $\phi_\parallel$ depends only implicitly on η and since all variables can be expressed through the density, $n(\eta)$, it is sufficient to consider only the last of the above equations and to solve for $\phi_\parallel$. This equation is readily transformed into a form containing the pseudo-potential. We chose to write it as

$$\frac{1}{2}\left(\frac{dn}{d\eta}\right)^2 = \frac{S(n;\alpha,\kappa_\parallel)}{\kappa_\perp^2 F^2(n;\alpha,\kappa_\parallel)} \tag{10.104}$$

where we have $F = (\alpha\kappa_\parallel^2/2)\Psi(n)$ and $\Psi(n) = (1 - 2/n^2\alpha\kappa_\parallel^2)/n$. Then we can write

$$S = [\Psi^2(n)\kappa_\perp^2\kappa_\parallel^2]^{-1} \int_1^n \frac{1-\xi}{\xi}(\xi - \kappa_\parallel^2)\Psi(\xi)\, d\xi \tag{10.105}$$

Solving the last integral, the pseudo-potential becomes explicitly

$$S = \frac{\alpha^2\kappa_\parallel^2}{4}\left[\frac{1-n}{n}(n+\kappa_\parallel^2) + (1+\kappa_\parallel^2)\ln n + \frac{(1-n)^2}{\alpha n^2}\left(\frac{1}{\kappa_\parallel^2} - \frac{n+2}{3n}\right)\right] \tag{10.106}$$

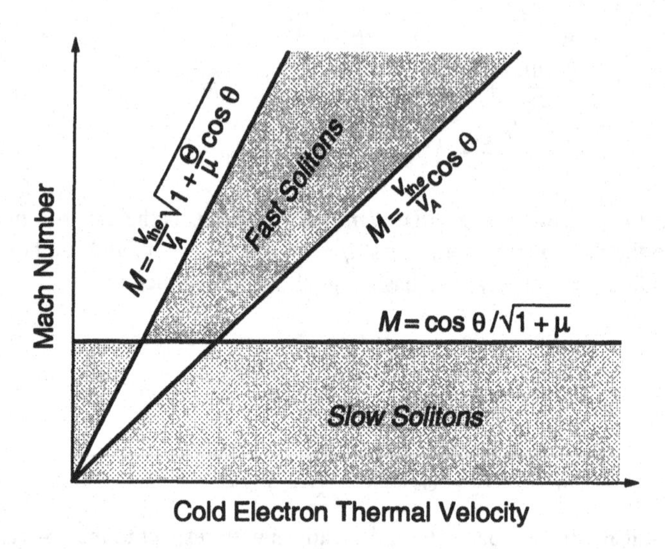

Fig. 10.12. Regions of existence of shear kinetic Alfvén solitons in parameter space.

For real solutions, $S \geq 0$ is required. There is a wide range where compressive solitons are possible, depending on the values of α and $\kappa_\|$. For sub-Alfvénic speeds only compressive solitons exist. But for super-Alfvénic Mach numbers also rarefaction solitons can exist. These are density holes which propagate across the plasma at high speeds. Usually, the holes are wide while compressions are very narrow in the super-Alfvénic case.

As for kinetic Alfvén solitons, it is possible to derive a Korteweg-de Vries equation for the small-amplitude shear kinetic solitons. Let us denote the right-hand side of Eq. (10.104) by $V(n)$. The Korteweg-de Vries equation is given as

$$n''' - \{V^{(3)}(1)n + [V^{(2)}(1) - V^{(3)}(1)]n' = 0 \tag{10.107}$$

Here the bracketed exponents stand for the order of derivations with respect to the density n. The solution of this equation is of the usual sech^2-type. Calculating the derivatives of $V(n)$, one obtains the two expressions

$$\begin{aligned} V^{(2)}(1) &= 4(\alpha\kappa_\perp^4\kappa_\|^6)^{-1}(\kappa_\|^2 - 1)(1 - 2/\alpha\kappa_\|^2)^{-3} \\ V^{(3)}(1) &= -\frac{8}{\alpha\kappa_\perp^4\kappa_\|^6}\left(1 - \frac{2}{\alpha\kappa_\|^2}\right)^3\left[\kappa_\|^2 + 5(\kappa_\|^2 - 1)\frac{(6/\alpha\kappa_\|^2) - 1}{1 - (2/\alpha\kappa_\|^2)}\right] \end{aligned} \tag{10.108}$$

Because in the solution for the soliton the second derivative appears under the root sign, these expressions impose some conditions on the existence of the solitons. In particular,

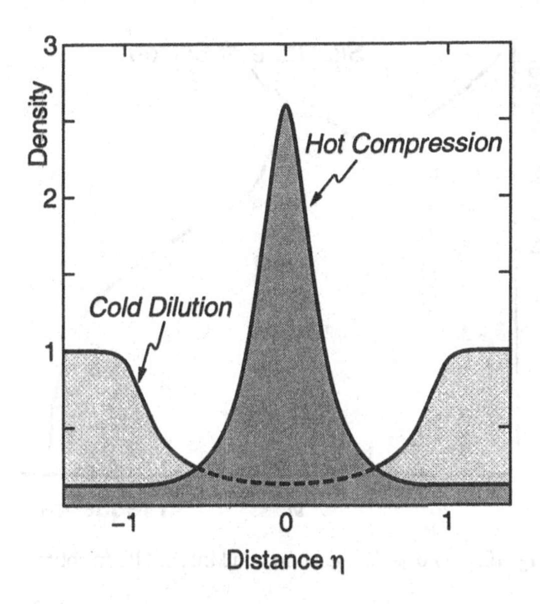

Fig. 10.13. Cold plasma depletion containing a hot electron sheet.

for compressive kinetic Alfvén solitons to exist one can derive the restriction

$$2(v_A/v_{\rm thc})^2 < \kappa_\parallel^2 < 7(v_A/v_{\rm thc})^2 \tag{10.109}$$

which must be satisfied simultaneously with $\kappa_\parallel^2 < 1.5$. Combining both we find the condition on the Mach number for shear kinetic compressive solitons (θ is the angle between the original wave and the ambient magnetic field)

$$\tfrac{1}{7}(v_A/v_{\rm thc})\cos^2\theta < M^2 < \tfrac{1}{2}(v_A/v_{\rm thc})\cos^2\theta \tag{10.110}$$

The corresponding condition for sub-Alfvénic rarefactive solitons to exist is

$$M^2 < \tfrac{1}{7}(v_A/v_{\rm thc})\cos^2\theta \tag{10.111}$$

Two-Electron Kinetic Alfvén Solitons

We now investigate one particularly interesting case when the plasma contains two electron components of different densities and temperatures. The cold and hot electron populations are indexed by the subscripts c and h, respectively. Quasineutrality requires that, in normalized coordinates,

$$n_i = (n_c + \mu n_h)/(1+\mu) \tag{10.112}$$

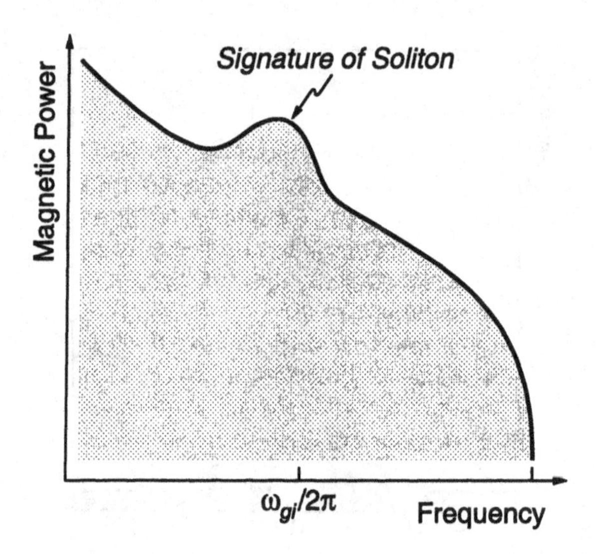

Fig. 10.14. Possible signature of a kinetic Alfvén soliton.

where $\mu = n_{0h}/n_{0c}$ is the ratio of the initial undisturbed densities of the hot to cold electron components. Applying the same methods as in the previous section, we obtain as the first integral of the stationary co-moving basic equations

$$\left(\frac{dn_c}{d\eta}\right)^2 = -\frac{\alpha_c n_c^6}{\kappa_\perp^2 (1 - n_c^2 \alpha_c \kappa_\parallel^2/2)^2} \int_1^{n_c} \frac{d\xi}{\xi^3} G(n; \mu, \kappa_\parallel) \left(1 - \frac{\xi^2 \alpha_c \kappa_\parallel^2}{2}\right) \quad (10.113)$$

Here $n = (n/n_c) = (1+\mu)^{-1}(n_c + \mu n_h)$ is the total electron density, n_h is expressed in terms of n_c as

$$n_h = n_c^{T_c/T_h} \exp\left[T_c(1 - n_c^2)/T_h \alpha_c n_c^2 \kappa_\parallel^2\right] \quad (10.114)$$

and the integrand is

$$G = (n^{-1} - 1)[(1+\mu)n - \kappa_\parallel^2] \quad (10.115)$$

The above integral replaces the pseudo-potential used in the previous section. Now the regions where the pseudo-potential is negative are modified by the additional two parameters, μ and $\Theta = T_h/T_c$.

An analytical treatment yields two types of solitons as shown in Fig. 10.12. There are slow and fast solitons possible, which can be either density compressions or dilutions, depending on the values of the parameters. The two-electron solitons have one special property. The cold component is more strongly affected than the hot component. In particular, density depletions appear only in the cold component. In the particular cases when such density depletions are observed, the density of the hot component peaks in the center of the cold depletion, forming a compressive hot electron structure inside the diluted cold plasma. Figure 10.13 shows one example of such a soliton. The cold plasma depletion contain a hot electron layer, with the peak density of the hot electrons raised by more than a factor ten in order to compensate for the cold electron decrease. It is the concentration of the hot electrons which pushes the cold plasma out of the center of the soliton.

Two-electron plasmas are frequently encountered in space plasma physics. In the plasmasphere it is the mixture of the ring current plasma and the low energy plasmaspheric plasma of atmospheric origin. At auroral latitudes the mixture of cold ionospheric and warm plasma sheet plasma leads to conditions where the present theory of kinetic Alfvén solitons is applicable. In the region of the low-latitude boundary layer adjacent to the magnetopause the warm magnetospheric plasma blends into the cold magnetosheath plasma on a scale where kinetic Alfvén waves and solitons may form and contribute to the dynamics. An example of the presumable observation of the signature of a kinetic Alfvén soliton in the low latitude boundary layer with frequency close to the ion-cyclotron frequency is given in Fig. 10.14.

10.6. Drift Wave Turbulence

Weak drift wave turbulence is described by the Hasegawa-Mima equation (9.40) in the two-dimensional approximation, which is based on wave-wave interactions. When the drift waves evolve nonlinearly, the weakly turbulent wave-wave interaction approximation breaks down. The drift waves may decay into large-amplitude solitons. When such solitons appear in large numbers, they constitute a state of turbulence which is very different from weak turbulence and deserves the name of *strong turbulence*.

Drift Wave Solitons

We consider the simplest case of drift waves when the electrons, for low wave frequencies, can be considered as Boltzmann-distributed, and the ions as cold. Then the ions perform a simple $E \times B$-drift in the crossed ambient magnetic and wave electric fields, a process which provides the nonlinearity, because when the waves are driven by the ions, as in the case of the lower-hybrid drift instability, this coupling couples the wave back to the motion of the ions. The waves, with their perpendicular and parallel electric field components, propagate obliquely to the magnetic field. We introduce the density and temperature scales L_n and L_T and scale all lengths with respect to $r_{gi} = c_{ia}/\omega_{gi}$.

Because of the obliqueness of the problem the nonlinear wave is two-dimensional. It is governed by the *Kadomtsev-Petviashvili equation*, which is the two-dimensional generalization of the Korteweg-de Vries equation. But for $k_\perp^2 r_{gi}^2 < 1$ and $k_y k_\perp^2 r_{gi}^3 \ll$

r_{gi}/L_T this equation can be reduced to the Korteweg-de Vries equation proper

$$\frac{\partial \phi}{\partial \tau} + \nabla_\perp \phi + \nabla_\perp^3 \phi - \phi \nabla_\perp \phi = 0 \tag{10.116}$$

for the normalized electrostatic potential, $\phi(x, t)$. Here time τ is measured in units of r_{gi}/v_{de}, where the drift velocity is $v_{de} = r_{gi}c_{ia}/L_n = k_B T_e / eB_0 L_n$.

In the linear approximation the system evolves into drift modes which are eigenmodes of the system considered. Because the nonlinear evolution of these drift waves is governed by the Korteweg-de Vries equations, we expect that solitons will be generated in large number for sufficiently large initial wave amplitudes, and the system will become turbulent consisting of non-interacting solitons of frequency, $\omega(k) \approx kv_{de}$, and dispersion relation $\omega_s = ku$, where $u > v_{de}$ is the soliton speed. The drift wave field is then assumed to be composed of a large number N of solitons

$$\phi(x, \tau) = \sum_{l=1}^{N} \phi_s(x, \tau, x_l, u_l) \tag{10.117}$$

each having a different central position, x_l, and speed, u_l, with the speed satisfying the soliton dispersion relation. Since the single solitons are nearly independent solutions of the Korteweg-de Vries equation, their analytical representation is of the form

$$\phi_s(x, \tau, x_0, u) = -3(u-1)\mathrm{sech}^2[\tfrac{1}{2}(u-1)^{1/2}(x - x_0 - u\tau)] \tag{10.118}$$

Collisions are elastic and produce only an unimportant phase shift of the soliton position.

The inverse scattering method gives us the possibility to define a distribution function of the soliton amplitudes and velocities, given in Eq. (10.55). The soliton amplitude is $A = 3(u-1)$. With $\beta = \ell = 1$ and $3u_0 \rightarrow A$, we have $\sigma = 1/3$ and

$$F[u] = \frac{1}{12\pi} \int_{\phi < -A/2} \frac{dx}{[-2\phi(x) - A]^{1/2}} \tag{10.119}$$

a formula which is valid for the negative-potential drift wave solitons with $\phi < 0$.

Spectrum of Drift Wave Solitons

Equation (10.119) may serve as starting point for a statistical description through an average over initial configurations. One may assume that the final turbulent state is reached as the most probable state of a number of initial configurations, such that the system containing many solitons will saturate at some randomly phased spectrum, $\langle|\phi(k)|^2\rangle$, with the angular brackets indicating the averaging procedure, and $\phi(k)$ being the Fourier transformed potential amplitude. The spread of the random-phased field, ϕ, can be taken as given by some mean square amplitude, $\langle|\phi|^2\rangle = \phi_0^2$.

Given such an initial amplitude, knowing the distribution (10.119) of soliton amplitudes resulting from the solution of the Korteweg-de Vries equation, the distribution of solitons can be obtained by calculating the average over $F[u]$ with random-phased amplitudes. For a broad wavenumber spectrum the initial spectrum can be taken as white noise. Then the average of $F[u]$ is a Gaussian integral written in discrete form as

$$f_{\rm sol}(u) = \frac{1}{Z}\prod_{j=1}^{l}\int d\phi_j \exp\left[\frac{-\phi_j^2}{2\phi_0^2}\right] F[u;\phi] \tag{10.120}$$

Normalization requires that the integral over u-space just gives the total number of solitons in the system

$$\int_{v_{de}}^{\infty} f_{\rm sol}(u)du = N \tag{10.121}$$

Moreover, the indexed potentials are the soliton potential functions at positions x_j. It is now possible to convert the integral in the functional $F[u]$ into a sum and to perform the Gaussian integrations. Using a representation of the parabolic cylinder function $D_{-1/2}(x)$ one finds

$$\boxed{f_{\rm sol}(u) = \frac{1}{24\pi}\frac{L}{\sqrt{\phi_0}}\exp\left[-\frac{9(u-1)^2}{16\phi_0^2}\right] D_{-1/2}\left[\frac{3(u-1)}{2\phi_0}\right]} \tag{10.122}$$

L is the length of the system, defined as $L = \sum_j^l x_j$, where $x_j = (j/l)L$. This is the distribution function for drift wave solitons, which is provided in drift wave turbulence arising from the random-phased initial condition with amplitude ϕ_0. The distribution of the soliton velocities turns out to be not purely Gaussian even in this simple case of strong turbulence. The parabolic cylinder function introduces a skewing into the distribution which is caused by the concentration of the wave energy in the solitary wave solutions of the Korteweg-de Vries equation. As one of the consequences the turbulent spectrum gets harder than a spectrum of randomly distributed waves.

It is not very difficult to calculate the spectral density of the soliton field knowing the distribution function. The spectral density, $S(k,\omega)$, is defined as the Fourier transform of the two-point correlation function. Taking $\phi_{\rm sol}$ from Eq. (10.117) and calculating the Fourier transform of $\langle\phi_{\rm sol}(x+\zeta, t+\tau)\phi_{\rm sol}(x,t)\rangle$ one finds for the spectral shape

$$\boxed{S(k,\omega) = \frac{96k^2}{\sqrt{3}L} f_{\rm sol}\left(\frac{\omega}{k}\right) \operatorname{cosech}^2\left[\pi k r_{gi}\left(\frac{kv_{de}}{\omega - kv_{de}}\right)^{1/2}\right]} \tag{10.123}$$

Note that k, v_{de}, L are all normalized quantities. This result has been obtained by replacing the sum over the solitons in $\phi_{\rm sol}$ with an integral using the soliton distribution, $f_{\rm sol}$, as weight, and $u = \omega/k$, which is justified for a large number of solitons. This number is defined as

$$N = \int_{v_{de}}^{\infty} du\, f_{\rm sol}(u) = QL\phi_0^{1/2} \tag{10.124}$$

where $Q \approx \Gamma(3/4)/119$. The first important property of the spectrum of solitons is that it vanishes for frequencies $\omega < kv_{de}$, simply because there are no solitons in this range with speeds smaller than $u < v_{de}$. Secondly, the spectrum peaks at frequencies $\omega \approx kv_{de}[1 + O(\phi_0)]$ above the lower frequency cut-off. If one expands the parabolic cylinder function for small and for large arguments these peaks are found at

$$\omega_{\rm max} \approx \begin{cases} kv_{de}(1 + \sqrt{2}\phi_0) & \text{for} \quad \pi^2k^2 \ll \phi_0 \\ kv_{de}[1 + \sqrt{2}(\pi k\phi_0^2)^{2/5}] & \text{for} \quad \pi^2k^2 \gg \phi_0 \end{cases} \tag{10.125}$$

The interesting conclusion is thus that the calculation of the soliton spectrum verifies the suspicion that the turbulent spectrum of a soliton gas, in this case the drift-soliton gas, is harder than the spectrum of drift waves themselves from which it developed.

Application of this kind of turbulence to waves in equatorial spread-F and to instabilities of the kind of the lower-hybrid drift instability in the Earth's plasma sheet and low-latitude boundary layer seems promising, but has not yet been tried. Spectra in these regions look relatively structureless and thus do point on the existence of evolved scale-independent or self-similar turbulence.

Similar considerations may also apply to kinetic Alfvén soliton turbulence, to ion acoustic turbulence and electron acoustic turbulence. For these waves it is known that the turbulence follows the Korteweg-de Vries equation as long as the amplitudes do not become too large. One may expect that when this happens, particle reflection at the many solitary structures in the turbulent plasma will produce dissipation and introduce irreversibility, and the second-order derivative will become non-negligible, small potential differences will arise, and the plasma will make the transition to micro-double layer turbulence. The existence of this kind of turbulence manifests itself in localized holes in phase space and may be very important in particle acceleration.

Concluding Remarks

A number of equations has been derived in this chapter, which lead to localized large-amplitude wave solutions travelling across the plasma. The important insight is that these solitons evolve as nonlinear dynamical equilibria between nonlinearity and dispersivity of single waves, but that the equations allow a given initial condition to decay into a large number of such solitons, travelling at their own and well-defined speeds with negligible interaction, thus comprising a turbulent state of the plasma which can be described in a statistical way.

A few examples are given in this chapter, e.g., ion-acoustic, electron-acoustic, kinetic Alfvén, and drift wave solitons. Their application to space plasmas is still open. The reason is that in spite of the well-known importance of soliton turbulence only few observations allow for an unambiguous interpretation in terms of solitons. In this situation the link between theory and observation in space plasma physics is still missing.

Further Reading

There is a small number of excellent books on nonlinear plasma theory like [1], [5], and [11], which all were written in the seventies. Meanwhile nonlinear plasma physics has grown into a wide field, but no comprehensive modern text is available. The reason is that analytically not so much has been achieved during the last twenty years while the activities and interests have turned to numerical simulations. A contemporary summary of numerical plasma simulations is given in [7].

Derivation of the Burgers and Korteweg-de Vries equations from the general hydrodynamics set of Navier-Stokes equations is given in [4]. Derivation from plasma fluid equations are contained in [8]. The inverse scattering method has found wide application in the theory of one-dimensional nonlinear wave equations. It is given in [4], [8], and many other texts. Drift wave turbulence is treated in [3]. The double layer theory is taken from Mace and Hellberg, *J. Plasma Phys.* **49** (1993) 283. For the Kadomtsev-Petviashvili equation we followed Meiss and Horton, *Phys. Rev. Lett.* **48** (1982) 1362. Kinetic Alfvén soliton theory is included in [2]. Double layer potentials were measured by Mälkki et al., *J. Geophys. Res.* **99** (1994) 6045. Shear kinetic Alfvén soliton theory for one- and two-electron fluids is taken from Treumann et al., *Astron. Astrophys.* **236** (1990) 242. Observational information on shocks is found in [6] and [10]. Laminar and turbulent shocks are discussed in [9].

[1] R. C. Davidson, *Methods in Nonlinear Plasma Theory* (Academic Press, New York, 1972).

[2] A. Hasegawa and C. Uberoi, *The Alfvén Wave* (U.S. Dept. of Energy, Springfield, 1982).

[3] B. B. Kadomtsev, *Plasma Turbulence* (Academic Press, New York, 1965).

[4] V. I. Karpman, *Nichtlineare Wellen in dispersiven Medien* (Vieweg Verlag, Braunschweig, 1977).

[5] R. Z. Sagdeev and A. A. Galeev, *Nonlinear Plasma Theory* (W. A. Benjamin, New York, 1969).

[6] R. G. Stone and B. T. Tsurutani (eds.), *Collisionless Shocks in the Heliosphere: A Tutorial Review* (American Geophysical Union, Washington, 1985).

[7] T. Tajima, *Computational Plasma Physics* (Addison-Wesley Publ. Co., Redwood City, 1989).

[8] T. Taniuti and K. Nishihara, *Nonlinear Waves* (Pitman Advanced Publ., Boston, 1983).

[9] D. A. Tidman and N. A. Krall, *Shock Waves in Collisionless Plasmas* (Wiley Interscience, New York, 1971).

[10] B. T. Tsurutani and R. G. Stone (eds.), *Collisionless Shocks in the Heliosphere: Reviews of Current Research* (American Geophysical Union, Washington, 1985).

[11] V. N. Tsytovich, *Nonlinear Effects in Plasmas* (Plenum Press, New York, 1970).

11. Strong Turbulence

The previous chapter attempted the first step into strong turbulence theory. We considered the evolution of large-amplitude waves when the dispersion competes with nonlinearity and obtained that plasmas may in this case evolve into turbulent states containing large numbers of quasi-stable solitary waves, so-called topological solitons, which constitute turbulence.

Another important effect arising in a plasma is a consequence of the fact that plasma waves are carriers of energy. The energy density stored in a plasma wave corresponds to a real pressure exerted on the plasma. Any inhomogeneity developing on the wave spectrum by localized reflection, for instance, by absorption or interaction with other waves and particles or simply by focussing or spreading of a wave, causes a wave-pressure force, the *ponderomotive force*. This force appears as an ordinary force on the particles or fluid and may act on the plasma and change the propagation properties of the wave itself or may affect the propagation of other waves. It also may lead to interchanges between the waves. We expect that under such conditions another type of turbulence will develop and call it *strong turbulence*, a term whose meaning will have to be clarified in this chapter.

Strong turbulence has a number of different aspects. In a first step we are going to derive the ponderomotive force which is exerted by the plasma wave pressure onto the background plasma. This can be done with the elementary knowledge of basic plasma physics using a simple high-frequency wave model. The simple theory can then be extended to a much more general case, but for the purposes of understanding the elementary theory is sufficient. Hereafter, we consider the basic equations which in the presence of nonlinearity and wave pressure govern strong turbulence. Solution of these equations leads to caviton formation instead of solitons. Cavitons behave differently from solitons. Their dynamics will be discussed in detail for Langmuir waves uncovering some entirely new effects.

11.1. Ponderomotive Force

The wave pressure force exerted by a large-amplitude wave on the plasma background is the ponderomotive force of the inhomogeneous wave field experienced by the particle components. Such forces can be generated by any wave transporting energy, i.e., moving at a certain group velocity and thus being spatially inhomogeneous.

Radiation Pressure Force

The simplest example of such a force is the radiation pressure of a high-frequency electromagnetic field experienced by an isotropic medium of dielectric function given in Eq. (I.9.36) of our companion book

$$\epsilon(\omega) = 1 - \omega_{pe}^2/\omega^2 \tag{11.1}$$

The pressure force in this most simple case can be estimated as follows. The electric field energy density is known to be given by

$$W_w = \frac{\partial(\omega\epsilon)}{\partial\omega} W_E \tag{11.2}$$

with $W_E = (\epsilon_0/2)|\delta\mathbf{E}|^2$. This energy density is equal to the wave field pressure which, from the point of view of the background medium carrying the wave, behaves exactly like an ordinary gas pressure acting on a hydrodynamic fluid. The radiation pressure or *ponderomotive force* is then simply defined as

$$\mathbf{F}_{\mathrm{pm}} = -\frac{1}{2n_0}\nabla W_w \tag{11.3}$$

where we divide by the number density in order to obtain the force acting on one particle. There is an additional factor of 1/2 in this expression, the origin of which will become clear below. Inserting the expression for the dielectric function (11.1) and carrying out the differentiations with respect to the frequency, ω, yields

$$\mathbf{F}_{\mathrm{pm}} = \frac{\epsilon - 1}{2n_0}\nabla W_E(\mathbf{x}, t) + \frac{W_E(\mathbf{x}, t)}{2n_0}\nabla\epsilon \tag{11.4}$$

For approximately constant ϵ the last term disappears, and the ponderomotive pressure force assumes the particularly simple form

$$\boxed{\mathbf{F}_{\mathrm{pm}} = \frac{\epsilon - 1}{2n_0}\nabla W_E(\mathbf{x}, t) = -\frac{\epsilon_0\omega_{pe}^2}{4n_0\omega^2}\nabla|\delta\mathbf{E}(\mathbf{x}, \mathrm{t}|^2} \tag{11.5}$$

In the present case this force is a potential force, with potential given by

$$\phi_{\rm pm} = -\frac{\epsilon(\omega) - 1}{2n_0} W_E = \frac{e^2}{4m_e\omega^2}|\delta\mathbf{E}|^2 \tag{11.6}$$

Because the Debye length is much smaller than the characteristic length of the field inhomogeneity, the electrons obey the Boltzmann law in the ponderomotive force potential

$$n_e = n_0 \exp[-\phi_{\rm pm}/k_B(T_e + T_i)] = n_0 \exp[-|\delta\mathbf{E}(\mathbf{x}, t)|^2/E_c^2] \tag{11.7}$$

where the ponderomotive critical electric field is given by

$$E_c^2 = 4m_e\omega^2 k_B(T_e + T_i)/e^2 \tag{11.8}$$

Therefore, for wave electric field intensities coming close to the critical electric field, the nonlinear pressure force effects become large and the distortions of the plasma density may become of the same order as the background density.

Density Variation

The nonlinear ponderomotive effect may result in a violent modification of the plasma background caused by the waves, in which case the plasma looses its homogeneous character, but changes on a comparably short spatial scale. This is a typical property of turbulence. Strong turbulence may thus be caused by the ponderomotive effects of large amplitude waves. The nonlinear distortion of the density for moderate turbulence is obtained from the Boltzmann law by small amplitude expansion

$$\delta n_{nl}/n_0 \approx -|\delta\mathbf{E}(\mathbf{x}, t)|^2/E_c^2 \tag{11.9}$$

This variation in the density is a negative modulation. For the model under consideration the expected modulation of the background plasma density in presence of a large-amplitude electromagnetic wave, which exerts a radiation pressure onto the plasma, the expected density variation will always result in a local decrease of the plasma density in the region of large amplitudes. Such depletions of density are usually called *cavitons* and should be distinguished from the ordinary or topological solitons discussed in the previous chapter, because of the entirely different physics involved in their production and evolution.

It should, however, be mentioned that the consideration of the ponderomotive force in this section is not restricted to the radiation pressure of the electromagnetic wave. Our only assumption was that the wave frequency should be high enough to neglect other effects. Therefore the same argument also applies to Langmuir waves. Their response function is the same, since their high frequencies radiation pressure works in exactly the same way as discussed above. But because Langmuir waves have very low phase and group velocities, their ponderomotive effect may even be stronger than that of electromagnetic waves when their energy accumulates locally.

Fluid Theory

The intuitive approach to the ponderomotive force given in the previous section must be based on safer grounds and at the same time must be extended to arbitrary large-amplitude waves. It is easiest to start from the hydrodynamic set of equations of a plasma and to average over the fast time scale of the large-amplitude wave. Depending on the fluid model used one obtains different results.

Here we only sketch the derivation of the ponderomotive force. The simplest model is that of a magnetohydrodynamic fluid. We assume that the fluid is subject to a nonlinear large-amplitude magnetohydrodynamic or electromagnetic wave

$$[\delta\mathbf{B}(\mathbf{x},t), \delta\mathbf{E}(\mathbf{x},t)] = [\delta\mathbf{B}_0(\mathbf{x},t), \delta\mathbf{E}_0(\mathbf{x},t)] \exp(\mathbf{k}\cdot\mathbf{x} - \omega t) \tag{11.10}$$

where the amplitudes are slowly varying functions of space and time. The nonlinear terms in the magnetohydrodynamic equations, i.e., the convective derivative, $\mathbf{v}\cdot\nabla\mathbf{v}$, the flux term, $n\mathbf{v}$, and the Lorentz force term, $\mathbf{j}\times\mathbf{B}$, contain products of the amplitudes and after fast-time and fast-space averaging will contribute to a wave force term.

The easiest way is to calculate the components of the wave stress tensor, $\boldsymbol{\Sigma}$, under such an averaging. The ponderomotive force density is then defined as

$$\mathbf{f}_{\rm pm}(\mathbf{x},t) = \nabla\cdot\boldsymbol{\Sigma} \tag{11.11}$$

When the wave amplitude depends on space and time as a slowly varying function, i.e., when the wave field is inhomogeneous and varying, the ponderomotive force will also be a function of space and time. Performing the calculations, one finds that the ponderomotive force density obtained is composed of two terms according to

$$\mathbf{f}_{\rm pm}(\mathbf{x},t) = \mathbf{f}_{{\rm pm},\nabla}(\mathbf{x},t) + \mathbf{f}_{{\rm pm},t}(\mathbf{x},t) \tag{11.12}$$

The first term results from the spatial dependence of the wave amplitude and is given by

$$\mathbf{f}_{{\rm pm},\nabla} = \frac{\epsilon_0}{4}\left[\nabla\left(\delta E_l^*\delta E_m\, n\frac{\partial\epsilon_{lm}}{\partial n}\right) - \delta E_l^*\delta E_m\left(\nabla\epsilon_{lm} - \frac{\partial\epsilon_{lm}}{\partial B_{0j}}\nabla B_{0j}\right)\right] \tag{11.13}$$

Here we have written the tensor $\boldsymbol{\epsilon} = \epsilon_{lm}$ in index notation. It depends on the plasma density, n, and on the ambient magnetic field, $\mathbf{B}_0$.

The second term in (11.12) comes from the slow time variation of the wave amplitude and is given by

$$\mathbf{f}_{{\rm pm},t} = \frac{\epsilon_0}{4\mu_0}\left\{\frac{\partial}{\partial t}[(\boldsymbol{\epsilon} - \mathsf{I})\cdot\delta\mathbf{E}\times\delta\mathbf{B}^*] + \left[\left(\omega\frac{\partial\boldsymbol{\epsilon}}{\partial\omega}\cdot\frac{\partial\delta\mathbf{E}}{\partial t}\right)\times\delta\mathbf{B}^*\right] + \text{c.c.}\right\} \tag{11.14}$$

where c.c. stands for the conjugate complex part. These expressions simplify for plane circularly polarized waves of parallel wavenumber, $k_\parallel$, and frequency, ω. In this case one simply adds the two components of the ponderomotive force

$$\begin{aligned} f_{\mathrm{pm}\parallel} &= \frac{\epsilon-1}{2}\nabla_{\parallel} W_E + \frac{k_{\parallel}}{2\omega^2}\frac{\partial\omega^2(\epsilon-1)}{\partial\omega}\frac{\partial W_E}{\partial t} \\ \mathbf{f}_{\mathrm{pm}\perp} &= -\frac{1}{2}\frac{\partial\omega(\epsilon-1)}{\partial\omega}\nabla_{\perp} W_E - \frac{B_0}{\mu_0}\nabla_{\perp}\delta B_{\parallel} \end{aligned} \tag{11.15}$$

to the stationary Lorentz force on the right-hand side of the equation of motion. For vanishing time dependence and otherwise homogeneous conditions and for an incident electromagnetic wave one recovers the simple wave pressure force of the previous section. But already Alfvén or whistler waves in the stationary case give different expressions for the ponderomotive force than the simple pressure force.

When the magnetohydrodynamic ponderomotive force is extended to the two-fluid case, it turns out that the ponderomotive force is different for each of the species. Thus it may drive currents and not only give rise to motions as in the magnetohydrodynamic case. Another possible effect is that the ponderomotive force may contribute to new terms in Ohm's law and affect plasma transport or accelerate particles.

11.2. Nonlinear Wave Equation

Imagine an intense but localized wave entering a low density plasma with a plasma frequency just above the electron-cyclotron frequency. If the wave pressure force pushes the plasma out of the region of the wave and dilutes the plasma below the electron-cyclotron frequency, the properties of the plasma change drastically, and wave propagation becomes very different from what it was before the wave was injected into the plasma. This simple example demonstrates the importance of wave pressure effects.

Effects of this kind are common in plasmas where the number of the possible modes is large. One immediately realizes that these effects must have to do with the amplitude of the wave instead of the wavenumber. In this section we investigate the importance of the wave amplitude on the dispersive properties and propagation of the wave.

Nonlinear Dispersion Relation

Let us assume that the wave dispersion relation is not only a function of the wavenumber, k, but also depends on the wave amplitude. In linear dispersion theory and in the consideration of the topological effects leading to the Burgers and Korteweg-de Vries equations any explicit dependence of the frequency on the wave amplitude had been neglected. Abandoning this neglect, introduces an entirely new view into wave theory.

Let us denote the wave function of the large-amplitude disturbance symbolically by $\tilde{A}$, where $\tilde{A}$ can be either a scalar or a column vector. A quite general ansatz for $\tilde{A}$ is to extract a plane wave rapidly variable phase factor from its space and time dependence and to assume that this plane wave is modulated by a possibly slower varying amplitude factor, $A_0(\mathbf{x}, t)$, which is also a function of space and time

$$\tilde{A} = A_0(\mathbf{x}, t) \exp[i(k_0 x - \omega_0 t)] \tag{11.16}$$

The wavenumber, k_0, and frequency, ω_0, belong to the fast carrier wave part of the wave field. The amplitude itself may now be written as the product of another unspecified phase factor and an amplitude which is again variable in space and time as

$$A_0(\mathbf{x}, t) = A(\mathbf{x}, t) \exp[i\varphi(\mathbf{x}, t)] \tag{11.17}$$

Clearly, in the linear approximation the wave function, $\tilde{A}$, satisfies a linear equation with $\omega_0(k_0)$ being the solution of the linear dispersion relation. In the general nonlinear case the frequency of the nonlinear wave which we denote by ω will become a function of the wave energy as well.

Let us derive an equation which takes into account this energy dependence of the frequency. The energy of a wave is proportional to the square of its amplitude, A, since when taking the product $\tilde{A} \cdot \tilde{A}^*$, the complex phase factors cancel. Letting the frequency depend on energy, the *nonlinear dispersion relation* can formally be written as

$$\omega = \omega(\mathbf{k}, A^2) \tag{11.18}$$

For simplicity we supposed an isotropic case and real A. Extensions to non-isotropic cases and complex A are straightforward. In the latter case one replaces $A^2 \rightarrow |A|^2$.

Finite-Amplitude Equations

Let us assume that the dependence on the energy is weak. In this case the nonlinear dispersion relation or frequency can be expanded with respect to the energy

$$\omega = \omega_0(\mathbf{k}) + \nabla_A \omega|_0 \, A^2 + \mathrm{O}(A^4) \tag{11.19}$$

where $\nabla_A \omega$ is defined as

$$\nabla_A \omega = \partial \omega / \partial (A^2) \tag{11.20}$$

and the index 0 implies that the indexed quantity has to be taken at zero wave amplitude. Thus ω_0 is the usual dispersion relation with its full dependence on the wavenumber, $\mathbf{k}$, which has been used in the previous sections. It contains the nonlinearities described there. Taking into account the dependence on the wave amplitude therefore introduces another nonlinearity which we so far has not been aware of.

In order to illuminate the nature of this kind of nonlinearity, let us assume that the wave is nearly stationary. In this case one applies the eikonal approximation, assuming that the phase of the wave varies slowly under the influence of its evolution, while its amplitude may be a function of space and time, $A(\mathbf{x}, t)$. We write the total phase as

$$\phi(\mathbf{x}, t) = \mathbf{k}_0 \cdot \mathbf{x} - \omega_0 t + \varphi(\mathbf{x}, t) \tag{11.21}$$

with φ the small disturbance of the phase which is introduced by the amplitude dependence. Geometrical optics prescribes the following relations between the frequency, wavenumber and the phase

$$\begin{aligned} \omega(\mathbf{x}, t) &= -\partial\phi(\mathbf{x}, t)/\partial t = \omega_0 - \partial\varphi(\mathbf{x}, t)/\partial t \\ \mathbf{k}(\mathbf{x}, t) &= \nabla\phi(\mathbf{x}, t) = \mathbf{k}_0 + \nabla\varphi(\mathbf{x}, t) \end{aligned} \tag{11.22}$$

These definitions can be used in Eq. (11.19) to derive an equation for the variation in the phase function. Restricting to terms of second order in A and in $\nabla\varphi$ gives

$$\boxed{\frac{d\varphi}{dt} + \frac{g_{gr0}}{2}(\nabla_x\varphi)^2 + \frac{v_{gr0}}{2k_0}(\nabla_\perp\varphi)^2 + \nabla_A\omega|_0\, A^2 = 0} \tag{11.23}$$

where the convective derivative

$$d/dt = (\partial/\partial t) + \mathbf{v}_{gr0} \cdot \nabla \tag{11.24}$$

is defined with the help of the zero-order group velocity

$$\mathbf{v}_{gr0} = \nabla_{\mathbf{k}}\omega(\mathbf{k}, A^2)\big|_{\mathbf{k}_0, A=0} \tag{11.25}$$

and the coordinate x points in the direction of $\mathbf{v}_{gr0}$. Moreover, the derivative of the group velocity is defined by

$$g_{gr0} = \nabla_{\mathbf{k}}^2\omega(\mathbf{k}, A^2)\big|_{\mathbf{k}_0, A=0} = \nabla_{\mathbf{k}} \cdot \mathbf{v}_{gr0} = 3v_{gr0}/k + \nabla_{\mathbf{k}} v_{gr0} \tag{11.26}$$

Equation (11.23) is an equation for the variation in the phase of the wave. The underlying assumption is that the wave constitutes a wave packet with well-defined group velocity, whose evolution can be described by following its phase variation. But another equation is needed for the wave amplitude. This equation is provided by the energy equation, which can be written in the following form

$$\partial A^2/\partial t + \nabla \cdot (A^2\mathbf{v}_W) = 0 \tag{11.27}$$

In this equation energy dissipation is suppressed. The velocity of energy convection, $\mathbf{v}_W(A^2, \mathbf{k})$, is itself a function of the wavenumber as defined in Eq. (11.22) and the square of the wave amplitude. Since we restrict ourselves to second-order expressions only, the A^2-dependence in the convective term in the energy transport equation contains only the zero-order energy convection velocity which is the zero-order group velocity

$$\mathbf{v}_W(0, \mathbf{k}) = \mathbf{v}_{gr0} = v_{gr0}\mathbf{k}/k \tag{11.28}$$

with $\mathbf{k}$ given in Eq. (11.22). Written explicitly the two components are

$$\begin{aligned} v_{W\parallel} &= v_{gr0} + g_{gr0}\nabla_x\varphi \\ \mathbf{v}_{W\perp} &= (v_{gr0}/k)\nabla_\perp\varphi \end{aligned} \tag{11.29}$$

and the equation for the wave amplitude follows from the energy transport equation

$$\boxed{\left(\frac{\partial}{\partial t} + v_{gr0}\nabla_x\right)A^2 + \left[g_{gr0}\nabla_x(A^2\nabla_x) + \frac{v_{gr0}}{k_0}\nabla_\perp(A^2\nabla_\perp)\right]\varphi = 0} \tag{11.30}$$

The two equations (11.23) and (11.30) describe the nonlinear evolution of the wave amplitude and phase variation up to second order in the amplitude. The amplitude is assumed to be small but finite.

The important conclusion which can be drawn from these two fundamental equations for the evolution of finite-amplitude nonlinear waves is that it is sufficient to determine the nonlinear dispersion relation in order to obtain the full set of nonlinear equations, from which the further behavior of the wave amplitude and phase can be determined. The problem is therefore reduced to finding the nonlinear dispersion relation before solving the above set of equations.

Nonlinear Parabolic Equation

The set of nonlinear equations (11.23) and (11.30) derived above can be combined into one single equation, which is known under the name *nonlinear parabolic equation.* Under some simplifying conditions it can be written in the form of a *nonlinear Schrödinger equation*, which we will use in later sections.

Let us return to the initial definition of our wave function, $\tilde{A}(\mathbf{x}, t)$. Formally introducing a response function, $F(\omega)$, the linearized equation for $\tilde{A}$ can be written as

$$\left[-\nabla^2 - F_0\left(i\frac{\partial}{\partial t}\right)\right]\tilde{A}(\mathbf{x}, t) = 0 \tag{11.31}$$

With the help of $-i\nabla = \mathbf{k}$ and $i\partial/\partial t = \omega$ one obtains the linear dispersion relation

$$k^2 = F_0(\omega) \tag{11.32}$$

The function $F_0(\omega)$ is the inverse of our usual linear dispersion relations which we wrote in the form $\omega = \omega(\mathbf{k})$, here written down for the isotropic case which is easily generalized to non-isotropic conditions.

Since we are dealing with the nonlinear case, the function $F(\omega)$ is an operator function acting on the complex wave amplitude, $\tilde{A}(\mathbf{x}, t)$. But if the change in the amplitude proceeds slow enough that the wave performs a number of oscillations before the amplitude has changed remarkably, we can expand the nonlinear equation corresponding to Eq. (11.31) with the full function $F(\omega)$ instead of $F_0(\omega)$, around $F_0(\omega_0)$ to obtain

$$F_0(\omega)\tilde{A} = \left[F_0(\omega_0) + i\left.\frac{dF_0(\omega)}{d\omega}\right|_0 \frac{\partial}{\partial t} - \frac{1}{2}\left.\frac{d^2F_0(\omega)}{d\omega^2}\right|_0 \frac{\partial^2}{\partial t^2}\right] A_0(\mathbf{x}, t)\mathrm{e}^{-i(k_0 x - \omega_0 t)} \tag{11.33}$$

From the linear dispersion equation we have $k_0^2 = F_0(\omega_0)$ and the resulting two relations for the derivatives of the response function are

$$\begin{aligned} dF_0(\omega)/d\omega|_0 &= 2k_0/v_{gr0} \\ d^2F_0(\omega)/d\omega^2|_0 &= 2(v_{gr0} - k_0 g_{gr0})/v_{gr0}^3 \end{aligned} \tag{11.34}$$

With the help of these three expressions we can rewrite the general dispersion relation in terms of an equation for the complex amplitude

$$\left[i\frac{d}{dt} + \frac{g_{gr0}}{2}\nabla_x^2 + \frac{v_{gr0}}{2k_0}\nabla_\perp^2\right] A_0(\mathbf{x}, t) = 0 \tag{11.35}$$

In the derivation of this equation we took into account only the nonlinearity of the response function, but still neglected the amplitude dependence. The coefficient of the missing term is proportional to $A^2 = |A_0|^2$, as we found before in the last term of Eq. (11.23). Moreover, it is proportional to the derivative of the frequency with respect to the amplitude. Hence, adding it to the above equation (11.35) gives the nonlinear equation we are looking for

$$\boxed{\left[i\frac{d}{dt} + \frac{g_{gr0}}{2}\nabla_x^2 + \frac{v_{gr0}}{2k_0}\nabla_\perp^2 - (\nabla_A\omega|_0)\,|A_0(\mathbf{x}, t)|^2\right] A_0(\mathbf{x}, t) = 0} \tag{11.36}$$

This is the *nonlinear parabolic equation* for the slowly varying complex wave amplitude, $A_0(\mathbf{x}, t)$. It is third order in the amplitude, but the nonlinearity is only in the modulus and not in the phase of the amplitude, which simplifies the equation a little.

The physical meaning of the nonlinear equation (11.36) or the equivalent system of equations for the square of the amplitude, A^2, and its phase, φ, is that the large-amplitude wave affects its own temporal and spatial evolution. This *self-modulation* of the wave is caused by the reaction of the background medium to the wave energy flow and therefore to the wave pressure exerted by the wave on the medium. The pressure variations change the local conditions of wave propagation by affecting the background plasma density and temperature, and in magnetized plasmas also the magnetic field pressure and stresses. When these changes become susceptible, they cause the plasma dielectric response function to deviate from its linear form. As a consequence, the wave properties themselves are modulated, and the nonlinear wave may look very unlike the familiar linear waves. Famous examples are hydrodynamic and plasma turbulence, where the media are highly distorted and the wave modes present cannot in a simple way be identified with the well-known linear hydrodynamic or plasma modes.

Nonlinear Schrödinger Equation

For electromagnetic waves with the dielectric constant given by Eq. (11.1), the nonlinear parabolic equation assumes a particularly simple form. As we already know, large-amplitude waves exert a dynamic pressure on the plasma which leads to a ponderomotive potential and ponderomotive force. Using a very simple quantum-mechanical argument, one can obtain an equation for the wave amplitude which is a generalization of the Schrödinger equation to include nonlinear effects, the *nonlinear Schrödinger equation.*

Before presenting this argument, we will derive the nonlinear Schrödinger equation from the nonlinear parabolic equation (11.36). The dielectric (11.1) allows to write the refraction index $N^2(\omega, A^2) = \epsilon(\omega, A^2)$ as function of the wave intensity. The square of the wave amplitude is $A^2 = |\delta E|^2$. Expanding with respect to $|\delta E|^2$, one obtains

$$N^2(\omega, |\delta E|^2) = N_0^2(\omega)(1 + \alpha|\delta E|^2) \tag{11.37}$$

where $N_0^2(\omega)$ is the linear refraction index. With $g_{gr0} = (-v_{gr0}^3/c)[\partial^2(\omega N_0)/\partial\omega^2]$, and $\nabla_A\omega = -(v_{gr0}k_0/2N_0^2)\nabla_A N^2$ along the wave ray path, the nonlinear parabolic equation can be rewritten into the following equation for the amplitude of the electric field

$$\left[2i\left(\frac{\partial}{\partial t} + v_{gr0}\nabla_x\right) - v_{gr0}^3\frac{\partial^2 k_0}{\partial\omega_0^2}\nabla_x^2 + \frac{v_{gr0}}{k_0}\nabla_\perp^2 + v_{gr0}k_0\alpha|\delta E|^2\right]\delta E = 0 \tag{11.38}$$

This is one form of the nonlinear Schrödinger equation. It is now trivial to see that this equation actually has the form of a quantum-mechanical wave equation if the amplitude, $E(\mathbf{x}, t)$, of the electric field is interpreted as being proportional to the wave function, $\psi(\mathbf{x}, t)$. Schrödinger's equation of a particle in a potential, U, reads

$$\left[i\hbar\frac{\partial}{\partial t} + \frac{\hbar^2}{2m}\nabla^2 - U(\mathbf{x})\right]\psi(\mathbf{x}, t) = 0 \tag{11.39}$$

Except for the linear spatial derivatives in Eq. (11.38), which can be absorbed by a suitable transformation of the coordinates, the two equations are formally identical. The nonlinear dependence of the ponderomotive potential on the wave amplitude is the main difference between both equations. It describes the self-modulation of the plasma wave.

The peculiar nonlinearity of the nonlinear Schrödinger equation (11.38) is of third order in the wave amplitude. In this respect the nonlinear Schrödinger equation differs from the other nonlinear wave equations, which contained first-order nonlinearities only but higher derivatives. Instead, the Schrödinger equation is second-order in space but third-order in the wave function equation. In the version given here it is free of any dissipation and has localized solutions similar to the Korteweg-de Vries equation.

But the present solutions are physically very different from those of the Korteweg-de Vries equation. The localized solutions of the latter equation are topological deformation of the wave profile. In contrast, the solutions of the nonlinear Schrödinger equation are holes in the background plasma density which is the carrier of the wave. These *cavities* are caused by the wave pressure force, and the wave itself is trapped inside the cavities with the cavities forming *envelopes* around the region of high wave intensity. In a stationary state the localized solution is produced by the equilibrium between the plasma and wave pressures or, more precisely, between the total pressure, including plasma and magnetic pressure, and the wave pressure.

11.3. Modulational Instability

Before proceeding to find a method to solve the nonlinear parabolic or nonlinear Schrödinger equations (11.36) and (11.38) for a number of particular cases, we investigate the stability of Eq. (11.36) or the equivalent set of equations (11.23) and (11.30).

Linearization Around Initial Wave

The idea is the following. Given a finite-amplitude wave of amplitude $A^2(0)$, it is asked if the above nonlinear equations allow for an instability to arise and what are the conditions of instability. This means that we are going to investigate the evolution of deviations from the initial large amplitude of the wave caused by the nonlinear character of the interaction between the wave and the plasma. Rewriting Eq. (11.23) for this particular case, we obtain

$$\frac{d\varphi}{dt} + \frac{g_{gr0}}{2}(\nabla_x\varphi)^2 + \frac{v_{gr0}}{2k_0}(\nabla_\perp\varphi)^2 + (\nabla_A\omega|_0)\left[A^2 - A^2(0)\right] = 0 \qquad (11.40)$$

Note that d/dt is the convective derivative. The formal replacement of $A^2 \to A^2 - A^2(0)$ is justified by Eq. (11.19). It shows that in an external initial wave field of amplitude $A(0)$ the frequency is Doppler-shifted by $-\nabla_A\omega|_0 A^2(0)$, which implies that the operator $\partial\varphi/\partial t$ is to be replaced by $(\partial\varphi/\partial t) - \nabla_A\omega|_0 A^2(0)$.

As taken from Eq. (11.40), the stationary wave has the solution $A = A(0), \varphi = 0$. The one-dimensional non-stationary case obeys the nonlinear equations

$$\begin{aligned} \frac{\partial\varphi}{\partial t'} + \frac{1}{2}[\nabla_{x'}\varphi]^2 + \frac{1}{2g_{gr0}}(\nabla_A\omega|_0)[A^2 - A^2(0)] &= 0 \\ \frac{\partial A^2}{\partial t'} + \nabla_{x'}[A^2\nabla_{x'}\varphi] &= 0 \end{aligned} \tag{11.41}$$

These equations have been written in the new coordinates, $t' = g_{gr0}t$ and $x' = x - v_{gr0}t$, in order to get rid of the linear derivative with respect to x. They are similar to the hydrodynamic set of equations if A^2 is interpreted as a density, and $\nabla_{x'}\varphi$ as a velocity. The first equation is then a Hamilton-Jacobi equation, and the velocity of pseudo-sound can be read from the pseudo-pressure term as

$$c_{s0}^2 = (A^2(0)/g_{gr0})\,\nabla_A\omega|_0 \tag{11.42}$$

Instability Criteria

It follows immediately that instability arises if the pseudo-sound speed becomes negative, because then the linear pseudo-sound dispersion relation

$$\tilde{\omega} = \pm i\tilde{k}|c_{s0}| \tag{11.43}$$

has complex conjugate imaginary solutions for the frequency, $\tilde{\omega}$. The condition for parallel instability, the *Lighthill condition*, is therefore

$$\boxed{\nabla_A\omega|_0/g_{gr0} < 0} \tag{11.44}$$

Note that g_{gr0} may be positive or negative and that parallel is meant with respect to the direction of the zero-order group velocity. Similarly, linearizing the above equations one easily finds that transverse disturbances become unstable under the simpler condition

$$\boxed{\nabla_A\omega|_0 < 0} \tag{11.45}$$

Again it is sufficient to know the nonlinear dispersion relation in order to decide whether or not the finite amplitude wave will become unstable. The initial behavior of the large-amplitude wave is entirely determined by its dispersive properties which, in contrast to the Korteweg-de Vries equation, in this case depends also on the wave amplitude.

The instability found is a result of the presence of the large-amplitude wave and the change of the properties of the medium on top of which the wave propagates caused by the pressure of the wave. This change modifies the dispersion properties, and instability of the wave amplitude may arise. As result of this instability the amplitude of the wave is modulated locally. Instabilities of this kind are therefore called *modulational instabilities*. They do not exist in the linear theory of waves and are a very peculiar dynamical response of the plasma to the presence of finite amplitude waves.

The modulational instability can formally be considered as a four-wave interaction which could also be described in the terms of the weakly turbulent wave-wave interaction theory. The four waves are the high-frequency carrier mode of the initial finite-amplitude spatially inhomogeneous wave, its slowly varying envelope wave and the two end products, the large-amplitude envelope and the modified carrier wave. It must, however, be emphasized that such a description, sometimes called the *oscillating two-stream instability* is valid only in the initial linear phase of the modulational instability. Later evolution cannot be described anymore by wave-wave interaction theory because of the strong coupling of the interaction which is implied by the dependence of the nonlinear dispersion relation on the large amplitude of the wave.

Modulational instabilities constitute the initial step of the nonlinear evolution of large-amplitude waves. Further evolution must take into account the dynamical effect of the nonlinearity, which in many cases leads to saturation and modification of the wave, formation of solitary structures and self-focussing. In some particular cases, however, further instability sets in leading to plasma collapse. Some of these problems will be briefly investigated below.

Nonlinear Schrödinger Equation

The above theory can be applied to the nonlinear Schrödinger equation (11.38), which in terms of the initial large-amplitude wave and in co-moving coordinates

$$\begin{aligned} \beta &= -\nabla_{|\delta E|^2}/(\partial^2\omega/\partial k_0^2) \\ \xi &= x - t(\partial\omega/\partial k_0) \\ \tau &= t(\partial^2\omega/\partial k_0^2) \end{aligned} \tag{11.46}$$

reads

$$\boxed{\left[i\frac{\partial}{\partial\tau} + \frac{1}{2}\nabla_\xi^2 + \beta\left(|\delta E|^2 - |\delta E_0|^2\right)\right]\delta E(\xi,\tau) = 0} \tag{11.47}$$

where $|\delta E_0|^2$ is the large-amplitude electric wave amplitude, which is injected into the plasma and tends to modulate itself.

We again introduce new variables, the modulated amplitude, a, and the modulated phase, ϕ, of the wave according to $\delta E = a\exp i\phi$, and find after substitution

$$\begin{aligned} \frac{\partial a^2}{\partial \tau} + \nabla_\xi (a^2 \nabla_\xi \phi) &= 0 \\ \frac{1}{4a^2}\nabla_\xi^2 a^2 - \frac{1}{8a^4}(\nabla_\xi a^2)^2 + \beta(a^2 - a_0^2) - \frac{\partial \phi}{\partial \tau} - \frac{1}{2}(\nabla_\xi^2 \phi)^2 &= 0 \end{aligned} \tag{11.48}$$

We linearize this set of equations around the initial amplitude factor, a_0^2, and phase, $\phi_0 = 0$, and solve the linearized set with the usual plane wave ansatz, $\exp i(\kappa\xi - \varpi\tau)$, for a and ϕ. This yields linearly oscillating wave solutions of wavenumber, κ, and frequency, ϖ, resulting from the dispersion relation

$$\varpi^2 = (\kappa^2 - 2\beta a_0^2)^2/4 - \beta^2 a_0^4 \tag{11.49}$$

This dispersion relation allows for purely growing or damped solutions. With $\beta > 0$ instability becomes possible for small κ. The maximum growth rate is obtained for $\kappa = (2\beta a_0^2)^{1/2}$ with maximum growth rate

$$\gamma_{\max} = \mathrm{Im}\varpi_{\max} = \beta a_0^2 \tag{11.50}$$

Instability sets on for wavelengths $\lambda > \lambda_{cr}$ where the critical threshold wavelength is

$$\boxed{\lambda_{cr} = \pi / a_0 \beta^{1/2}} \tag{11.51}$$

The nonlinear Schrödinger equation provides an example how a large-amplitude wave under the action of its own pressure force begins to self-modulate. Self-modulation of its amplitude commences with a simple linear phase, which can be described as a modulational instability. As mentioned before, this instability can be envisioned as a four-wave process where the initial large-amplitude wave, a low-frequency sound wave, the modulated wave and the growing sound wave are involved. But this description is valid only for the initial short linear state. Later on the modulation becomes strong and generates a very large number of waves in both modes, the sound and the modulated wave mode which must be described by a broad spectrum nonlinear theory. Here the modulational instability picture fails and we deal with what we call strong turbulence. In the next section we give a brief account of strong turbulence in one particular mode, the Langmuir wave in a non-magnetized plasma.

Envelope Cavitons

We now proceed to solve the *stationary nonlinear Schrödinger equation* (11.38). For simplicity we write it in one-dimensional form

$$\left[i\left(\frac{\partial}{\partial t} + \frac{\partial \omega}{\partial k_0}\nabla_x \right) + \frac{1}{2}\frac{\partial^2 \omega}{\partial k_0^2}\nabla_x^2 - \left(\nabla_{|\delta E|^2}\omega\right)|\delta E|^2 \right] \delta E = 0 \tag{11.52}$$

and use the explicit expressions for the coefficients. In co-moving normalized coordinates the linear term in the nonlinear Schrödinger equation disappears

$$\boxed{\left(i\frac{\partial}{\partial\tau}+\frac{1}{2}\nabla_\xi^2+\beta|\delta E|^2\right)\delta E=0} \tag{11.53}$$

For localized solutions with vanishing fields and derivatives at infinity we can write

$$\delta E(\xi,\tau)=a(\xi,\tau)\exp i\phi(\xi,\tau) \tag{11.54}$$

The two coupled equations obtained for real a and ϕ are $a^2\nabla_\xi\phi=\text{const}$, and

$$\frac{d}{da^2}\left[\frac{1}{4a^2}\left(\frac{\partial a^2}{\partial\xi}\right)^2+\beta a^4\right]-\left(\frac{\partial\phi}{\partial\xi}\right)^2=0 \tag{11.55}$$

Since ϕ can be expressed through a^2, this equation becomes an ordinary differential equation for a^2 which is of the form of the energy conservation equation $(da^2/d\xi)^2=-S(a^2)$ with pseudo-potential

$$S(a^2)=4\beta a^6-8c_1a^4-c_2a^2+4c_3 \tag{11.56}$$

For $\beta>0$ the localized solution requires $c_2=c_3=0$. Introducing $a_{\rm m}^2=2c_1/\beta$ for the nonlinear maximum amplitude, the conservation equation can be written as

$$(da^2/d\xi)^2=-4\beta a^4(a^2-a_{\rm m}^2) \tag{11.57}$$

Its solution is, like for the Korteweg-de Vries equation, a hyperbolic function

$$\boxed{a=a_{\rm m}\,\text{sech}\left(a_{\rm m}\xi\beta^{1/2}\right)} \tag{11.58}$$

In order to exist, the amplitude of the initial wave must be positive, $a_{\rm m}>0$. In contrast to the Korteweg-de Vries soliton solution the envelope caviton solution is proportional to the first power of the hyperbolic secans of the wave amplitude. Cavitons filled with trapped waves are wider than the corresponding topological solitons. Another difference is that the caviton moves at the group velocity of the waves, and this is independent of the wave amplitude. In the case of the topological soliton the velocity depends on the amplitude, leading to faster speeds for larger amplitudes, but cavitons move all at the same group velocity and therefore do not overtake each other.

For $\beta<0$, we have $c_3\neq 0$, and the basic equation can be written with $\tilde{a}=\text{const}$

$$(da^2/d\xi)^2=4|\beta|(a^2-a_{\rm m}^2)(a^2-\tilde{a}^2)^2 \tag{11.59}$$

which when integrated yields the inverse caviton solution

$$a = \tilde{a}\left[1 - b^2\mathrm{sech}^2(\xi b|\tilde{a}\beta|^{1/2})\right]^{1/2} \tag{11.60}$$

where we used the normalized quantity $b^2 = (\tilde{a}^2 - a_{\mathrm{m}}^2)/\tilde{a}^2 < 1$. This solution corresponds formally to the accumulation of density in a region where the wave intensity is very low. It is another question if such solutions exist at all and if they are stable.

11.4. Langmuir Turbulence

Strong Langmuir turbulence is concerned with the formation of cavitons under the action of spatially inhomogeneous radiation pressure of large-amplitude Langmuir waves. For the purposes of the dynamics of the Langmuir wave field it is not important to ask how the Langmuir waves can be excited. The most frequent generation mechanism for Langmuir turbulence will be the gentle bump-in-tail instability of an electron beam. The linear theory of this instability has been discussed in Sec. 4.1. A certain problem arises insofar as we have shown that on a comparably short time scale this instability will quasilinearly saturate with the electron escaping from resonance towards lower speeds. The quasilinear saturation level may not be strong enough to drive modulational instability and to initiate strong turbulence. But when modulational instability sets in on a time scale short compared with the quasilinear time scale, the system will necessarily evolve towards strong turbulence. We will follow such a philosophy before justifying our assumptions a posteriori.

Zakharov Equations

Consider a large-amplitude Langmuir wave. The energy density of the wave is assumed to be inhomogeneously distributed over the plasma. In such a case the radiation pressure of the wave will exert a ponderomotive force (11.5) on the background plasma, which will drive low-frequency waves. In an unmagnetized homogeneous plasma there is only one other eigenmode, the ion-acoustic wave at low frequency below the ion plasma frequency, ω_{pi}. Its frequency is much less than the plasma frequency. Hence, it is the mode which is driven by the radiation pressure of the high frequency Langmuir wave.

In a two-fluid model, with ions and electrons as separate fluids, the wave equation for the ion-acoustic wave is given by

$$\boxed{\left[\frac{\partial^2}{\partial t^2} - c_{ia}^2\nabla^2\right]\frac{\delta n}{n_0} = \frac{\epsilon_0}{4m_i n_0}\nabla^2|\delta\mathbf{E}|^2} \tag{11.61}$$

This equation follows from a combination of the ion continuity equation and the ion-momentum conservation equation if the ponderomotive pressure-force term is included into the latter equation. Quasineutrality has been imposed, and the ion-acoustic speed is given by $c_{ia}^2 = (\gamma_e k_B T_e + \gamma_i k_B T_i)/m_i$. The ion-acoustic oscillations described by the wave equation part on the left-hand side of Eq. (11.61) are driven by the ponderomotive source term on the right-hand side of Eq. (11.61). Hence, they are by no means free oscillations, but exist only because the high-frequency Langmuir wave-field feeds energy into the low-frequency waves and excites low-frequency density fluctuations.

In the presence of these self-excited low-frequency density fluctuations, which may reach large amplitudes, the dispersion relation

$$\omega = \omega_{pe}\left(1 + \tfrac{3}{2}k^2\lambda_D^2\right) \tag{11.62}$$

of the high-frequency Langmuir waves will change, because the plasma switches from the originally homogeneous to an inhomogeneous state. Let us assume that the density fluctuations are slow enough and any variations in the electron temperature can be smoothed out within one oscillation. Then the electron temperature can be considered constant, and the variation of the frequency in the dispersion relation is entirely determined by the density variation

$$\delta\omega_1 = (\delta n/2n_0)\omega_{pe0} \tag{11.63}$$

where $\omega_{pe0}^2 = e^2 n_0/\epsilon_0 m_e$ is the plasma frequency with respect to the undisturbed density, n_0. The total variation, $\delta\omega_{\mathrm{tot}}$, of the frequency in the long-wavelength region (small k) is the sum of the above variation in the plasma frequency, caused by the density modulation, plus the thermal correction term, $3k^2\lambda_D^2/2$, which is of the same order as $\delta\omega_1$

$$\delta\omega_{\mathrm{tot}} = \frac{\omega_{pe0}}{2}\left(\frac{\delta n}{n_0} + 3k^2\lambda_D^2\right) \tag{11.64}$$

We interpret this expression as an operator acting on the electric field amplitude and obtain the following equation

$$\boxed{\left[i\frac{\partial}{\partial t}\mathsf{I} + \frac{3}{2}\omega_{pe0}\lambda_D^2\nabla\nabla\cdot\right]\delta\mathbf{E}(\mathbf{x},t) = \frac{\omega_{pe0}}{2}\left(\frac{\delta n}{n_0}\right)\delta\mathbf{E}(\mathbf{x},t)} \tag{11.65}$$

Equations (11.61) and (11.65) are the *Zakharov equations* of strong Langmuir turbulence. They describe the coupling between the high-frequency Langmuir wave and the low-frequency ion-acoustic wave via the density fluctuation caused by the ponderomotive force and the reaction on the amplitude of the electric Langmuir wave field. The latter of the two Zakharov equations has the form of the nonlinear Schrödinger equation.

Modulation Threshold

We have already found that Langmuir waves may exert a kind of radiation pressure on the plasma. Assume that we have a large number of Langmuir plasmons whose number density obeys a Boltzmann distribution with respect to their momenta, $\hbar k$, around an average momentum, $\hbar k_0$

$$N(k) \approx N_0(2\pi k_0^2)^{-1/2} \exp(-k^2/2k_0^2) \tag{11.66}$$

In the presence of a density perturbation, δn, we have shown that the dispersion relation of the Langmuir plasmons, i.e., is their energy law, includes the density variation of Eq. (11.64), and the plasmon distribution becomes

$$N_\ell(k) \approx N_0(2\pi k_0^2)^{-1/2} \exp\left\{-\left[\tfrac{1}{2}k^2\lambda_D^2 + \tfrac{1}{3}(\delta n/n_0)\right]/k_0^2\lambda_D^2\right\} \tag{11.67}$$

The total plasmon density is then the integral over all wavenumbers, k, of this expression

$$N_{\ell,\mathrm{tot}} = \int dk\, N_\ell(k) = N_0 \exp\left(-\frac{1}{3k_0^2\lambda_D^2}\frac{\delta n}{n_0}\right) \tag{11.68}$$

From this we can calculate the variation of the density of plasmons, which is simply given by the argument of the exponential. The radiation pressure turns out to be proportional to this plasmon density variation

$$\delta p_{\mathrm{pm}} \approx -\frac{p_{\mathrm{pm}0}}{3k_0^2\lambda_D^2}\frac{\delta n}{n_0} \tag{11.69}$$

and we obtain the threshold for instability when we require that this variation exceeds the variation in the plasma pressure, $k_B T_e \delta n$. So the instability criterion becomes simply that the initial radiation pressure satisfies

$$p_{\mathrm{pm}0} > 3k_0^2\lambda_D^2 n_0 k_B T_e \tag{11.70}$$

which tells us that for small k_0 modulational instability will set on. This is exactly the same conclusion as drawn above in the general discussion of modulational instability, but now specialized to Langmuir wave turbulence. Using Eq. (11.5), the instability condition is written as

$$\boxed{\frac{\epsilon_0|\delta E|^2}{2n_0 k_B T_e} > 6k_0^2\lambda_D^2} \tag{11.71}$$

which shows that it is sufficient to know the initial wavenumber, k_0, and the plasma parameters to decide if a Langmuir wave of a certain initial amplitude will undergo modulational instability. The above condition tells that the wavelength of the Langmuir wave must be sufficiently long for modulational instability to evolve. We are now going into the more interesting effects which arise when this is the case.

Subsonic Cavitons

The exact solution of the system of Zakharov equations is a formidable task. So far only numerical solutions have been obtained. But there are two limiting cases which allow a qualitative discussion of the behavior of its solutions.

The first limit is obtained when the time-derivative term in the ion equation (11.61) can be neglected. Formally this case corresponds to changes with velocity $\varpi/\tilde{k} \ll c_{ia}$, where ϖ and $\tilde{k}$ are the frequency and wavenumber of the driven ion-acoustic wave. When we assume that this wave will be localized, the ratio $\varpi/\tilde{k} = u$ is interpreted as the speed of the ion-acoustic wave. Therefore this case is known as the *subsonic approximation* to the Zakharov equations. We find that in this approximation

$$\boxed{\delta n \approx -\frac{\epsilon_0}{4m_i c_{ia}^2}|\delta E|^2} \tag{11.72}$$

Inserting for $\delta n/n_0$ into the right-hand side of Eq. (11.65) just reproduces the nonlinear Schrödinger equation for the Langmuir wave amplitude

$$\left[i\frac{\partial}{\partial t}\mathbf{I} + \frac{3}{2}\omega_{pe0}\lambda_D^2\nabla\nabla\cdot\right]\delta\mathbf{E}(\mathbf{x},t) = -\frac{\epsilon_0\omega_{pe0}}{8m_i n_0 c_{ia}^2}\left[|\delta\mathbf{E}(\mathbf{x},t)|^2\delta\mathbf{E}(\mathbf{x},t)\right] \tag{11.73}$$

We have shown in Eq. (11.58) that the one-dimensional version of this equation, transformed to the co-moving system of coordinates centered on the localized caviton, has a solution $\delta E \propto (L\beta^{1/2})^{-1}\mathrm{sech}(x/L)$, which is a caviton of width L. The shape of this caviton is preserved during its evolution, because in this one-dimensional state the nonlinearity and the dispersion just balance each other. One immediately observes that the wave intensity in the caviton changes with the inverse square of its width

$$|\delta E|^2 \propto L^{-2} \tag{11.74}$$

Intense one-dimensional Langmuir cavitons will be narrower than weak cavitons. This resembles Korteweg-de Vries solitons. Remember, however, that here all cavitons move at the same speed. The one-dimensional nonlinear Schrödinger equation can also be solved by the inverse scattering method, in which case it generates a large number of cavitons from a given initial large-amplitude Langmuir wave in close similarity to the behavior of the solutions of the Korteweg-de Vries equation. In the purely one-dimensional case subsonic Langmuir turbulence will decay into a large number of cavitons which may fill the plasma volume and generate small-scale large-amplitude density fluctuations with all the cavitons being density depletions filled with Langmuir plasmons.

Let us check the dependence of stable caviton formation on the dimensionality, d, of the system. The number of Langmuir plasmons trapped in one caviton is

$$N_\ell \propto \int d^d x |\delta E|^2 \tag{11.75}$$

Because N_ℓ is finite and conserved, we immediately find instead of Eq. (11.74) the general dependence of the wave intensity on the characteristic dimension of the caviton

$$|\delta E|^2(t) \propto L^{-d}(t) \tag{11.76}$$

We found from Eq. (11.64) that $\delta n/n_0 \approx k^2(t)\lambda_D^2$. Since $k(t) \approx L^{-1}(t)$ we have

$$\delta n^2 \propto L(t) \tag{11.77}$$

which is independent of the dimensionality d. The pressure of the expelled plasma is proportional to δn and thus is also independent of d, while the radiation pressure from Eq. (11.76) varies as L^{-d}. Therefore, for $d = 1$ the radiation pressure can be balanced by the plasma pressure. In other words, for $d = 1$ one obtains one well-defined width L from pressure balance for the width of the caviton during the evolution, and a stable caviton is formed. The pressure equilibrium condition just produces the above caviton solution with its relation between the amplitude and the width of the caviton.

However, for $d > 1$ shrinking of the caviton in the course of the modulation cannot be halted by the plasma pressure. The cavitons gets deeper and narrower with time progressing. This behavior is known as *plasma collapse*. Of course, shrinking will proceed only down to a width of the order of the Debye length, when the trapped plasmon wavelength becomes so short that the Langmuir waves are strongly Landau damped.

Caviton Collapse

Collapse of cavitons proceeds for $|E|^2 > L^{-2}$ (in normalized variables). In addition to the plasmon number density, there are two other invariants of the nonlinear Schrödinger equation for Langmuir waves, the total momentum $\mathbf{P}$, and the total energy, H

$$\begin{aligned} 2\mathbf{P} &= i\int d^dx(\delta\mathbf{E}\nabla\cdot\delta\mathbf{E}^* - \delta\mathbf{E}^*\nabla\cdot\delta\mathbf{E}) \\ 2H &= \int d^dx(|\nabla\cdot\delta\mathbf{E}|^2 - |\delta\mathbf{E}|^4) \end{aligned} \tag{11.78}$$

These quantities satisfy a virial theorem. Defining the root mean square spatial width of the cavitons as $\langle(\Delta L)^2\rangle = \langle(L - \langle L\rangle)^2\rangle$, where the average is $\langle f\rangle = \int f d^dx|\delta E|^2/N_\ell$, it can be shown that $\partial\langle L\rangle/\partial t = P/N_\ell = \text{const}$ and

$$\boxed{\partial^2\langle(\Delta L)^2\rangle/\partial t^2 = H/N_\ell - P^2/N_\ell^2 + (2-d)\langle|\delta\mathbf{E}|^2\rangle} \tag{11.79}$$

Integrating this virial theorem twice with respect to time yields

$$\langle(\Delta L)^2\rangle = c_1t^2 + c_2t + c_3 + (2-d)\int_0^t dt'\int_0^{t'} dt''\langle|\delta\mathbf{E}|^2\rangle \tag{11.80}$$

where $2c_1 = H/N_\ell - P^2/N_\ell^2$, and c_2, c_3 are constants of integration. One concludes that the average width of the cavitons will shrink to zero during a finite time in all cases when $d \geq 2$ and $c_1 < 0$, and the field, δE, will become singular in this case because N_ℓ is conserved.

Collapse Scaling

The above argument demonstrates that in a certain regime cavitons cannot be prevented from collapsing down to a few Debye lengths of width. Behavior like this is a typical property of strong plasma turbulence. It has a number of consequences, which we will briefly discuss below. But before doing so, let us quickly look into the dynamics of such a collapse. Returning to the ion wave equation, we observe that in the regime where the speed of the cavitons is larger than the ion acoustic speed, i.e., in the *supersonic* limit, one finds

$$\frac{\partial^2 \delta n}{\partial t^2} \approx -\frac{\epsilon_0}{4m_i}\nabla^2|\delta \mathbf{E}|^2 \qquad (11.81)$$

There is a continuous change in density with time under the action of the wave electric field in this case. Because $\delta n \propto k^2 \propto \nabla^2$, we find from a dimensional consideration of the last equation that the electric field scales as

$$|\delta \mathbf{E}|^2 \propto (t_c - t)^{-2} \propto L^{-d} \qquad (11.82)$$

where the second part of the equation has already been obtained above. Hence, the density varies with time according to

$$\delta n \propto L^{-2} \propto (t_c - t)^{-4/d} \qquad (11.83)$$

The constant time, t_c, appearing in these expressions is the finite instant when the collapse ends. Clearly, this time will never be reached in reality, because the field would become infinitely large. Before this happens, Landau damping has set in, and electron heating will cause the waves to dissipate their energy. This scaling suggests that it is possible to introduce a self-similar scaling according to $|E| \to \xi^{-1}|E(x/\xi)|$ and $\xi(t) = (t_c - t)^{1/2}$. In these quantities the shape of the caviton is conserved during the collapse.

Figure 11.1 sketches the process of caviton formation from a long-wavelength initial Langmuir wave. The thin line is the Langmuir wave intensity $|E|^2$. The radiation pressure force is strongest at the position where the gradient is steepest but the intensity already high. This is the region close but above the turning point on the intensity profile. As shown, at this place high intensities of short-wavelength Langmuir waves accumulate due to modulational instability. At the same time cavities form on the background density at these positions which trap the short Langmuir waves. For $d = 1$ these would become stable envelope cavitons. In higher dimensions they collapse, as shown by the

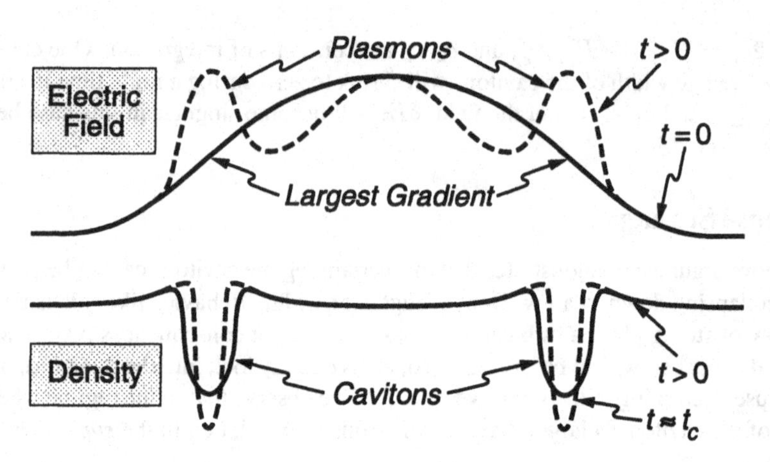

Fig. 11.1. Caviton formation and collapse in Langmuir turbulence.

deepening and narrowing of the cavitons near $t \sim t_c$. The slight enhancements in density at the borders of the cavitons in the figure schematically account for the expelled plasma density.

Returning briefly to the ion equation (11.61), we find that in the supersonic case, where the non-local Laplace term on the left-hand side is neglected, the density variation during collapse scales as

$$\delta n \propto \left[(m_e/m_i)|\delta E|^2\right]^{1/2} \tag{11.84}$$

From here we find that the fastest growing caviton has a wavenumber

$$k \propto |\delta n|^{1/2} \propto \left[(m_e/m_i)|\delta E|^2\right]^{1/4} \tag{11.85}$$

Because of the above scaling we can construct a collapse speed, $u_c \approx L/t_c$. When all terms in the second Zakharov equation are of the same order and thus of the same importance, we have the scaling $t_c^{-1} \approx L^{-2} \approx \delta n$. Then the collapse speed scales as

$$u_c = L/t_c \propto |\delta n|^{1/2} \tag{11.86}$$

The collapse speeds in the adiabatic subsonic and non-adiabatic supersonic limits therefore scale according to the following law

$$u_c \propto \begin{cases} |\delta E| & \text{for subsonic collapse} \\ [(m_e/m_i)|\delta E|^2]^{1/4} & \text{for supersonic collapse} \end{cases} \tag{11.87}$$

In the supersonic case the rate of collapse is therefore slower than in the subsonic case. The supersonic collapse takes longer time.

11.5. Lower-Hybrid Turbulence

Langmuir waves are only one example of modulational instability, caviton formation and collapse. A large number of other waves may experience the same or similar fate. Among those waves are kinetic Alfvén waves, magnetosonic waves, electromagnetic cavitons driven by intense laser and maser pulses, whistler waves in weakly magnetized plasmas, and, as a particularly important application, lower-hybrid waves.

Musher-Sturman Equations

Lower-hybrid waves have the particular property that they propagate perpendicular to the magnetic field and are intrinsically two-dimensional. As a consequence, the wave radiation pressure becomes an anisotropic tensor, and the ponderomotive forces parallel and perpendicular to the magnetic field are different. The equations describing strong lower-hybrid turbulence are not as simple as the Zakharov equations (11.65). In particular, the parabolic equation cannot be written in the simple form of a cubic nonlinear Schrödinger equation. This is obvious from a first glance at the linear dispersion relation of lower hybrid waves

$$\omega = \omega_{lh}\left(1 + \frac{R^2k^2}{2} + \frac{m_i}{2m_e}\frac{k_\parallel^2}{k^2} - \frac{\omega_{pe}^2}{k^2c^2}\frac{\omega_{pe}^2}{\omega_{pe}^2+\omega_{ge}^2}\right) \tag{11.88}$$

where the quantity R is a typical dispersion length of lower-hybrid waves

$$R^2 = \frac{3k_BT_i}{m_i\omega_{lh}^2} + \frac{2k_BT_e}{m_e\omega_{ge}^2}\frac{\omega_{pe}^2}{\omega_{ge}^2+\omega_{pe}^2} \tag{11.89}$$

which in the case of a dense plasma with $\omega_{pe} \gg \omega_{ge}$ is the electron gyroradius.

If one takes the variation of the frequency in the same way as we did in strong Langmuir turbulence, the resulting expression contains variations with respect to density, δn, magnetic field, $\delta\mathbf{B}$, and terms containing the parallel, perpendicular, and full wavenumbers. Interpreting the latter components as operators in space, the resulting equations contain complicated and mixed derivatives of the electric field vector, $\mathbf{E}$, in all directions. One may, however, introduce a number of approximations. The simplest one is the assumption that the wave field is purely electrostatic and that the electric field drifts contribute most. In this case the electric field obeys the following set of equations

$$\begin{aligned}\left(\frac{\partial^2}{\partial t^2} - c_{ia}^2\nabla^2\right)\delta n &= \frac{i\epsilon_0}{4m_i}\frac{\omega_{pe}^2}{\omega_{ge}\omega_{lh}}\nabla^2\left(\nabla\delta\phi^*\times\nabla\delta\phi\right)_\parallel \\ \left(\frac{2i}{\omega_{lh}}\nabla^2\frac{\partial}{\partial t} + R^2\nabla^4 - \frac{m_i}{m_e}\nabla_\parallel^2 - \frac{m_i}{m_e}\frac{\omega_{pe}^2}{c^2}\frac{\omega_{lh}^2}{\omega_{ge}^2}\right)\delta\phi &= \frac{i}{n_0}\frac{m_i}{m_e}\frac{\omega_{lh}}{\omega_{ge}}\left(\nabla\delta\phi\times\nabla\delta n\right)_\parallel\end{aligned} \tag{11.90}$$

These equations are written in terms of the electric wave potential, $\delta\phi$, instead of the electric field, $\delta\mathbf{E}$. They are called *Musher-Sturman equations* and govern strong lower-hybrid turbulence and lower-hybrid collapse.

Lower-Hybrid Collapse Scaling

The Musher-Sturman equations (11.90) are written in terms of the wave potential and the density variation in the background. The latter is simply the ion-acoustic wave equation as before, but takes into account the anisotropy. The cross-products appear because $E \times B$ motions and polarization motions are important in the lower-hybrid wave. Solving these equations is much more difficult than in the Langmuir caviton case. But it can also be shown that their solutions yield cavitons, which are formed by modulational instability. The cavitons turn out to be two-dimensional structures, which form cigars or pancakes of transverse and parallel lengths

$$\begin{aligned} L_\perp &\approx R\left[\frac{\alpha m_e}{2m_i}\frac{n_0 k_B(T_e+T_i)}{W_{lh}}\right]^{1/2} \\ L_\parallel &\approx \frac{L_\perp^2}{R}\left(\frac{m_i}{m_e}\right)^{1/2} \end{aligned} \tag{11.91}$$

Hence, these cavitons are elongated along the magnetic field but propagate at the ion-acoustic velocity. They also experience self-similar collapse which can be described by self-similar formulae. The threshold for modulational instability and caviton formation is given by

$$\boxed{\frac{W_E}{n_0 k_B(T_e+T_i)} \approx 2\Delta_0\left(1+\frac{k^2}{k_0^2}\right)\left(1+\frac{\omega_{pe}^2}{\omega_{ge}^2}\right)^{-1}} \tag{11.92}$$

The abbreviations used are

$$\Delta_0 = \frac{1}{2}\left(R^2k^2 - \frac{\alpha\omega_{pe}^2}{k^2c^2+k_0^2c^2} + \frac{\alpha\omega_{pe}^2}{k_0^2c^2} + \frac{m_i}{m_e}\frac{k_\parallel^2}{k^2+k_0^2}\right) \tag{11.93}$$

and $\alpha = (1+\omega_{ge}^2/\omega_{pe}^2)^{-1}$. This threshold condition shows that the threshold is fairly low for lower-hybrid waves and that they can easily reach the regime of modulational instability. Once the threshold given in Eq. (11.92) is exceeded, lower-hybrid waves will inevitably undergo modulational instability, caviton formation and collapse in a way similar to Langmuir turbulence. But this collapse is two-dimensional and anisotropic and shows a number of differences when compared with Langmuir collapse.

In case of collapse, the length scales of the caviton become functions of the wave intensity, and the time of caviton formation is equal to the inverse growth rate of the modulation of the lower-hybrid wave, $t_{\rm mod} \propto \gamma_{\rm mod}^{-1}$, where

$$\gamma_{\rm mod} \approx \omega_{lh} \frac{m_i}{m_e} \left(1 + \frac{\omega_{pe}^2}{\omega_{ge}^2}\right) \frac{W_E}{2n_0 k_B (T_e + T_i)} \tag{11.94}$$

If one considers the self-similar evolution, one finds from the basic equations that the density evolves as $\delta n \propto (t_c - t)^{-1}$, while the speed of the collapse, which has been defined above for Langmuir waves, is $u_c \propto (t_c - t)^{-1/2}$.

11.6. Particle Effects

So far we have neglected the effects on the particles and how the reaction of these particles affects the evolution of the collapse. Before performing some relevant estimates let us explain the physics of this interaction.

Langmuir Turbulence Effects

The Zakharov equations with the ion wave equation (11.61) and the nonlinear Schrödinger equation (11.65) describe the localized excitation of ion-acoustic waves forming cavitons and the trapping of short-wavelength Langmuir waves in these cavitons under the action of the Langmuir wave radiation pressure. The initial state is a long-wavelength Langmuir pump wave, excited presumably by a gentle beam instability.

As the linear theory of the gentle beam instability suggests, Langmuir modes of this kind are resonant with the positive slope part of the beam. Such waves may become scattered at background cold ions in the plasma, in which process they transfer a large part of their momentum to the ions. This scattering is symbolically described by

$$\ell + i \rightarrow \ell' + i' \tag{11.95}$$

where ℓ stands for the Langmuir wave, i stands for the ion, and a prime indicates scattered products. In particular, for the wavenumber this equation reads

$$\hbar \mathbf{k}_\ell + \mathbf{p}_i \rightarrow \hbar \mathbf{k}'_\ell + \mathbf{p}'_i \tag{11.96}$$

Since the momentum, $\hbar k_\ell$, of a wave is proportional to its wavenumber, k, loss of momentum implies increase of the Langmuir wavelength and decreasing wavenumber. The Langmuir spectrum, in the course of many scattering events, cascades down to low k.

This implies that the phase velocity of the scattered wave, ω/k, increases considerably, and the waves shift out of resonance to high velocities, where no particles exist which could Landau damp the waves. This scenario is depicted in Fig. 11.2. Clearly

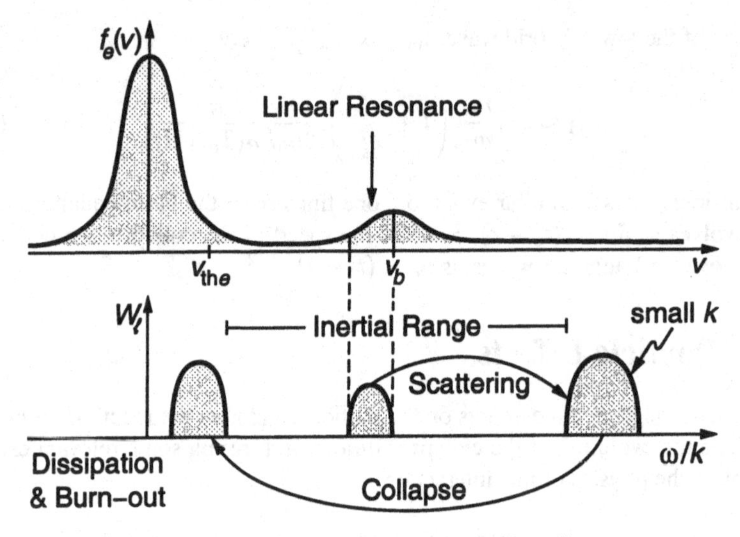

Fig. 11.2. Langmuir wave dynamics during collapse.

the Langmuir wave energy accumulates here at long wavelengths until it surpasses the threshold for modulational instability given in Eq. (11.71). When this happens, caviton formation takes over and breaks the wave off into smaller parts of intense but localized plasmon density.

Caviton Burn-Out

The process of modulational formation of cavitons is very fast so that the Langmuir wave readily divides into short-wavelength modes trapped in the cavities. When the wavelength of these modes becomes so short that the phase velocity decreases until it matches the thermal velocity of the plasma, the trapped waves become quickly Landau damped and disappear. One speaks of a *burn-out* of the cavitons. The burn-out leaves empty cavitons which cannot sustain the plasma pressure anymore and break off into propagating ion-acoustic holes which run away at the ion-acoustic speed until they resolve. These processes cause particle heating and acceleration and ultimately lead to nonlinear deformation of the particle as well as the Langmuir wave spectrum.

Inertial Range Spectrum

Here we are interested in the deformation of the Langmuir wave spectrum. We consider the inertial collapse range, when the Langmuir wave energy flows through the wavenumber space during collapse from the long-wavelength regime into the short-wavelength regime before becoming dissipated. In this range the energy flux is simply a constant, meaning that the energy flowing in at long wavelengths flows out at short wavelengths

$$W_\ell(k)[dk/dt(k)] = \text{const} \tag{11.97}$$

This equation corresponds to Kolmogorov's postulate. The velocity of energy flow in k-space is given by dk/dt, with the time depending on the wavenumber. From the above scaling we found for the collapse time $t_c \propto L^{d/2} \propto k^{-d/2}$. Inserting into Kolmogorov's condition, we immediately find that the spectral energy density of the Langmuir waves in the inertial range, where no dissipation is taken into account, scales as

$$\boxed{W_\ell(k) \propto k^{-(1+d/2)}} \tag{11.98}$$

In two dimensions the spectrum thus scales as $W_\ell \propto 1/k^2$, while in three dimensions it scales as $W_\ell \propto 1/k^{5/2}$. This shape of the spectral energy density of Langmuir waves in strong turbulence is sketched in Fig. 11.3. The spectral energy density is given on a logarithmic scale in dependence on the wavenumber, k. Long waves are injected at small k, from where the energy flows across a broad inertial range into the domain of dissipation of wave energy by Landau damping.

Dissipative Range Spectrum

The spectral shape in the dissipative range can be estimated assuming self-similar collapse and Landau damping. Self-similarity requires that $W_\ell = A\xi^{-3}F(r/\xi)$ for a spherically symmetric caviton of shape function F. The wave amplitude, A, and coordinate, ξ, during collapse obey

$$\begin{aligned} \partial A/\partial t &= -2\gamma(A, \xi^{-1})A \\ \partial \xi/\partial t &= -\nu(A, \xi)\xi \end{aligned} \tag{11.99}$$

where γ is the quasilinear Landau damping rate from Eq. (4.3), and the coefficient $\nu = (A/\xi^3)^{1/2}$ is the collapse rate expressed in the self-similar coordinates, since ξ is the self-similar size corresponding to wavenumber. If ξ and A are assumed to be random variables, the spectral energy density is defined as

$$W_\ell(k) \propto n_{\rm cav} \int dA d\xi\, f(A, \xi)\xi F(\mathbf{k}\xi) \tag{11.100}$$

with $n_{\rm cav}$ the spatial caviton density, $f(A, \xi)$ the probability distribution function, and $F(\mathbf{k}\xi)$ the shape function. The former satisfies the Liouville equation

$$\partial f/\partial t = \nabla_\xi(\nu\xi f) + \partial(2\gamma A f)/\partial A \tag{11.101}$$

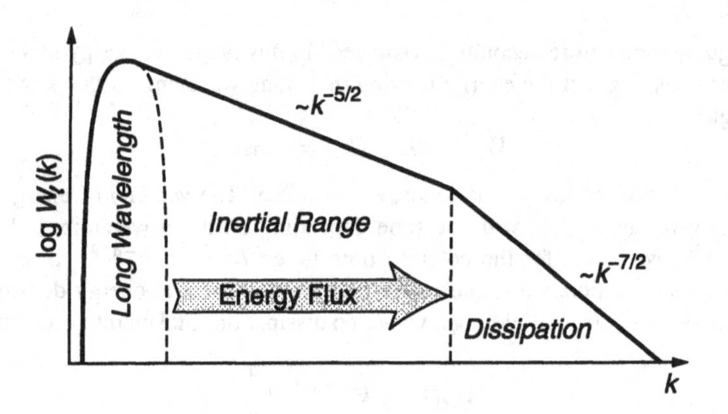

Fig. 11.3. Langmuir wave spectral energy in strong turbulence.

In exactly the same way one defines averages of the damping rate $\gamma(W_\ell)$ and of $\nu(W_\ell)$ as integrals over $f\xi g$. With these definitions, taking the time derivative of the spectral energy density integral and replacing the distribution function with the Liouville equation, one obtains the following kinetic equation for the spectral energy density

$$\partial W_\ell(k)/\partial t = -\alpha\nabla_k(k^5 W_\ell^2) - 2\gamma(k)W_\ell \tag{11.102}$$

Here α is a constant, which results from the various replacements of integrals and variables in arriving at this equation and whose exact form we do not need to know in this qualitative consideration. The first term on the right-hand side is the divergence of the energy flow in k-space, the second is the dissipation term.

In the inertial range the time derivative and damping rate are both zero. The first term on the right-hand side then immediately reproduces the $k^{-5/2}$ dependence of the inertial spectrum. In the dissipative range one can use the quasilinear set of equations consisting of the growth rate, the quasilinear diffusion equation for the electron distribution function, and the quasilinear diffusion coefficient which we write here as

$$\begin{aligned}
\gamma(k) &\approx \int d^3v\mathbf{k}\cdot\nabla_{\mathbf{v}} f_e\delta(\omega_l - \mathbf{k}\cdot\mathbf{v}) \\
\partial f_e/\partial t &= \nabla_v D(v)\nabla_v f_e \\
D(v) &\approx \int d^3k W_\ell(\mathbf{k})\delta(\omega_l - \mathbf{k}\cdot\mathbf{v})\mathbf{kk}/k^2
\end{aligned} \tag{11.103}$$

These equations have to be solved simultaneously with the above equation for W_ℓ.

Let us assume isotropy in velocity and wavenumber space. In this case one can find a steady state solution for both the spectral energy density and the particle distribution function, f_e, which holds for Landau damping in the dissipative domain of k-space

$$\boxed{\begin{aligned} W_\ell(k) &\propto k^{-2-d/2} \\ f_e(v) &\propto v^{1-3d/2} \end{aligned}} \tag{11.104}$$

This dependence yields in three-dimensional space the $k^{-7/2}$ spectrum of the dissipative short wavelength region. This kind of spectrum is shown at large k in Fig. 11.3. As a side product, the electron distribution function turns out to develop a tail, which decays much less steeply with velocity than the initial electron Maxwell distribution function.

Lower-Hybrid Turbulence Effects

Lower-hybrid collapse does also affect the particle distribution, but the interactions are even more subtle than in Langmuir turbulence. This is not only due to the anisotropy of the lower-hybrid wave spectrum, but also to the entirely different dynamics involved in these waves due to the magnetization of the plasma.

Ions behave unmagnetized and observe a wave field, which is a high-frequency field on their time scales. In such a field they can readily be heated. However, because the wave field is predominantly perpendicular to the magnetic field with only a small parallel electric field component, ion heating is predominantly transverse. This transverse heating implies transverse damping and is one of the main dissipation mechanisms of lower-hybrid turbulence. Electrons, on the other hand, are strongly magnetized in lower-hybrid waves and see a low-frequency field. Hence, their acceleration and heating proceeds only parallel to the magnetic field. Parallel Landau damping due to electrons is the main dissipation mechanism for the parallel wave component.

The collapse itself proceeds in an anisotropic way forming cigars. But at the same time it has been found in numerical simulations that the trapping of lower-hybrid waves in cavitons is not complete. Some of the lower-hybrid plasmons may leak out of the caviton and form escaping wave trains, which, as long as their intensity exceeds the threshold for modulational instability, may collapse themselves and produces a very broad wavenumber spectrum of low intensity lower-hybrid cavitons at wave energies close to threshold. Lower-hybrid collapse is therefore a very complicated process, which has not yet been understood satisfactory. Assuming that an inertial range exists as in Langmuir turbulence, the spectral energy density can be estimated as

$$\boxed{W_E \propto k^{-3}} \tag{11.105}$$

It decays with k at a slightly steeper power law than the Langmuir collapse spectrum. But by the argument given above one expects that the spectral energy density of lower-

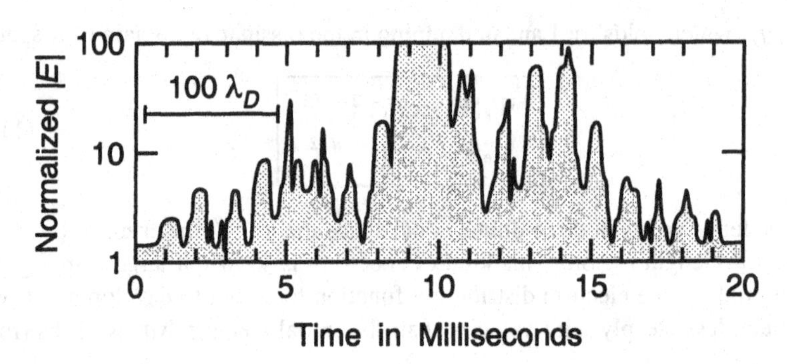

Fig. 11.4. Modulated and burnt-out Langmuir waves and cavitons in Jupiter's foreshock.

hybrid cavitons will settle close to the threshold intensity for caviton formation and modulational instability.

The nonlinear evolution of large-amplitude lower-hybrid waves is thus similar to that of Langmuir waves. Lower-hybrid waves evolve towards strong turbulence, creating a large number of cavitons with energies close to the threshold. The large-amplitude cavitons undergo collapse, accompanied by transversely heated ions, parallel accelerated electrons, and by burnt out cigar structures in the background density.

Collapse Observations

Jupiter's bow shock accelerates fast electron beams in the same way as the Earth's bow shock against the solar wind, giving rise to a gentle-beam situation and generation of intense Langmuir waves. These waves scatter off solar wind ions, condensate at long wavelengths, and undergo modulational instability and collapse. As result highly modulated Langmuir wave forms will be produced. Such large-amplitude narrow-modulated wave packets are shown in Fig. 11.4. Their amplitudes are more than four orders of magnitude above the normal level, and the modulation length is some ten λ_D, resembling the expectations. However, there is some difference between naive collapse theory and observation. Collapse theory predicts that the most narrow packets have the largest amplitudes, which is not the case. Hence, the narrow structures should be interpreted as burnt-out cavitons, having undergone dissipation and nucleation into smaller structures.

Measurements of narrow cavitons in the solar wind are very difficult, since the cavitons propagate at the speed of ion-acoustic waves, which is much lower than the undisturbed solar wind speed, $c_{ia} \ll v_{sw}$. Hence, direct measurements of structures only a few Debye lengths wide and blown across the spacecraft at the solar wind speed are impossible at the currently achievable time resolution. However, because the cavitons have a broad k-spectrum, broadband spikes of electrostatic low-frequency waves

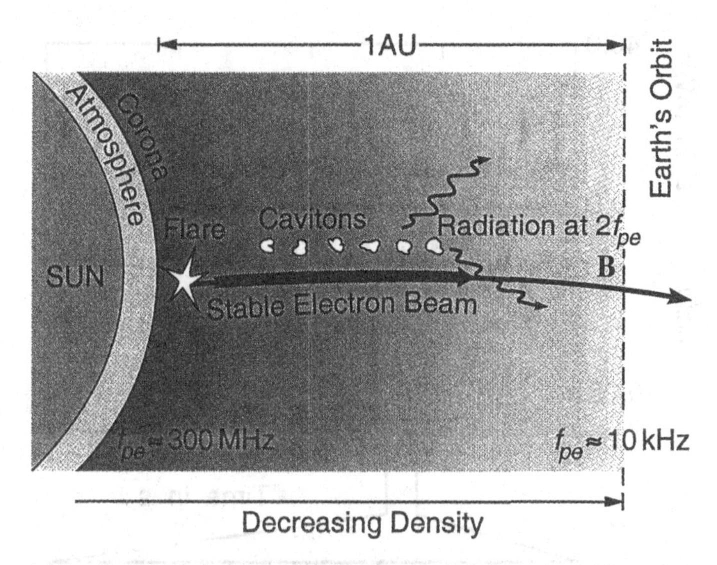

Fig. 11.5. Interplanetary type III scenario with flare ejected beam.

can be interpreted as the signature of cavitons in the solar wind plasma.

Stabilization of Interplanetary Electron Beams

Another important application of Langmuir collapse theory are the type III radio bursts. Here the production of cavitons provides the counter-propagating Langmuir waves inside the caviton, which are needed for the Langmuir wave-wave interaction producing the escaping radio wave (see Sec. 9.4). Actually, the strongly enhanced wave amplitudes inside the caviton lead to much more intense radio waves than predicted by weakly turbulent wave-wave interaction theory. Moreover, the modulation of the wave spectrum and its condensation into cavitons removes the Langmuir waves from resonance with the beam, where they otherwise would deplete the electron beam which causes the Langmuir waves by linear instability. This way the nonlinear modulation of the Langmuir wave spectrum self-stabilizes the electron beam and lets it survive over large distances, from the solar surface far into interplanetary space beyond the orbit of the Earth. This is sketched in Fig. 11.5 for an interplanetary type III radio burst as shown in Fig. 9.5.

Auroral Lower-Hybrid Cavitons

Spikelets of electrostatic wave emissions near the lower-hybrid frequency are observed above the aurora. They are correlated with localized density variations and have am-

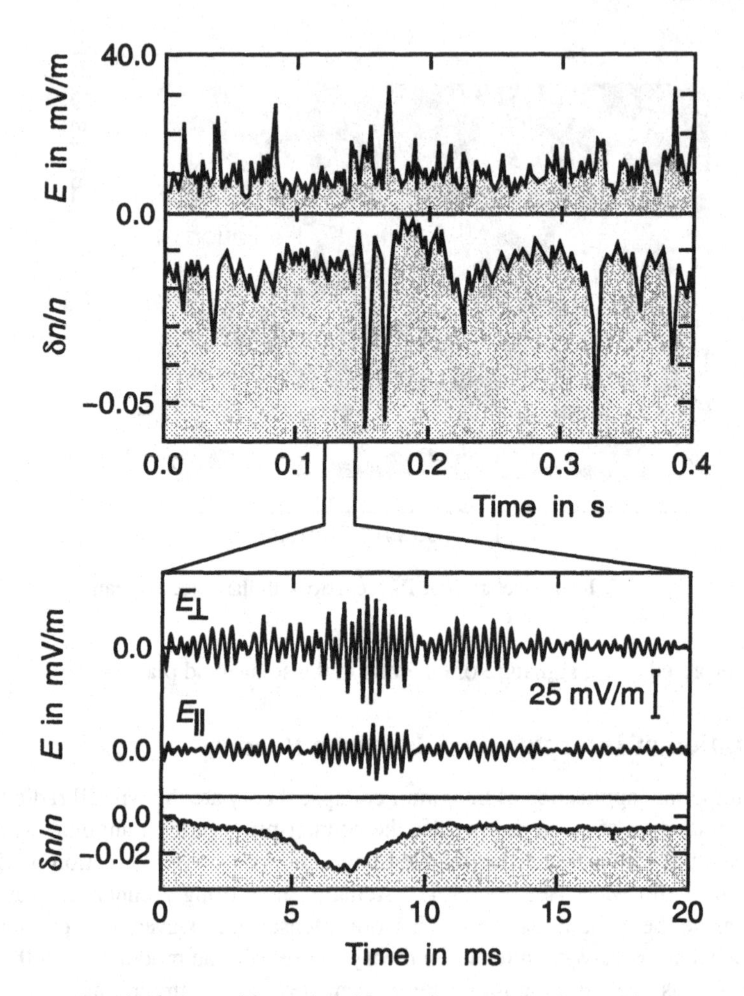

Fig. 11.6. Auroral lower-hybrid waveform and density cavities.

plitudes of up to 100 mV/m. The density variations are negative, indicating that the plasma is expelled from the region of high wave pressure. Such observations suggest caviton formation in lower-hybrid wave collapse.

Figure 11.6 shows a waveform measurement combined with the observed density variation. The association of the density depletions and the simultaneous excursions of the lower-hybrid wave amplitude is interpreted as strong indication for the existence of lower-hybrid cavitons and possible lower-hybrid collapse in the auroral plasma. Fast electron beams and transverse ion heating are also observed, but the association with

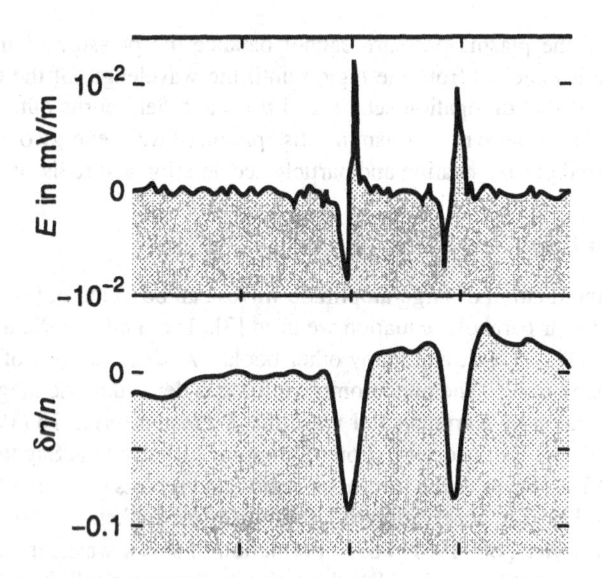

Fig. 11.7. Lower-hybrid cavitons as double layers in the auroral magnetosphere.

the collapse and caviton structure could not be established. Sometimes electric field measurements also show a double-peak structure across a density depletion (see Fig. 11.7), which may be interpreted as a double-layer (see Sec. 10.4) inside a caviton.

Concluding Remarks

The theory of strong turbulence is different from weak turbulence theory, i.e., more qualitative, since the basic equations cannot be solved analytically. Only when they can be reduced to, e.g., the Korteweg-de Vries or the nonlinear Schrödinger equation, closed solutions exist for the stationary state. These solutions describe solitary wave structures, which in the case of strong turbulence are driven by the wave pressure but balanced by the dispersive effect introduced by the background plasma pressure.

These solitary structures trap the high-frequency waves, but are themselves low-frequency waves. In the case of Langmuir and lower-hybrid turbulence they are ion-acoustic waves, but one can imagine that magnetosonic modes or kinetic Alfvén waves can do the same service. Little has been done in investigating these modes because of the mathematical difficulties involved and the presumably sparse applications. But low-frequency modes other than ion-acoustic waves might be important in understanding large-scale turbulence on scales close to the magnetohydrodynamic turbulence scale.

The nonlinear evolution of cavitons uncovers a new instability, *plasma collapse*.

During collapse the plasma pressure cannot balance the pressure of the wave field, and the plasma is expelled from the region until the wavelength of the trapped waves becomes so short that dissipation sets in and the wave field burns out. Collapse may therefore provide the basic mechanism for dissipation of wave energy out of resonance. It causes localized plasma heating and particle acceleration and redistributes energy.

Further Reading

The theory of modulation of large-amplitude waves can be found in [8]. In the derivation of the nonlinear parabolic equation we used [3]. The nonlinear Schrödinger equation is given in [1], [2], [4], and many other books. A short account of modulational instability is found in [2]. The ponderomotive force is derived in the magnetohydrodynamic approximation by Karpman and Washimi, *J. Plasma Phys.*, **18** (1977) 173. The briefest reference for the theory of strong Langmuir turbulence is Sagdeev, *Rev. Mod. Phys.*, **51** (1979) 1. Waves in the foreshock region and reports on strong turbulence are given in [5] and [7]. The observation of Langmuir wave packets is taken from Gurnett et al., *J. Geophys. Res.*, **86** (1981) 8833. Formation of shock waves in nonlinear wave evolution is found in [6]. In lower-hybrid collapse theory we followed Shapiro et al., *Phys. Fluids B*, **5** (1993) 3148. The measurements of lower-hybrid cavitons are taken from Eriksson et al., *Geophys. Res. Lett.*, **21** (1994) 1843, and Dovner et al., *Geophys. Res. Lett.*, **21** (1994) 1827.

[1] R. C. Davidson, *Methods in Nonlinear Plasma Theory* (Academic Press, New York, 1972).

[2] A. Hasegawa, *Plasma Instabilities and Nonlinear Effects* (Springer Verlag, Heidelberg, 1975).

[3] V. I. Karpman, *Nichtlineare Wellen in dispersiven Medien* (Vieweg Verlag, Braunschweig, 1977).

[4] D. R. Nicholson, *Introduction to Plasma Theory* (Wiley, New York, 1983).

[5] R. G. Stone and B. T. Tsurutani (eds.), *Collisionless Shocks in the Heliosphere: A Tutorial Review* (American Geophysical Union, Washington, 1985).

[6] D. A. Tidman and N. A. Krall, *Shock Waves in Collisionless Plasmas* (Wiley Interscience, New York, 1971).

[7] B. T. Tsurutani and R. G. Stone (eds.), *Collisionless Shocks in the Heliosphere: Reviews of Current Research* (American Geophysical Union, Washington, 1985).

[8] G. B. Whitham, *Linear and Nonlinear Waves* (Wiley, New York, 1974).

12. Collective Effects

In this last chapter we consider three special types of collective interactions, the generation of anomalous collisions and diffusion in an otherwise collisionless plasma, collisionless shock waves, and the acceleration of plasma particles to high energies. This selection of topics is guided by practical needs in space physics. Of course, all chapters of this book include collective effects. But the present chapter selects those effects which fall somewhat outside the main course of the theory-oriented presentation, which proceeded from linear instability to the nonlinear processes.

The first effect, anomalous collisions, is probably the most important macroscopic effect in a collisionless plasma. It is caused by close correlations between the particles, which replace the two-body collisions in ordinary collision-dominated systems. Binary collisions let collisional systems evolve towards relaxation and equilibrium and allow for a macroscopic description in terms of a fluid theory. In a collisionless plasma there are no binary collisions. Instead the collisions are replaced by interactions between the particles and various kinds of waves, which are responsible for relaxation processes in collisionless plasmas and ultimately justify the use of a fluid description. The immediate consequence of anomalous collisions is anomalous plasma resistivity, the appearance of finite relaxation times, and plasma diffusivity. These enter the phenomenological fluid equations of the plasma as transport coefficients and lead to its non-ideality.

Collisionless shock waves, the second collective effect selected, play a dominant role in many fast flows encountered in space. Since the magnetosphere including its dynamics is the result of the interaction of the fast solar wind flow with the geomagnetic field as an obstacle, the investigation of the processes at bow shock is of vital interest. Shock physics can be studied in situ by investigating its properties. Many of the results have been extrapolated to other shocks in the solar system and astrophysical objects.

The last collective effect treated in this chapter is particle acceleration in collisionless plasmas, leading to particle heating and particle beams. We discuss a number of conditions under which such acceleration can appear. This discussion is by no means exhaustive nor even complete. Only some basic acceleration processes are sketched.

12.1. Anomalous Resistivity

The determination of transport coefficients is one of the most important aspects of microscopic theory. Measured quantities, like density, flow velocity, and temperature, are usually of macroscopic nature. These quantities are mutually related through macroscopic conservation laws or evolution equations. The latter are derived from the microscopic equations, as demonstrated in Chap. 7 of our companion book, *Basic Space Plasma Physics*. In collisionless plasmas, the absence of collision and correlation terms in the Vlasov equation corresponds to a lack of transport coefficients and, hence, to the absence of any obvious irreversible effects. Transport of a macroscopic quantity in this case proceeds by convection. This means that only macroscopic motions like flow and wave propagation lead to transport.

In such a case there would be no energy exchange, and in reality this is not possible. We have learned that instabilities arise and that irreversible processes may exist in a collision-free plasma. The irreversible processes are a consequence of nonlinear interactions between the fields and particles. Viewed from a macroscopic position, they produce transport coefficients which are called *anomalous*, because they arise from these pseudo-collisions. These pseudo-collisions are a result of the property of the Vlasov equation to generate unstable and irreversible interactions between fields and particles, which can formally be reflected in the appearance of an average correlation term on the right-hand side of the general quasilinear equation (8.38).

Determination of anomalous transport coefficients requires to solve this equation and to derive the macroscopic evolution or transport equations by taking moments of Eq. (8.38). This is a difficult problem. In the present section we give a short account on how to calculate anomalous resistivities in a plasma under certain restrictive conditions.

Anomalous Collisions

Resistivity is caused by friction. Friction results from collisions between particles. The problem of finding the resistivity is thus reduced to the problem of finding the collision frequency. Instead of anomalous resistivity one may speak of anomalous collision frequencies. How collision frequencies are determined in the collisional case has been shown in Chap. 4 of our companion book. Let us restrict ourselves to the one-dimensional unmagnetized case and consider the electron equation of motion in an external electric field

$$m_e\, dv_e/dt = -eE - m_e \nu v_e \tag{12.1}$$

Under stationary conditions the current is defined as $j = -en_0 v_e = \sigma E$. The electrical resistivity, $\eta = 1/\sigma$, is given by

$$\eta = \nu/\omega_{pe}^2 \epsilon_0 \tag{12.2}$$

For a collisional plasma, with electron-ion Coulomb collisions dominating the dynamics, Eq. (I.4.9) from our companion book suggests that the Spitzer collision frequency can be expressed as

$$\nu_{ei} \propto \frac{\omega_{pe}}{n_0 \lambda_D^3} = \frac{\omega_{pe} W_{\rm tf}}{n_0 k_B T_e} \tag{12.3}$$

The classical collision frequency in a plasma thus turns out to be proportional to the energy density of thermal fluctuations, $W_{\rm tf}$ in Eq. (2.54), normalized to the thermal energy density of the plasma. One thus realizes that in Coulomb collisions the interaction between electrons and ions is mediated by the thermal fluctuation energy stored in the Langmuir wave field, assuming that the plasma is in thermal equilibrium. The colliding electrons are scattered by the oscillations in the thermal wave field.

This picture is somewhat surprising, but it explains the physical mechanism and may serve as starting point of a theory of anomalous collisions, since if one can increase the strength of the wave fluctuation field by some means, e.g., by instability, one also enhances the collision frequency. This is the basic idea of the mechanism of anomalous collisions. It is of particular importance in plasmas which are so dilute and hot that Coulomb collisions are spurious, as in the case of most space plasmas. Instabilities which saturate the wave fields on a high level, orders of magnitudes higher than the thermal fluctuation level, may be responsible for the observed enhanced collisions, dissipation, and transport.

Collisions do not act on electrons and ions in the same way. Electrons are typically affected most. Hence, collision frequencies will always cause currents and the problem of finding self-consistent anomalous collision frequencies becomes the problem of current instabilities. Initially a sufficiently strong current excites plasma oscillations which grow until saturation. At that time the current is either disrupted by strong anomalous resistivity or at least retarded, with the energy of the current transformed into irregular electron scattering and anomalous Joule heating of the plasma. Thus the anomalous resistivity will heavily depend on the kind of instability, on the instability threshold, and on the strength of the current.

A simple expression for the anomalous collision frequency arising from any linear current instability can be found in a way, which formally duplicates the above derivation of the classical collision frequency. There the collision frequency has been defined implicitly as the rate of momentum transfer between the electron fluid and the wave fluctuations. In the presence of an unstable current-driven wave spectrum and under stationary conditions, when both the instability has saturated and an equilibrium has been established between the friction the electrons experience and the electric force which accelerates them, the momentum transfer is determined by

$$m_e \nu_{an} \mathbf{v}_{de} = -e\mathbf{E} \tag{12.4}$$

The momentum on the left-hand side of this equation is transferred to the wave with energy density W_w. Remember that the wave momentum can be defined as the wave energy divided by frequency, ω, and multiplied by wavenumber, $\mathbf{k}$. Thus the change in wave momentum, $\Delta \mathbf{p}_w$, is

$$\Delta \mathbf{p}_w = \int \frac{d^3k}{8\pi^3} \frac{\mathbf{k}}{\omega(\mathbf{k})} \gamma(\mathbf{k}) W_w(\mathbf{k}) \tag{12.5}$$

In equilibrium between the change in electron momentum and wave momentum, this expression gives an estimate of the anomalous collision frequency

$$\nu_{an} \approx \frac{1}{n_0 m_e v_{de}^2} \int \frac{d^3k}{8\pi^3} \frac{\mathbf{k} \cdot \mathbf{v}_{de}}{\omega(\mathbf{k})} \gamma(\mathbf{k}) W_w(\mathbf{k}) \tag{12.6}$$

in terms of the phase velocity of the unstable spectrum, the wave spectral energy density, and the drift velocity of the electrons which provide the electric current. The problem is thus reduced to the problem of finding the saturated spectral density (2.19) of the waves, $W_w(\mathbf{k})$. The growth rate entering Eq. (12.6) is understood in the quasilinear sense as the growth rate before saturation, but depending on the saturation mechanism.

Let us assume, for example, that we are dealing with the current-driven ion-acoustic instability. The ion-acoustic instability is an electrostatic resonant instability with drifting electrons. The quasilinear equation for the electron distribution function responsible for an electrostatic resonant instability can be written as

$$\frac{\partial f_{0e}(t)}{\partial t} = \frac{e^2}{8\pi^2 m_e^2} \int d^3k \, \mathbf{k} \cdot \frac{\partial}{\partial \mathbf{v}} \left[|\delta\phi(\mathbf{k}, t)|^2 \, \delta(\omega - \mathbf{k} \cdot \mathbf{v}) \mathbf{k} \cdot \frac{\partial}{\partial \mathbf{v}} \right] f_{0e}(t) \tag{12.7}$$

were $\delta\phi(\mathbf{k})$ is the wave electric potential. We multiply this equation with $m\mathbf{v}$ and integrate over velocity space, using the dispersion relation for ion-acoustic waves. In the stationary state the integral over the term on the right-hand side gives the friction term

$$\nu_{an} m_e n_0 \mathbf{v}_{de} = \frac{1}{8\pi^3} \int d^3k \frac{\mathbf{k}}{\omega(\mathbf{k})} \gamma(\mathbf{k}) W_w(\mathbf{k}) \tag{12.8}$$

where we replaced $W_w = (\partial \omega \epsilon / \partial \omega) W_E$. This expression agrees with Eq. (12.6).

Sagdeev Formula

Estimates for the anomalous collision frequencies can be obtained in a particular case by calculating the saturated wave amplitudes and finding the growth rates. The simplest way is to use the linear maximum growth rates and the wave dispersion relations. The maximum growth rate for the ion-acoustic instability given in Eq. (4.27) can be used, for instance, to find an estimate for the ion-acoustic anomalous resistivity. As we know, the ion-acoustic wave becomes unstable for current drifts $v_{de} > c_{ia}$. The maximum unstable wavelength is of the order of the Debye length, such that $k_{\max} \approx 1/\lambda_D$. We approximate $\gamma \approx \omega v_{de}/v_{the}$ for large drifts and find the *Sagdeev formula* for the ion-acoustic anomalous collision frequency

$$\boxed{\nu_{ia,an} \approx \omega_{pe} W_w / n_0 k_B T_e} \tag{12.9}$$

which leaves us with the determination of the saturation level of ion-acoustic waves.

Buneman Collision Frequency

The Buneman instability is a very strong instability, which must switch off itself when evolving. Its phase velocity is $\omega/k \approx v_{the}\sqrt{m_e/m_i}$, and the wave energy density is $W_w \leq W_i$, less or equal to the ion thermal energy. Hence, the collision frequency turns out to be $\nu_{bun,an} \approx \omega_{pi}$, which is of the order of the ion plasma frequency, the highest possible frequency of ion-acoustic waves. For example, in the solar wind the ion plasma frequency is of the order of 3 kHz. If one compares this with the Spitzer collision frequency of roughly 0.5 collisions per 1 AU, or $1.4 \cdot 10^{-4}$ Hz, one realizes that the anomalous collision frequency caused by the Buneman instability is about $2 \cdot 10^7$ times larger than the Spitzer collision frequency.

Ion-Acoustic Collision Frequency

In order to find the ion-acoustic collision frequency, one must calculate the wave saturation level, which depends on the mechanism by which the waves saturate. There are several possibilities. The instability can saturate quasilinearly, by particle trapping in the wave and subsequent resonance broadening, by wave-wave interaction between the current-driven ion-acoustic wave and Langmuir waves, resulting in escaping radiation, by other kinds of wave-wave interactions with Langmuir or lower-hybrid waves, leading to the formation of envelope solitons or cavitons, by nonlinear evolution of the ion-acoustic waves into topological solitons, or, finally, by generation of holes in phase space and hole collisions. Few of these mechanisms have been explored and some of them may be unimportant, since the fastest saturation mechanism will in most cases dominate. In particular, single wave effects as particle trapping, topological soliton formation or envelope solitons will be of little importance in broadband ion-acoustic turbulence excited by currents at sufficiently large speeds above threshold.

In the weak turbulence approximation the evolution of the wave energy density is described in the simplest case by the evolution equation

$$\frac{\partial W_w(\mathbf{k})}{\partial t} = \left[2\gamma(\mathbf{k}) - A\omega(\mathbf{k})\frac{W_w}{n_0 k_B T_e} - B\omega(\mathbf{k})\left(\frac{W_w}{n_0 k_B T_e}\right)^2 + \ldots\right] W_w(\mathbf{k}) \tag{12.10}$$

The different terms in the expansion on the right-hand side of this equation correspond to the quasilinear, three-wave, four-wave interaction, etc. The coefficients A, B in this expression are in fact integral operators describing the different coupling terms contributing to the second and third orders in the expansion. Clearly, the mere expansion requires that the wave energy density is small compared to the plasma energy, thus excluding non-perturbative nonlinearities like solitons. The only process contributing to the quadratic term is scattering of ion-acoustic waves on ions with the resonance condition $\omega - \omega' = (\mathbf{k} - \mathbf{k}') \cdot \mathbf{v}_i$. Under stationary conditions the terms on the right-hand side balance. Usually one keeps only the two first terms balancing the linear growth rate and the scattering of waves by ions at very large electron temperatures, $T_e \gg T_i$, when the ion-acoustic instability can develop. It leads to the following anomalous ion-acoustic collision frequency

$$\boxed{\nu_{an,ia} = 0.01\, \omega_{pi} \frac{v_{de}}{c_{ia}} \frac{T_e}{T_i} \theta^{-2}} \tag{12.11}$$

where the small factor 0.01 is partially compensated by the large electron-to-ion temperature ratio. Also, for large current drift velocities, exceeding the ion-acoustic velocity by far, and for narrow wave scattering angle, θ, the anomalous collision frequency can become substantial and of the order of the Buneman collision frequency.

Critical Electric Field

The production of anomalous resistivity by scattering of current-carrying electrons off the self-excited ion-acoustic waves will reduce the current and thus lead to quasilinear effects and possible reduction of the current, such that the current speed is reduced until some equilibrium is reached. This equilibrium depends on the applied electric field. Quasilinearly the current will saturate at a certain velocity, $j_{\rm cr} = -env_{d,\rm cr}$.

But when the electric field is very large, larger than another critical value, $E = E_{\rm cr}$, the current will again increase, since then fast electrons, the *run-away electron* part of the distribution function, start to escape the collisions because of their large mean free paths. Actually, the fast electrons ignore the field fluctuations like a fast car ignores the short-length roughness of a street cover.

The electron-ion Spitzer collision frequency is given by

$$\nu_{ei} = n_0 \sigma_c \langle v_e \rangle \tag{12.12}$$

where the angular brackets indicate the ensemble average as well as the average over the electron deflection angle. The Spitzer collisional cross-section (I.4.9) is a function of the actual electron velocity, v_e. It can be written as

$$\sigma_c = \frac{\omega_{pe}^4}{16\pi n_0^2 v_e^4} \tag{12.13}$$

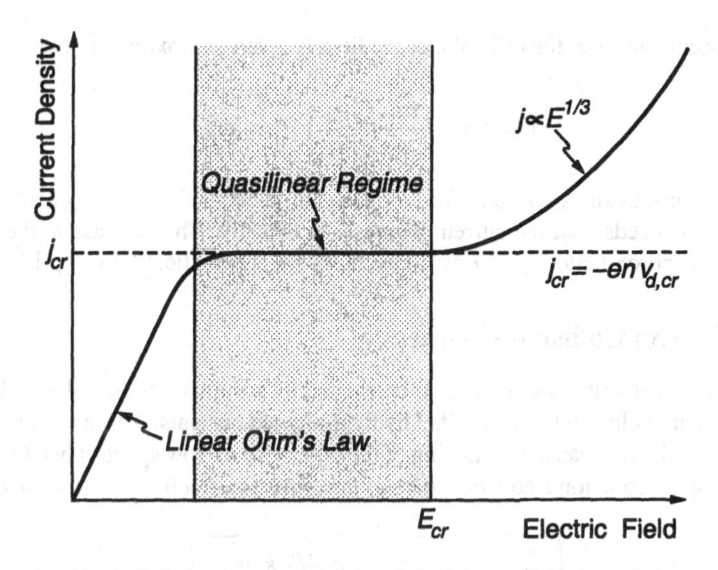

Fig. 12.1. Relation between current and field in a collisionless plasma.

All the electrons with velocities up to the thermal speed, $\langle v_e \rangle = v_{\text{the}}$, are confined by collisions. However, for the much faster tail electrons the collision frequency decreases as v_e^{-3}, such that these electrons become asymptotically free again.

For large electric fields and given collision frequency, ν, no equilibrium can be reached because the acceleration of the electrons by the field is faster than the collisional deceleration. The critical electric field above which the electrons become free is easily found from the one-dimensional non-relativistic electron equation of motion

$$m_e dv_e/dt = -eE - m_e \nu v_e \tag{12.14}$$

Hence, the marginal field amplitude for equilibrium is given by

$$\boxed{E_{\text{cr}} = m_e \nu v_e / e} \tag{12.15}$$

Inserting the Spitzer collision frequency yields the *Dreicer field* strength

$$\boxed{E_D = e/16\pi\epsilon_0 \lambda_D^2} \tag{12.16}$$

But Eq. (12.15) holds more generally for any anomalous collision frequency. It may be expressed through the wave intensity using the Sagdeev formula (12.9)

$$E_{\text{cr}} = W_w / 2en_0 \lambda_D \tag{12.17}$$

In the ion-acoustic case, the critical electric field is given approximately by

$$E_{\rm cr} \approx 0.01\frac{m_i}{e}\left(\frac{m_e}{m_i}\right)^{1/4}\omega_{pi}c_{ia} \tag{12.18}$$

The critical current stays constant for an increasing electric field as long as $E < E_{\rm cr}$. But when E exceeds $E_{\rm cr}$, the current increases drastically. The increase in the current above $E_{\rm cr}$ is approximately $j \propto E^{1/3}$. This is shown schematically in Fig. 12.1.

Ion-Cyclotron Collision Frequency

A number of other instabilities can also be excited by currents (see Sec. 4.4). Most important are ion-cyclotron waves excited by field-aligned currents. These waves saturate predominantly by resonance broadening of the otherwise narrow spectral band (see Sec. 8.4), causing the collision frequency to be of the order of a fraction of the ion-cyclotron frequency

$$\boxed{\nu_{ic,an} \approx 0.25\omega_{gi}\left(\frac{\pi}{2}\right)^{1/2}\frac{k_\| v_{\rm the}}{\omega_{gi}}} \tag{12.19}$$

This value is clearly much lower than that of the ion-acoustic instability. Nevertheless, anomalous collisions due to ion-cyclotron waves may sometimes be important when the threshold of the ion-acoustic instability is not exceeded by weak field-aligned currents and ion-cyclotron waves undergo instability. This may occur in strong magnetic fields when the plasma is underdense, $\omega_{pe} \ll \omega_{ge}$, like in the upper auroral ionosphere.

Electron-Cyclotron Collision Frequency

Similar to ion-cyclotron waves, electron-cyclotron waves may also become unstable in the presence of field-aligned currents. One can imagine that these waves are excited by the electron beam constituting the current, especially if the current is carried by energetic electrons. In this case the wave also saturates by resonance broadening.

If the current is carried by the bulk electrons, the current flow will Doppler-shift the frequency down to $\omega' = l\omega_{ge} - k_\| v_{de}$ in the rest frame of the ions, which do not participate in the current flow. For $\omega' \approx k_\| v_{\rm thi}$ the ions can interact with the electron-cyclotron waves, and a negative energy wave will be excited. This happens even for low electron temperatures, but saturation of the waves occurs at very small amplitudes due to trapping and broadening. The wavenumber of the growing waves is about $k_\| \approx \omega_{ge}/v_{de}$, and the collision frequency becomes

$$\boxed{\nu_{ec,an} \approx 0.1\omega_{ge}(v_{de}/v_{\rm the})^3} \tag{12.20}$$

Because the current drift speed is generally much less than the electron thermal velocity, this value is low. For instance, in the auroral magnetosphere, where this instability might be important, the drift velocity is of the order of the ion-acoustic speed, implying $\nu_{ec,an} \approx 10^{-6}\omega_{ge}$. The most important region for auroral processes is the excitation region of auroral kilometric radiation at about 3000 km altitude. There the electron-cyclotron frequency is about 300 kHz. Multiplying with 2π one finds that the collision frequency becomes not more than $\nu_{ec,an} \approx 1$ Hz, too low to be of any importance.

Lower-Hybrid Drift Collision Frequency

The last instability important for the excitation of anomalous collisions is the lower-hybrid drift instability. The quasilinear saturation level of this instability is given in Eq. (8.75). We recall that the lower-hybrid drift instability is caused by transverse currents, not by the field-aligned currents discussed so far. This instability will therefore be most important for interrupting perpendicular currents, which exist in extended current layers subject to reconnection. One can expect that it will provide the diffusivity and anomalous resistivity needed to initiate reconnection in such current sheets and to disrupt magnetic fields and release magnetic field energy stored in transverse currents.

Lower-hybrid drift instabilities require density and temperature gradients but are stabilized by magnetic shear which counteracts the density gradients. On the other hand, plasma streaming may help to destabilize the waves by coupling it to the Kelvin-Helmholtz mode and low-frequency turbulence. However, it is generally believed that low-β conditions are needed to keep this instability growing while high-β conditions should inhibit its growth. This is true for the excitation of the instability, but if a guide field is present and the current causes an additional transverse shear in the field, this argument does not apply, because the guide field makes up for the low-β condition while the transverse field may in the region of anomalous resistivity merge and reconnect.

The anomalous collision frequency caused by the lower-hybrid drift instability in a dense plasma, $\omega_{pe} > \omega_{ge}$, can be represented as

$$\nu_{lh,an} \approx \left(\frac{\pi}{2}\right)^{1/2} \frac{\omega_{pe}^2}{\omega_{lh}} \frac{W_E(k_\perp, \infty)}{n_0 k_B T_i} \tag{12.21}$$

Using Eq. (8.75) to express the saturation level of the lower-hybrid drift instability under the assumption that saturation proceeds via quasilinear plateau formation, one gets

$$\boxed{\nu_{lh,an} \approx \left(\frac{\pi}{2}\right)^{1/2} \left(\frac{r_{gi}}{4L_n}\right)^2 \omega_{lh}} \tag{12.22}$$

This value is of the order of the lower-hybrid frequency itself. In a strong magnetic field with steep density gradients of the order of the ion gyroradius, it becomes equal to the ion plasma frequency. In weaker magnetic fields it is the geometric mean of the electron- and ion-cyclotron frequencies.

That $\nu_{lh,an}$ becomes proportional to the cyclotron frequency implies that the collision frequency is proportional to the magnetic field. The lower-hybrid drift anomalous collision frequency will be very important whenever the plasma is able to evolve short scale density gradients. Such situations are encountered in the auroral magnetosphere, near the magnetopause and the bow shock, and in some regions of the magnetotail.

The critical electric field for the lower-hybrid drift instability, above which electrons will escape, can be estimated from the collision frequency, if one assumes that the waves propagate at a large angle to the magnetic field. In this case $k_\parallel/k_\perp \ll 1$, only the parallel field plays a role in the confinement, and the critical electric field is given by

$$E_{\rm cr} \approx \frac{m_i r_{gi}}{e} \frac{k_\parallel^2}{k_\perp^2} \left(\frac{\pi T_e}{2T_i}\right)^{1/2} \left(\frac{r_{gi}}{4L_n}\right)^2 \omega_{lh}^2 \tag{12.23}$$

Electric fields exceeding this value will readily accelerate the electrons parallel to the magnetic field. However, due to the small angular factor these parallel fields can be quite small. On the other hand, regions of strong lower-hybrid drift wave turbulence may serve as sources of parallel electron fluxes. Below we will show that in such regions electrons may be accelerated to substantially high energies fitting into the range of auroral electron energies.

Strongly Turbulent Collisions

Anomalous collision frequencies are dominated by the level of wave turbulence. The lowest level is the thermal fluctuation level, which in the case of Langmuir fluctuations just leads to the Spitzer collision frequency given in Eq. 12.12. In the former sections we estimated fluctuation levels based on saturation levels of various plasma instabilities in weak turbulence. However, in many cases the state of weak turbulence is not reached, and the plasma undergoes modulational instability and switches over to strong turbulence. In strong turbulence the wave spectrum breaks off into a large number of solitons and cavitons with heavily modulated background density. Electron-ion collisions become considerably more frequent and violent under these conditions, and the collision frequencies may exceed the quasilinear and weakly turbulent values.

In order to determine the strongly turbulent collision frequencies, one can follow the same line of arguments as before and substitute the strong turbulence level of the corresponding waves into the Sagdeev formula (12.9). Let us write this formula for a strong turbulence wave level, $W_{\rm st}$

$$\nu_{an} \approx \alpha\omega_{pe}\left(W_{\rm st}/n_0 k_B T_e\right) \tag{12.24}$$

with an unknown numerical proportionality factor, α, of the order of 0.1–10, which must be determined from numerical simulations. The wave power dissipated in the plasma due to the presence of this anomalous collision frequency will be

$$dW_E/dt \approx \nu_{an} W_E \tag{12.25}$$

At the same time the power entering the inertial range by modulational instability from the injection site in k-space will be

$$dW_{\rm st}/dt \approx \gamma_{\rm mod} W_{\rm st} \tag{12.26}$$

where $\gamma_{\rm mod}$ is the modulational growth rate. This growth rate is found from the modulational dispersion relation, $\omega^2 \approx k^2(k_B T_e - p_{pm0}/n_0 k^2 \lambda_D^2)/m_i$, as

$$\gamma_{\rm mod} \approx (W_{\rm st}/n_0 m_i \lambda_D^2)^{1/2} \tag{12.27}$$

where we used the Langmuir strong turbulence theory. Equating the absorbed with the generated power, one finds that in equilibrium

$$\frac{W_{\rm st}}{n_0 k_B T_e} \approx \alpha^2 \frac{m_i}{m_e} \left(\frac{\epsilon_0 |\delta \mathbf{E}|^2}{2 n_0 k_B T_e} \right)^2 \tag{12.28}$$

The large mass ratio in this formula reflects the increase in the wave intensity in the collapsing inertial range in strong turbulence. But the wave energy density is proportional to the fourth power of the pump wave field $|\delta \mathbf{E}|$, i.e., it is proportional to the square of the Langmuir wave energy density to thermal energy density ratio, $W_E/n_0 k_B T_e$, after scattering off ions and condensating in the long-wave range just before modulational instability commences. Hence, unless the pump wave density is high, the wave energy density will be moderate in the inertial regime.

Substituting Eq. 12.28 into Eq. 12.24 for the anomalous collision frequency, one finds for strongly turbulent collisions

$$\boxed{\nu_{an} \approx \alpha^3 \omega_{pe} \frac{m_i}{m_e} \left(\frac{\epsilon_0 |\delta \mathbf{E}|^2}{2 n_0 k_B T_e} \right)^2} \tag{12.29}$$

This collision frequency is enhanced over the Spitzer collision frequency. But as long as the numerical factor has not been precisely determined, it is not certain how far it exceeds the quasilinear collision frequencies derived above, e.g., for the ion-acoustic mode. Moreover, the dependence of the strongly turbulent collision frequency on the pump wave energy density is quadratic. Therefore, if the pump wave energy density is not very large, the anomalous collision frequency may be even reduced with respect to the quasilinear value. The reason for such a reduction lies in the fact that dissipation

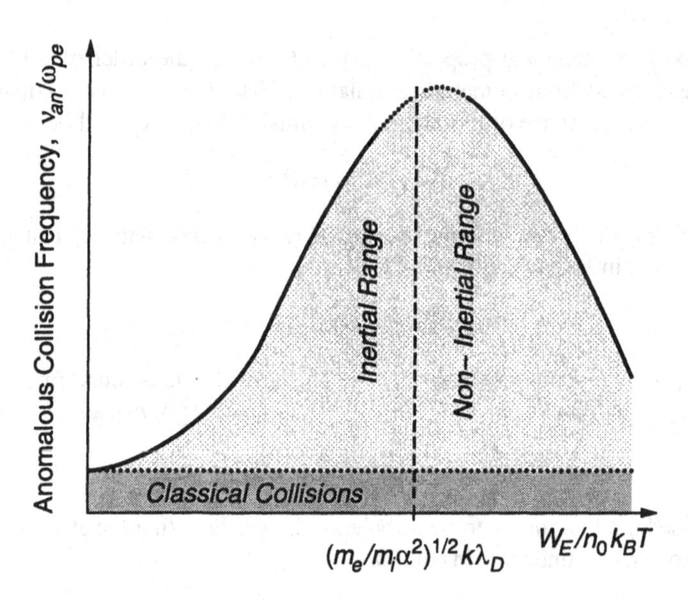

Fig. 12.2. Collision frequency in strong turbulence.

in the strongly turbulent regime is not necessarily restricted to Joule dissipation and heating, but proceeds via transport of the wave energy into the dissipative range across the inertial range and resonant generation of run-away electrons.

Close to the threshold of the modulational instability, but nevertheless well inside the inertial range, the collision frequency is

$$\nu_{an} \approx 6\alpha^3(m_i/m_e)\omega_{pe}k_0^2\lambda_D^2 \tag{12.30}$$

Like the quasilinear collision frequency, the nonlinear collision frequency found here is based on interaction with ion-acoustic modes, but in strong turbulence this mode is driven by the plasma wave pressure force. The interaction between the wave field and the plasma is stronger only in the localized regions, where the streaming electrons are possibly slowed down more than by quasilinear interaction. This corresponds to locally higher Joule dissipation and heating. But at the same time the concentration of wave energy in cavitons reduces the frequency of interaction of the streaming particles with the wave spectrum and thus counteracts the Joule dissipation, leading to an overall decrease of the collision frequency in strong interactions compared to the quasilinear value.

The above theory breaks down when the wave intensity becomes too strong. Then the source region overlaps with the dissipation region of Landau damping and the inertial range disappears. In this case the anomalous collision frequency becomes reduced.

Figure 12.2 shows the dependence of the anomalous collision frequency on the initial pump wave energy density. The collision frequency increases in the inertial range and presumably decays when the inertial range disappears for very strong initial wave amplitudes. The absolute scale can be determined only after the unknown factor α has been determined by numerical simulations.

12.2. Anomalous Diffusion

It is natural to extend the discussion of anomalous collision frequencies to other processes where collision frequencies play a role. Such processes are thermal conduction and diffusion of matter. Thermal conductivities are known to be inversely proportional to the collision frequency. Hence, substantial increases in the collision frequency as suggested in the previous section will tend to inhibit thermal conduction and decrease the heat flux, supporting storage of heat in regions bounded by zones of high anomalous collision frequencies. When anomalous collisions arise in regions of strong current flow, the heat flux from this region will become strongly inhibited, the region will thermally decouple from the environment, and will heat up to substantial temperatures. In regions of field-aligned currents the zone of anomalous collision frequency and resistivity along the magnetic field lines will always become a region of strong plasma heating and accumulation of heat.

The other important effect is plasma diffusion. Anomalous collision frequencies cause diffusion coefficients to change. Parallel diffusion behaves similar to heat conductivity. Since it is defined as

$$D_{\|} = k_B T / m\nu \tag{12.31}$$

use of the anomalous collision frequency in the place of ν indicates that diffusion parallel to the magnetic field is strongly inhibited in regions of high anomalous collision frequency. Particles are trapped in these regions due to the large number of collisions they experience. Moreover, ambipolar effects set on and limit the diffusion of both particle components to about the diffusivity of the lesser mobile species, thereby maintaining plasma quasineutrality.

For perpendicular diffusion with a diffusion coefficient defined as

$$D_{\perp} = D_{\|} / (1 + \omega_{ge}^2 / \nu^2) \tag{12.32}$$

no ambipolar diffusion arises, because a transverse electric field can exist in magnetized plasmas. They simply reflect plasma motions. In strong magnetic fields perpendicular diffusion is proportional to the anomalous collision frequency and is strongly enhanced due to the scattering of particles in the wave fields in all directions, which provides a transverse motion of the particles. Classically, the diffusion is known to be inversely proportional to the square of the magnetic field.

Lower-Hybrid Drift Diffusion

Two kinds of anomalous collisions do change this dependence, i.e., the cyclotron instabilities of ions and electrons. However, the electron-cyclotron collision frequency is restricted by the very low saturation level of the instability, and the ion-cyclotron instability also produces only low collision frequencies. The most important instability for diffusion is the lower-hybrid drift instability with an anomalous collision frequency given in Eq. (12.22). Using its dependence on the cyclotron frequency one finds that it not only increases the transverse diffusion rate, but it also reproduces the Bohm-scaling of the diffusion coefficient

$$\boxed{D_{\perp lh} \approx \frac{k_B T_e}{16eB}\left(\frac{\pi m_e}{2m_i}\right)^{1/2}\frac{r_{gi}^2}{L_n^2}} \tag{12.33}$$

In all regions, where the density gradient is steep enough to become comparable to the ion gyroradius, anomalous diffusion due to the lower-hybrid drift instability starts scaling as $D_{\perp lh} \propto k_B T_e/B$, which is Bohm diffusion. It will thus cause fast diffusion and losses of plasma from the corresponding gradients.

Kinetic Alfvén Diffusion

Kinetic Alfvén waves may also be efficient for plasma diffusion. These waves are low-frequency waves and it can be assumed that the particles will behave adiabatically in the wave field. However, kinetic Alfvén waves have a field-aligned electric wave component and carry field-aligned currents. Therefore, electrons may be in resonance with these waves and may be scattered in the wave field. Their average distribution function, $\langle f_e \rangle$, in the presence of a density gradient transverse to the magnetic field will evolve according to a resonant quasilinear diffusion equation

$$\frac{\partial \langle f_e \rangle}{\partial t} = \nabla_x \left[\sum_{\mathbf{k}} \frac{k_\perp^2 v_\parallel^2 |\delta E_\parallel(\mathbf{k})|^2 \pi \delta(k_\parallel v_\parallel - \omega)}{2\omega^2 B_0^2} \nabla_x\right] \langle f_e \rangle \tag{12.34}$$

which results from the quasilinear random-phase ensemble average of the drift-kinetic equation (I.6.27) for the electron distribution function, f_e

$$\left\{\frac{\partial}{\partial t} + v_\parallel \nabla_\parallel + \nabla_\perp \cdot \mathbf{v}_{de} - \frac{e}{m_e}\left[\delta E_\parallel + (\mathbf{v}_{de} \times \delta \mathbf{B}_\perp) \cdot \hat{\mathbf{e}}_\parallel\right] \frac{\partial}{\partial v_\parallel}\right\} f_e = 0 \tag{12.35}$$

with $\mathbf{v}_{de} = \mathbf{v}_E + v_\parallel(\delta \mathbf{B}_\perp / B_0)$ and using Ampère's law to replace the magnetic field component by the wave electric field. The averaging procedure is performed in the usual resonant manner, replacing the resonant denominator by a delta function.

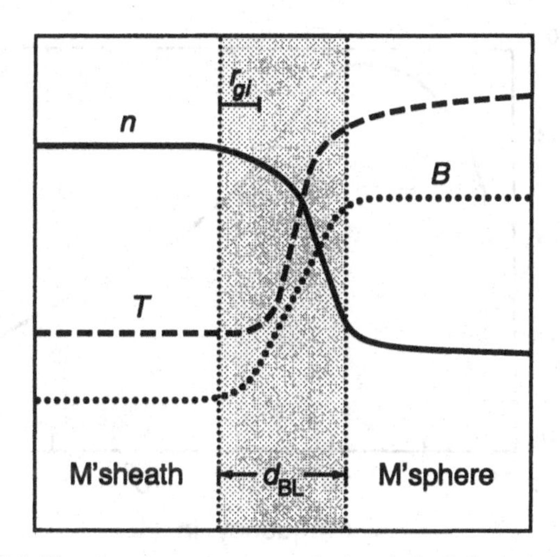

Fig. 12.3. Plasma parameters across the low-latitude boundary layer.

Integrating Eq. (12.34) with respect to velocity and assuming that the undisturbed distribution function is a Maxwellian, the diffusion equation for the electrons becomes

$$\partial n_e/\partial t = \nabla_x(D_e \nabla_x n_e) \tag{12.36}$$

By comparison the diffusion coefficient is defined as

$$D_e = \left(\frac{\pi}{8}\right)^{1/2} \sum_{\mathbf{k}} \frac{1}{k_\parallel v_{\rm the}} \frac{k_\perp^2 |\delta E_\parallel(\mathbf{k})|^2}{k_\parallel^2 B_0^2} \tag{12.37}$$

This diffusion coefficient holds for any low-frequency electromagnetic wave with a non-vanishing parallel electric field. Specifying the waves as kinetic Alfvén modes allows to write it in the form

$$D_e = \left(\frac{\pi}{8}\right)^{1/2} \sum_{\mathbf{k}} \frac{\zeta_{ia} v_A^2}{|k_\parallel| v_{\rm the}} \frac{|\delta B_x(\mathbf{k})|^2}{B_0^2} \frac{1 - \Lambda_0(\zeta_i)}{1 + [1 - \Lambda_0(\zeta_i)]T_e/T_i} \frac{T_e}{T_i} \tag{12.38}$$

where $\Lambda_0(\zeta) = I_0(\zeta)\exp(-\zeta)$ and $\zeta = k_\perp^2 v_{\rm ths}^2/\omega_{gs}^2$ with $s = i, e$ for ions and electrons, respectively. The value of $\zeta_{ia} = k_\perp^2 c_{ia}^2/\omega_{gi}^2$ is based on the ion-acoustic speed.

Low-Latitude Boundary Layer

The most important application of this theory is magnetopause diffusion and the formation of the low-latitude boundary layer. To put the problem into the right context, let

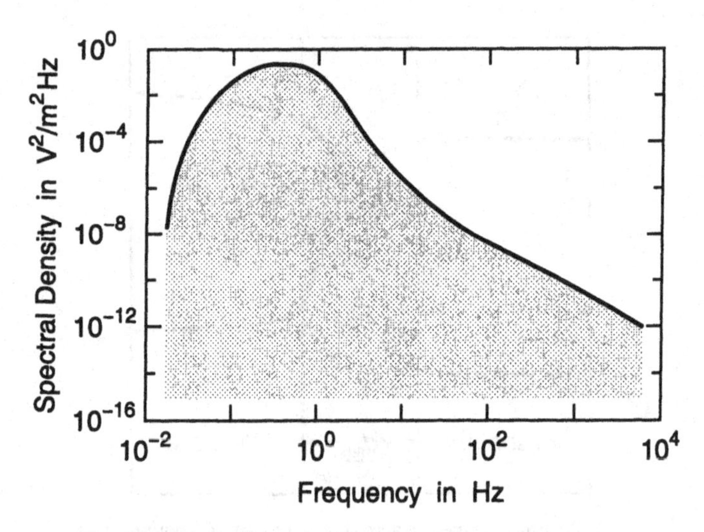

Fig. 12.4. Typical boundary layer electric wave spectrum.

us recall the physical conditions at the magnetopause. Figure 12.3 shows a schematic collection of the variation of magnetic field, density, and temperature during a transition from the magnetosheath across the low-latitude boundary layer of thickness $d_{\rm BL}$ into the magnetosphere. For northward interplanetary magnetic field, the magnetopause is a non-permeable sharp discontinuity between the turbulent magnetosheath plasma and the magnetosphere. The width of the boundary transition layer should be of the order of the ion gyroradius, roughly 100 km. But observations demonstrate that the average width of the transition is 1000–1200 km.

Estimates of the diffusion required for maintenance of such a thick boundary layer show that it must reach the Bohm limit, which at the magnetopause is of the order of $D_B \approx 10^9\,\mathrm{m^2/s}$, a comparably large value. It is hardly possible to reproduce this value with any of the anomalous collision frequencies except for the lower-hybrid drift anomalous collision frequency. In order to compare with measured wave intensities it is convenient to use

$$W_w = \left(1 + \frac{\omega^2}{\omega_{ge}^2}\right)\left(1 + \frac{k^2 v_{\mathrm{th}i}^2}{\omega_{lh}^2}\right) W_E \tag{12.39}$$

Using $\gamma_{\max} \approx 0.3(v_{de}/v_{\mathrm{th}i})^2\omega_{lh}$, the collision frequency can be expressed as

$$\boxed{\nu_{an,lh} \approx \omega_{lh}\left(\frac{\pi}{8}\right)^{1/2}\frac{m_i}{m_e}\left(1 + \frac{\omega_{pe}^2}{\omega_{ge}^2}\right)\frac{W_E}{nk_BT_i}} \tag{12.40}$$

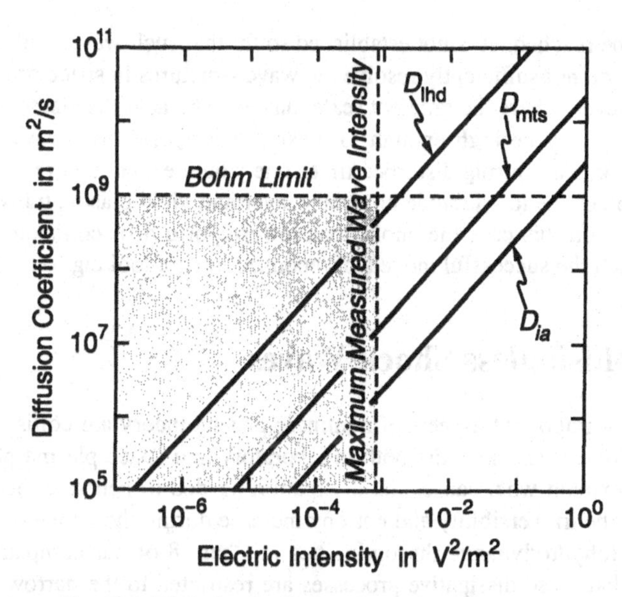

Fig. 12.5. Diffusion coefficients in the boundary layer.

which serves as an expression that contains the measurable total wave energy, W_E. Measurements show that the wave spectra in the boundary layer are nearly featureless, with possible peaks near the lower-hybrid frequency (Fig. 12.4). Integrating over frequency yields the value of W_w. Using this value and applying it to the magnetospheric low-latitude boundary layer, with measured values of $n_0 \approx 10\,\mathrm{cm}^{-3}$, $k_BT_e \approx 25\,\mathrm{eV}$, and $k_BT_i \approx 1\,\mathrm{keV}$, one obtains Fig. 12.5, which shows the dependence of the transverse diffusion coefficient on the measured value of the squared electric field amplitude.

In this figure the horizontal line at $D_\perp \approx 10^9\,\mathrm{m}^2/\mathrm{s}$ is the Bohm limit required for maintenance of the low-latitude boundary layer by scattering magnetosheath particles in the lower-hybrid field generated in the density gradient at the magnetopause, and subsequent diffusion perpendicular to the magnetopause into the magnetosphere. The vertical line is the uppermost measured electric wave intensity. As one observes, the lower-hybrid drift instability is marginally capable of providing the required diffusion rate for maintenance of the low-latitude boundary layer. We have included into this figure also the diffusion coefficients of the ion-acoustic, D_{ia}, and modified two-stream instabilities, D_{mts}. These coefficients turn out to be much smaller than the lower-hybrid drift diffusion coefficient for all reasonable wave intensities. They give considerably lower diffusion and presumably do not contribute to the formation of a low latitude boundary layer.

But one should take this figure with care. It assumes that very high electric field strengths can be reached. It is not established so far that such wave fields are realistic. Measurements do not sufficiently resolve the wave structures in space or time. In local gradients, however, where the gradient scale may even be near the electron gyroradius, wave intensities may be high enough to make lower-hybrid drift waves an efficient diffusion mechanism. During disturbed times one would expect that reconnection will cause formation of the low-latitude boundary layer and yield plasma transport into the magnetosphere. In this case the anomalous lower-hybrid drift collisions provide the diffusivity which the successful models of reconnection are looking for.

12.3. Collisionless Shock Waves

One of the most important aspects of nonlinear plasma theory are collisionless shock waves. Collisionless shocks exist both in laboratory and space plasma physics. The problem is to explain what causes the dissipation needed to generate the shock front and to provide the irreversibility and entropy increase during the transition. When discussing magnetohydrodynamic discontinuities in Chap. 8 of our companion book, it was assumed that these dissipative processes are restricted to the narrow shock front transition layer. Hence, anomalous transport coefficients must be generated therein by some instability mechanism. Most probable are ion-acoustic anomalous collisions, electron-acoustic anomalous collisions, though these are known to be low, Buneman anomalous collisions, though these can act only temporarily, and lower-hybrid drift anomalous collision, which need low plasma β. Related to the latter are modified two-stream anomalous collisions, which survive in high-β plasmas but have lower collision frequencies.

Because of the irreversibility of the shock transition process, a shock wave cannot be described as a purely steepened nonlinear wave. While wave steepening is a transitory process which, when not stopped, inevitably leads to wave breaking, shock waves are stationary states with the breaking prevented by some other process, in general by dissipation. Any fast, slow, or intermediate shock is, from a macroscopic point of view, a steepened fast, slow, or intermediate magnetosonic wave, with breaking inhibited due to the balance of steepening and dissipation. Hence, nonlinear wave theory is applicable to the shock process. The steepening of a fast magnetosonic wave is the ignitor of the follow-up microscopic processes, which provide the dissipation inside the shock front and cause irreversibility. In principle, such a picture will be correct for sufficiently slow shocks, where the flow speed leaves sufficient time for the steepening and the subsequent evolution of the microscopic processes.

Low Mach number shocks will follow this scenario. But in high Mach number shocks there will be not enough time for the anomalous process to provide the required collision frequencies. When a critical Mach number is surpassed, new effects like particle reflection from the shock front arise. Such shocks are called supercritical. In addition, the shock character depends strongly on the magnetic field direction. Quasi-perpendicular shocks behave more regularly than quasi-parallel shocks.

Rankine-Hugoniot Conditions

The Rankine-Hugoniot conditions do neither provide any information about the formation of shocks nor about the intrinsic dissipative processes which are at work in the shock transition layer. Moreover, they are based on the one-fluid model of the plasma, which is a very rough approximation to reality. Therefore, they hold only sufficiently far away from the shock front itself, deep in the ideal magnetohydrodynamic regions to both sides of the shock front. Prescribing the inflow parameters the serve as conditions from which the outflow parameters can be calculated.

Denoting the upstream and downstream values by the indices 1 and 2, and defining the following new variables $N = n_2/n_1$, $u_n = v_{n2}/v_{n1}$, $u_{t2} = v_{t2}/v_{n1}$, $u_{\mathrm{th}1} = v_{\mathrm{th}1}/v_{n1}$, $u_{\mathrm{th}2} = v_{\mathrm{th}2}/v_{n1}$, $\mathbf{b}_{t1} = \mathbf{B}_{t1}/(\mu_0 m_i n_1 v_{n1}^2)^{1/2}$, $\mathbf{b}_{t2} = \mathbf{B}_{t2}/(\mu_0 m_i n_1 v_{n1}^2)^{1/2}$, and $\mathbf{b}_n = \mathbf{B}_n/(\mu_0 m_i n_1 v_{n1}^2)^{1/2}$, the whole set of the shock jump conditions with non-vanishing flow across the shock may be written as

$$
\begin{aligned}
Nu_n &= 1 \\
2N(u_{\mathrm{th}2}^2 + u_n^2) + b_{t2}^2 &= 2(u_{\mathrm{th}1}^2 + 1) + b_{t1}^2 \\
u_n b_{t2} - u_{t2} b_n &= b_{t1} \\
b_{t2} b_n - u_{t2} &= b_{t1} b_n \\
5u_{\mathrm{th}2}^2 + u_n^2 + u_{t2}^2 + 2b_{t2} b_{t1} &= 5u_{\mathrm{th}1}^2 + 1 + 2b_{t1}^2
\end{aligned}
\tag{12.41}
$$

From the third and fourth equation one finds for the tangential magnetic and velocity fields behind the shock

$$
\begin{aligned}
b_{t2} &= b_{t1}(1 - b_n^2)/(u_n - b_n^2) \\
u_{t2} &= b_{t1} b_n (1 - u_n)/(u_n - b_n^2)
\end{aligned}
\tag{12.42}
$$

The rest of the equations can be reduced to a cubic equation for the normal velocity, u_n, behind the shock. But it is more interesting at this stage to consider the limiting case of a perpendicular shock. This case is defined as $b_n = u_{t2} = 0, b_{t1} \neq 0$. The cubic equation becomes quadratic in this case

$$
u_n^2 - \left(\tfrac{1}{4} + \tfrac{5}{4} u_{\mathrm{th}1}^2 + \tfrac{5}{8} b_{t1}^2\right) u_n - \tfrac{1}{8} b_{t1}^2 = 0 \tag{12.43}
$$

This equation yields the values of the quantities behind the shock in terms of those in front of the shock. In particular, the density, velocity and magnetic field jumps are all given by the same expression

$$\boxed{\frac{1}{N} = u_n = \left(\frac{b_{t2}}{b_{t1}}\right)^{-1} = \frac{1}{8}\left\{1 + 5u_{\rm th1}^2 + b_{t1}^2 + \left[\left(1 + 5u_{\rm th1}^2 + \tfrac{5}{2}b_{t1}^2\right)^2 + 8b_{t1}^2\right]^{1/2}\right\}} \tag{12.44}$$

and the jump in the temperature or the thermal velocity follows as

$$\boxed{u_{\rm th2}^2 = u_{\rm th1}^2 + \tfrac{1}{5}[1 - u_n^2 + 2b_{t1}^2(1 - u_n^{-1})]} \tag{12.45}$$

For the shock to exist one requires that the medium behind the shock is slowed down, and therefore $u_n < 1$. This condition leads to the additional condition on the magnetosonic Mach number of the incident flow

$$M = \left(\tfrac{5}{3}u_{\rm th1}^2 + b_{t1}^2\right)^{-1/2} > 1 \tag{12.46}$$

The last condition simply means that the flow is super-magnetosonic for fast perpendicular shocks to develop.

The jump conditions of the perpendicular shock given in Eq. (12.44) depend on the upstream ratio of thermal-to-magnetic field energy density, i.e., on the plasma beta parameter in the upstream flow, β_1. For $\beta_1 \ll 1$ the influence of the magnetic field is strong. In this case the Mach number becomes the upstream Alfvénic Mach number, $M_A = v_{n1}/v_{A1}$. In the opposite case, it is negligible, and the magnetohydrodynamic shock resembles a simple hydrodynamic shock. But one should be careful with this conclusion, since the magnetohydrodynamic approximation readily fails in reality, and particle and wave processes may become extremely important in shock dynamics.

The most important conclusion drawn from the above jump conditions is that for relatively high Mach numbers $M \gg 1$ the jumps simplify to some expressions which can easily be used for estimating the strongest possible changes of the field quantities across the shock transition in perpendicular shocks. For the original unormalized quantities these are written as

$$\begin{aligned} n_1/n_2 = B_{t1}/B_{t2} = v_{n2}/v_{n1} &\approx 1/4 \\ k_B T_2/m_i &\approx 3v_{n1}^2/16 \end{aligned} \tag{12.47}$$

These conditions show that the density and magnetic fields in a perpendicular shock will not increase by more than a factor of 4.

Theoretically, parallel shocks should behave like ordinary gasdynamic shocks, because the magnetic field inside the shock is aligned exactly along the inflow and the shock normal, and thus drops out of the magnetohydrodynamic Rankine-Hugoniot conditions. But, again, one must be careful in drawing such a conclusion, since for fast shocks with strong magnetic fields, $\beta_1 < 1$, the fast magnetosonic velocity parallel to the field becomes the shear Alfvén velocity, and the fast wave couples over to the shear Alfvén wave, which has a transverse magnetic field component. At the shock front the magnetosonic wave steepens, and this perpendicular field component increases. Hence, strictly parallel shock waves do not exist in a plasma. Near the shock front these shocks contain a transverse magnetic field component, and the shock wave becomes necessarily oblique and exhibits a complicated magnetic structure.

Critical Mach Numbers

The limit of extremely high Mach numbers given above is unrealistic. When the Mach number of a collisionless shock becomes too large, the character of the shock transition changes from laminar to turbulent. This change has to do with the process which dissipates the inflowing kinetic and magnetic energies in order to increase the entropy during shock transition and to generate irreversibility.

Laminar shock waves are steepened magnetohydrodynamic waves with breaking of the wave inhibited by dissipative effects. Two kinds of such dissipative effects may work in laminar shock fronts. The first is Joule heating due to the generation of anomalous collisions and resistivities in the current layer inside the shock, the second is viscous interaction in the shock front. Both processes are mutually related. In wave steepening both processes account for the production of different flow, density, and potential levels on both sides of the shock front. Therefore, neglecting dissipation and considering a solitary wave state should give an estimate of the maximum possible Mach number in laminar shocks.

Figure 12.6 shows a soliton-shock configuration with B_1 the undisturbed upstream tangential magnetic field, B_m the soliton, and B_s the shock amplitude. Consider the one-dimensional two-fluid model of magnetosonic waves at a perpendicular shock with the magnetic field parallel to the front

$$\begin{aligned} d(nv_{sx})/dx &= 0 \\ dB_z/dx &= \mu_0 env_{ey} \\ m_e v_{ix} dv_{ix}/dx &= eE_x + ev_{iy}B_z \\ m_e v_{ix} dv_{iy}/dx &= eE_y - ev_{ix}B_z \\ m_e v_{ix} dv_{iz}/dx &= 0 \end{aligned} \tag{12.48}$$

which is to be completed by the condition for the continuity of the tangential electric field, $dE_y/dx = 0$, and quasineutrality, $n_e \approx n_i$. The x direction is normal to the shock front, z and y are tangential. The first of these equations is the continuity equation, which holds for electrons and ions. The second equation is Ampère's law with the current carried exclusively by the electrons, $j_y = -env_{ey}$. The three remaining equations are the equations of motion of the ions.

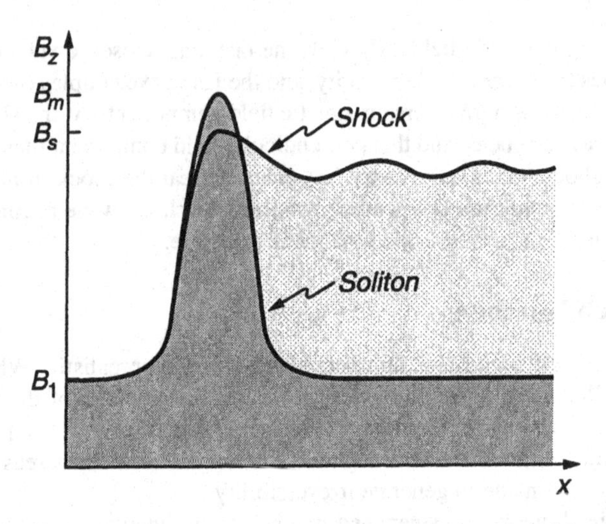

Fig. 12.6. Magnetosonic soliton-shock model to estimate M_{cr}.

The continuity equation implies constancy of the flux for both plasma components such that $n = n_1 v_1/v_{sx}$, where the index 1 refers to the values far upstream of the shock at $x = -\infty$. There the electron and ion densities and speeds are the same. Hence, for the assumed quasineutrality the ion and electron perpendicular speeds are approximately the same at the shock as well, $v_{ix} \approx v_{ex} = v_n$. It follows that $E_y = v_1 B_1$, $n = n_1 v_1/v_n$, and is $E_x = -B_z v_{ey}$. This equation results from neglecting the very small contribution of the ions to the y component of the velocity or from the fact that only the electrons are frozen-in. Remember that we consider perpendicular shocks with the magnetic field only in the z direction and changing in value only across the shock transition. In addition, summing the species in the equation for B_z and integrating once produces for the normal component of the velocity

$$v_n = v_{ix} = v_{ex} = v_1 - (B_z^2 - B_1^2)/2\mu_0 m_i n_1 v_1 \tag{12.49}$$

which shows that the normal fluid velocity decreases in the shock. This form can be used to express the current speed

$$v_{ey} = \frac{1}{\mu_0 e n_1}\frac{dB_z}{dx}\left(1 - \frac{B_z^2 - B_1^2}{2\mu_0 m_i n_1 v_1^2}\right) \tag{12.50}$$

With the help of these expressions one derives an equation for the magnetic field component tangential to the shock, B_z, which can be integrated once with respect to x in order to be written in the form of the energy conservation of a pseudo-particle

$$(dB_z/dx)^2 = -2S(B_z) \tag{12.51}$$

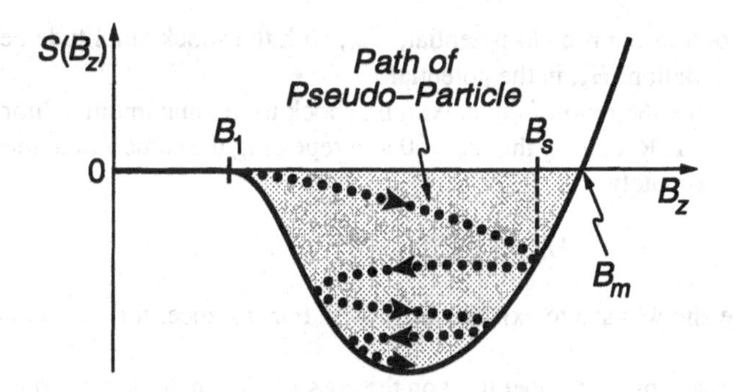

Fig. 12.7. Pseudo-potential of a laminar shock.

with the pseudo-potential, $S(B_z)$, defined through

$$S(B_z) = \frac{\omega_{pe}^2}{2c^2}(B_z - B_1)^2 \left[1 - \frac{(B_z + B_1)^2}{4\mu_0 m_i n_1 v_1^2}\right] \left(1 - \frac{B_z^2 - B_1^2}{2\mu_0 m_i n_1 v_1^2}\right)^{-1} \tag{12.52}$$

where we assumed that at infinity all derivatives vanish. In order to permit for solutions, this potential must be negative in a certain region, $B_1 \le B_z \le B_m$ and $dB_z/dx = 0$ at B_m. From $S(B_m) = 0$ one immediately finds that

$$B_m/B_1 = 2(v_1/v_{A1}) - 1 \tag{12.53}$$

where v_{A1} is the upstream Alfvén speed. The requirement that the normal velocity component from Eq. (12.49) after transition is positive, gives an upper limit on the Mach number. This limit is found by combining Eqs. (12.53) and (12.49)

$$\boxed{M_A = v_1/v_{A1} < M_{\text{cr}} = 2} \tag{12.54}$$

When inserted into Eq. (12.53), one obtains for the maximum magnetic field $B_m < 3B_1$, i.e., a compression factor of maximum 3.

The fast magnetosonic soliton solutions considered here correspond to dissipation-free laminar shocks and are therefore the extreme case of laminar shocks. For larger Mach numbers these waves overturn and cannot sustain stationary states anymore. Including dissipation into the above equations means that the pseudo-particle moves down in time in the pseudo-potential trough towards the minimum of the potential in a way similar as discussed for the solution of the Burgers equation. During this falling-down, the pseudo-particle encounters a gradually narrower potential. Hence, the maximum amplitude, $B_s < B_m$ of the shock will be smaller than the soliton amplitude. This behavior is shown in Fig. 12.7, where the amplitude of the solitary solution corresponds to the zero-point of the pseudo-potential, B_m, with the shock amplitude being the first downward oscillation, B_s, in the potential.

Let us define the ratio of the maximum shock to the maximum soliton amplitude, $B_s/B_m = \alpha \leq 1$. Requiring that $v_n > 0$ and repeating the same calculation as before, we have approximately

$$M_{\rm cr} \approx 2\alpha/(2\alpha - 1) \qquad \alpha > 1/2 \tag{12.55}$$

for a resistive shock wave to exist in fast flows. For instance, for $\alpha = 0.9$ one obtains $M_{\rm cr} = 2.25$.

However, an absolute upper limit on the possible Mach number for resistive shocks is obtained from the simple argument that the downstream flow velocity, $v_n = v_{n2}$, should be less than the relevant downstream wave speed for information transport between the obstacle and the shock. If this condition is not satisfied, the information about the obstacle cannot reach the shock in a flow time, resistivity cannot sustain the shock, and resistive shocks cannot evolve. The condition imposed is related to the *evolutionary condition* of the shock. For fast shocks the wave speed for information transport from the obstacle to the shock is the fast mode velocity given in Eq. (I.9.100) of our companion book. Heating of the plasma in the shock and behind increases the sound speed. It is thus sufficient to require that the downstream flow speed should be less than the downstream sound speed, $v_{n2} < c_{ia2}$.

Evaluating this condition, by using the temperature jump and velocity drop across a perpendicular shock, it can be shown that the critical dissipative Mach number up to which shocks can resistively be sustained is about

$$\boxed{M_{\rm cr1} \approx 2.7} \tag{12.56}$$

This limit is valid for strictly perpendicular shocks with a magnetic field to shock normal angle $\theta_{Bn} = 90°$. For oblique shocks with smaller angles, the critical Mach number decreases, and the shock becomes supercritical already at moderate Mach numbers. Enhancing the resistivity will cause the shock transition to widen. But resistivity cannot prevent supercritical shocks to break and to become turbulent. The most common and most efficient way is the generation of shock feet and foreshock.

The critical Mach number, $M_{\rm cr1}$, is only approximate and is obtained on the assumption that resistivity is the basic dissipation mechanism. A second, slightly larger critical Mach number, $M_{\rm cr2}$, can be defined if anomalous viscosity is added. Anomalous viscosity heats the ions, increasing their momentum when passing through the shock. Accordingly, the second critical Mach number is defined as the Mach number at which the downstream speed equals the downstream heated ion thermal velocity, $v_n = v_{{\rm th}i2}$. $M_{\rm cr2}$ depends on the downstream electron-to-ion temperature ratio, the incident plasma beta, β_1, and the angle θ_{Bn1}. It can be several times the first critical

Mach number. Below M_{cr2} the shock is sustained by resistivity and viscosity. Above M_{cr2} it can only be supported by ion reflection into the upstream region as the basic dissipation mechanism.

But for extremely strong shocks with very high Mach numbers even ion reflection cannot support the shock anymore, and a third critical Mach number will exist, because even a very fast flow into the shock cannot reflect more ions than a certain percentage of the incident ions. Such *super-supercritical* shocks will heat the plasma to such high temperatures that the magnetic field can be neglected and the shock will become gas-dynamic. Actually, compression of the magnetic field is restricted. Hence, the plasma may indeed become super-heated. The third critical Mach number, M_{cr3}, would be the solution of the balance equation between the downstream flow and sound speeds, $v_n = c_s$, where c_s is to be calculated by accounting for the unknown ion dynamics, reflection, and heating processes.

12.4. Shock Wave Structure

Collisionless shocks have a well-defined substructure, which depends on the angle between the upstream magnetic field and the shock normal (see Sec. 8.5 of the companion volume). For a given upstream magnetic field structure, this angle often depends on the form and nature of the obstacle. For obstacles of finite extension around which the incident matter can flow, it changes with location along the shock. Shocks can have a totally different character at different locations and may thus have different structure.

Shock Foot

Mach numbers exceeding one or more of the the critical Mach numbers are not unusual in space and astrophysical plasmas. The solar wind has variable Mach numbers in the range $1.5 < M < 15$, with the lower and upper bounds depending on the state of solar activity. Clearly, even during times of high Mach numbers the Earth's bow shock wave exists. Astrophysical shock waves may have Mach numbers up to several thousands. Such shocks are *supercritical* or even *super-supercritical* in contrast to the *subcritical* shocks with low Mach numbers. How can such shocks be sustained and how do they manage to generate the required dissipation?

Within the framework of the cold plasma theory used in the previous section, the width of the soliton transition is determined by the electron inertial length, c/ω_{pe}. This assumption is unrealistic for a shock. Widening due to dissipation increases the width to v_1/ν_{an}, and one expects the transition to be of the order of the ion gyroradius in the compressed magnetic field of the shock ramp. The main property of supercritical shocks is that they need to reflect particles from the shock front into the undisturbed flow region upstream of the shock. For some part of the incident particle population,

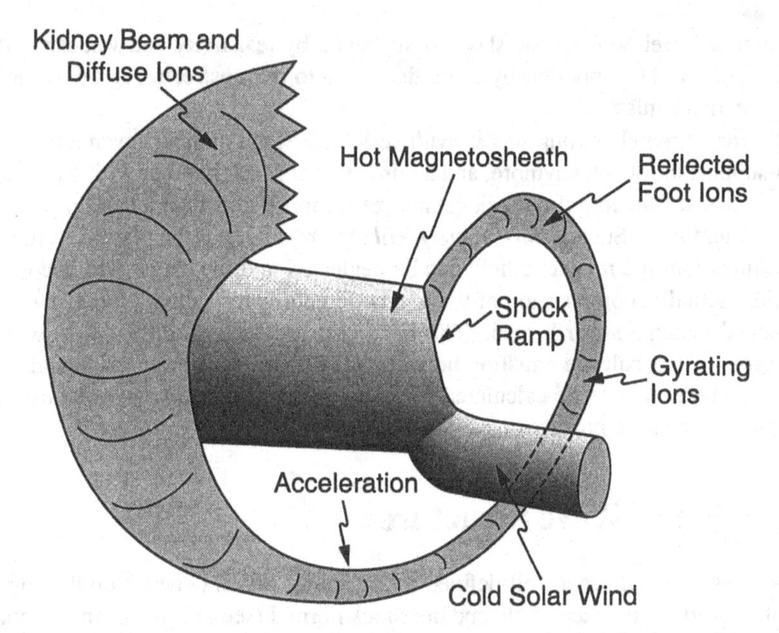

Fig. 12.8. Ion phase space components in a quasi-perpendicular shock transition.

the shock itself assumes the property of a rigid obstacle at which these particles are specularly reflected. The purpose of this reflection is to inform the incident flow about the existence of the obstacle-shock system and, before reaching the shock transition, to slow it down to a speed which allows to satisfy the evolutionary condition. In addition, the crest of the shock starts breaking and generates turbulence behind the shock which helps slowing down the flow further.

At a perpendicular shock the reflection of particles produces a *foot region* in front of the shock ramp. This foot region has the width of an energetic ion gyroradius. Two kinds of ions may form the foot. The first component is the lowest energy part of the incident ions, which have energies less than the shock potential and are reflected from the shock potential well. Note that the shock generates a potential leading to a non-zero normal electric field component pointing upstream, as described in Sec. 8.5 of the companion volume, *Basic Space Plasma Physics*.

The other component consists of ions with large gyroradii. To them the shock ramp appears as a thin steep wall and the ions are again reflected. The condition of reflection is that the adiabatic motion of these ions is broken at the shock ramp, or simply that $r_{gi} > d$, where d is the width of the ramp which is about a thermal ion gyroradius. In addition, some perpendicularly heated ions from the ramp may escape upstream and contribute to the foot region. All these ions are further accelerated by

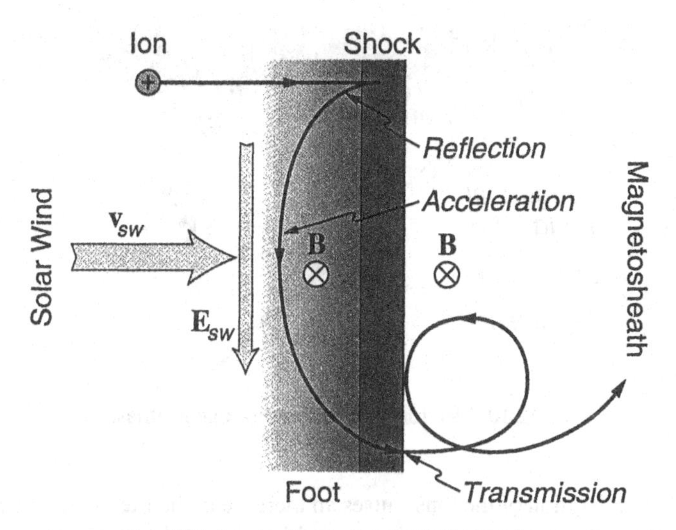

Fig. 12.9. Ion motion in the shock foot region.

the solar wind electric field. The ion phase space representation sketched in Fig. 12.8 shows all these components. The incident cool but dense solar wind ion beam hits the shock transition. It splits into a broad heated and slowed-down downstream distribution and a dilute nearly monochromatic gyrating reflected energetic ion component which after gyration and acceleration becomes more diffuse.

Electrons contribute very little to the foot. Since their small gyroradii let them behave adiabatically, most of them will not be reflected. In addition, the normal shock electric field points sunward and accelerates electrons into the shock instead of reflecting them. The foot region is thus produced solely by reflected ions, which constitute a non-compensated perpendicular current whose magnetic field forms the foot.

A large part of the gyratory orbit of the foot-region ions is parallel to the electric field in the incident flow, $\mathbf{E}_{sw} = -\mathbf{v}_{sw} \times \mathbf{B}_{sw}$. This field accelerates the foot region ions further (Fig. 12.9), to about twice the incident solar wind velocity, thereby increasing the foot current and magnetic field at the expense of the flow energy. After sufficient acceleration the angle of incidence of such ions onto the shock front may change in such a way that the reflection condition does not hold anymore, and the ions finally pass through the shock. In the shocked region behind the shock ramp these ions still have a temperature anisotropy and generate anisotropy-driven wave turbulence, until they are quasilinearly scattered and merge into the background.

Foot-region ions constitute a fast ion beam in the solar wind flow and, in addition, cause instability and wave activity by counterstreaming beam interaction. Hence, the foot region contains an enhanced level of low-frequency magnetic fluctuations. Fur-

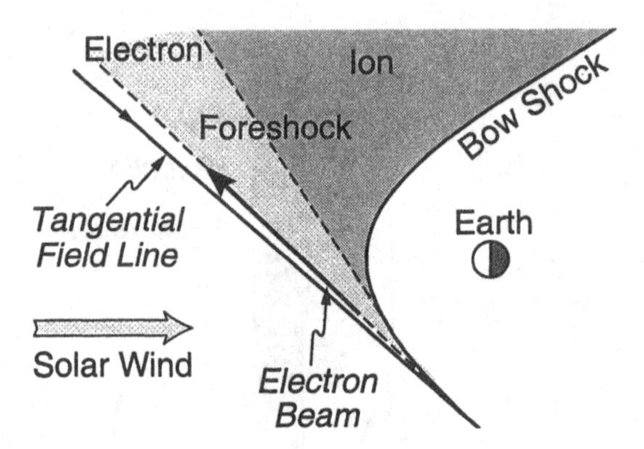

Fig. 12.10. Electron and ion foreshock geometries.

thermore, the acceleration of the ions causes an increase in their temperature anisotropy and contributes to the excitation of anisotropy-driven instabilities, plasma heating, and further dissipation of flow energy in front of the shock.

Foreshock

A behavior like described above is characteristic for all quasi-perpendicular supercritical shocks ($\theta_{Bn} > 45°$). But for shocks with angles $\theta_{Bn} < 90°$, the ions may escape further upstream along the magnetic field into the incident flow, where they generate waves by beam- and anisotropy-driven instabilities, which quasilinearly scatter them into a hot halo distribution. This distribution is convected downstream by the incident flow towards the shock front. The foot region of a strictly perpendicular shock thus transforms into a broad ion foreshock for quasi-perpendicular shocks. Quasi-parallel supercritical shocks have even broader foreshock regions. Hence, a foreshock is a very general property of any supercritical shock which is not strictly perpendicular. Its purpose is to warn the incident flow about the existence of an obstacle, to dissipate part of the incident energy, to raise the incident temperature, and to slow down the flow. In principle, a foreshock already belongs to the shock transition region.

Curved shocks like the Earth's bow shock wave can always be divided into regions of perpendicular, quasi-perpendicular and quasi-parallel shocks (see Fig. 8.10 in the companion volume). Such shocks always have extended foreshock regions. These are further divided into two zones, the *electron foreshock* and the *ion foreshock* (see Fig. 12.10). The electron foreshock is a narrow downstream region, bounded on one side approximately by the magnetic field line tangential to the shock. It contains electrons which have been reflected at or have escaped from the shock. The most energetic electrons appear at the tangential point. They are either specularly reflected or heated in the

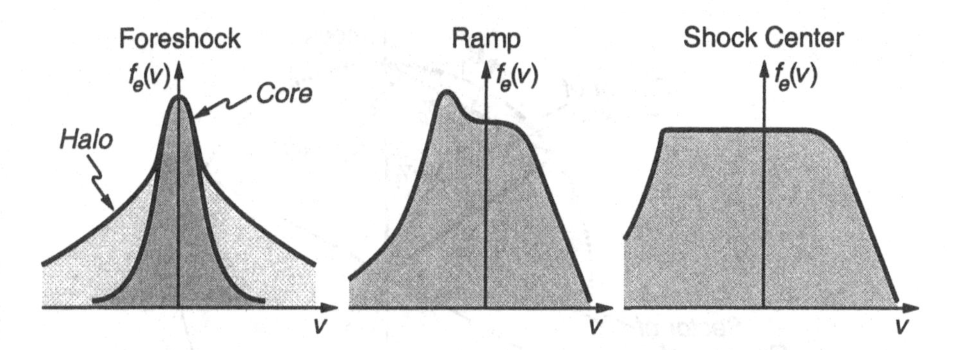

Fig. 12.11. Electron distribution functions in foreshock, ramp, and shock center.

shock ramp. Some electrons have sufficiently large field-aligned velocities to escape upstream into the solar wind, where they may travel far along the tangential field line, excite Langmuir and upper-hybrid waves, become slowed down and scattered into an isotropic halo distribution (see left-hand panel of Fig. 12.11), and are ultimately swept down again by the solar wind flow into the foreshock region. In contrast, the ion foreshock (see Fig. 12.10) forms a much larger angle with the tangential field line than the electron foreshock, because the velocity of the reflected ions is much lower than the speed of the reflected electrons and the upstream solar wind flow velocity cannot be neglected.

De Hoffmann-Teller Frame

Because the foreshock is formed by particle reflection, the mechanism of reflection and upstream injection is of interest. The simplest assumption is that the particles, mainly ions of large gyroradii but also a few energetic electrons, are specularly reflected from the narrow shock front. This mechanism is most easily visualized in the *de Hoffmann-Teller frame*, a reference frame moving parallel to the shock surface with a velocity which transforms the upstream inflow velocity of the solar wind into a velocity component which is entirely parallel to the incident field

$$\mathbf{v}_{sw} = \mathbf{v}_{\mathrm{HT}} + \mathbf{v}_{sw\|} \tag{12.57}$$

Since the de Hoffmann-Teller velocity is parallel to the shock front, it can be expressed by the shock normal unit vector, $\hat{\mathbf{n}}$. Noting that $\hat{\mathbf{n}} \cdot \mathbf{v}_{\mathrm{HT}} = 0$, one obtain a

$$\boxed{\mathbf{v}_{\mathrm{HT}} = \hat{\mathbf{n}} \times (\mathbf{v}_{sw} \times \mathbf{B}_{sw})/\hat{\mathbf{n}} \cdot \mathbf{B}_{sw}} \tag{12.58}$$

as the general expression for the de Hoffmann-Teller velocity. This velocity is parallel to the shock and at the same time transverse to the magnetic field, which implies that

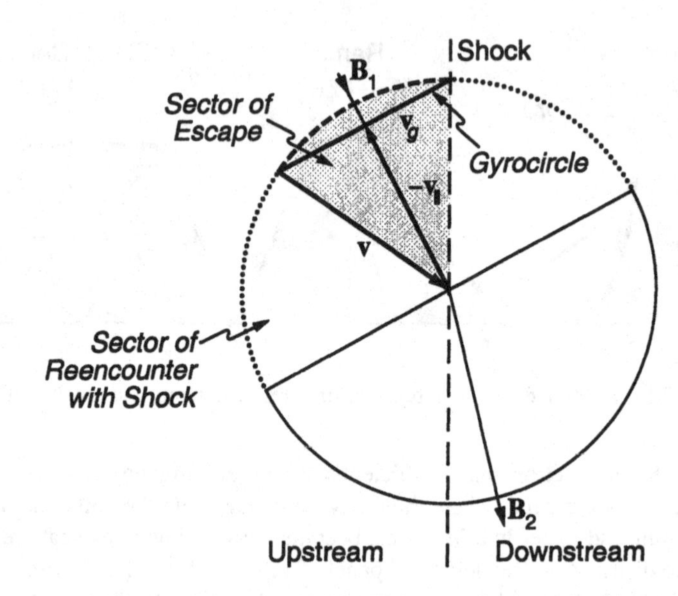

Fig. 12.12. Specular reflection in HT-frame and sector of escape.

the magnetic field is moving with just this velocity along the shock front. In the de Hoffmann-Teller frame any particle incident on the shock front will have only a gyromotion around the moving magnetic field line in addition to its parallel bulk speed along the magnetic field. Of course, rapid variations and accelerations by outer forces are not affected by this transformation, which merely reverses the Lorentz transformation in order to get rid of stationary perpendicular electric fields at the shock.

Shock Potential and Non-Coplanarity

Each shock has a significant cross-shock potential which is generated by the charge separation built up in the shock front when particles of different mass and gyroradii encounter the magnetic shock ramp (see Sec. 8.5 of the companion volume). Ions are reflected by this potential while electrons are accelerated downstream. This shock potential is frame-dependent. The difference between the electric fields in the normal incidence and de Hoffmann-Teller frames is related to a magnetic component at the shock front which is not coplanar (the coplanarity theorem derived by the ideal magnetohydrodynamics treatment in Sec. 8.4 of the companion volume does not necessarily hold in two-fluid or kinetic theory.)

The normal incidence frame is the frame in which the plasma stream (e.g., the solar wind) flows into the shock along its normal vector. Denoting components in this frame by the index NI, the Lorentz transformation gives the following relation between the normal components of the electric fields in the two frames

$$E_{n\mathrm{HT}} - E_{n\mathrm{NI}} = v_n B_{\mathrm{nc}} \tan\theta_{Bn} \tag{12.59}$$

where v_n is the normal incident plasma speed, B_{nc} is the non-coplanar component of the magnetic field, and the angle θ_{Bn} is the angle between the shock normal and the incident magnetic field. Hence, the shock potential difference between the two frames becomes

$$\Delta\phi = \phi_{\mathrm{NI}} - \phi_{\mathrm{HT}} = v_n \tan\theta_{Bn} \int B_{\mathrm{nc}} dn \tag{12.60}$$

the integral of the non-coplanar magnetic field component across the shock transition. Whenever the above potential difference is non-zero, the magnetic field possesses a non-coplanar component.

Specular Reflection

During specular reflection the normal component of an incident particle is reversed from downstream to upstream. Though the reflection mechanism is not known and should in principle be non-adiabatic, in ideal specular reflection the inversion of the normal component of the particle velocity does not change the particle magnetic moment. This is the simplest model of upstream injection. No definite mechanism has been given up to date. Figure 12.12 demonstrates what happens to a particle in the de Hoffmann-Teller frame when its incident normal velocity is reversed. The particle velocity consists of the sum of the gyration and the parallel velocity. Its total velocity in the plane containing the shock normal and the magnetic field is $\mathbf{v}$. Decomposition into gyrovelocity, $\mathbf{v}_g$, and sign-reversed parallel component, $-\mathbf{v}_{\|}$, yields escape from the front upstream along the magnetic field line only if the reflected parallel velocity points upward and, in addition, the gyration of the particle does not intersect the shock front.

Specular reflection may not be the correct mechanism to produce the upstream particle component. However, at least for ions the properties of the measured distributions are in relatively good qualitative agreement with the specular reflection mechanism. Actual reflection is based on a combination of reflection in the shock potential, which works only for the low-energy component of the incident particles, and non-adiabaticity. The latter poses an unresolved problem in which the shock-generated plasma turbulence is strongly involved. In a qualitative picture one can assume that small-scale, much shorter than the ion gyroradius, but large-amplitude magnetic fluctuations in the shock front prevent that ions pass through the shock and keep them at the shock front. Here they are accelerated by the electric field and by reflection from magnetic waves convected downstream towards the shock from the foreshock region, until they reach sufficiently high energies to escape along the magnetic field into the upstream direction (Fig. 12.13). Recent numerical simulations suggest that most of the

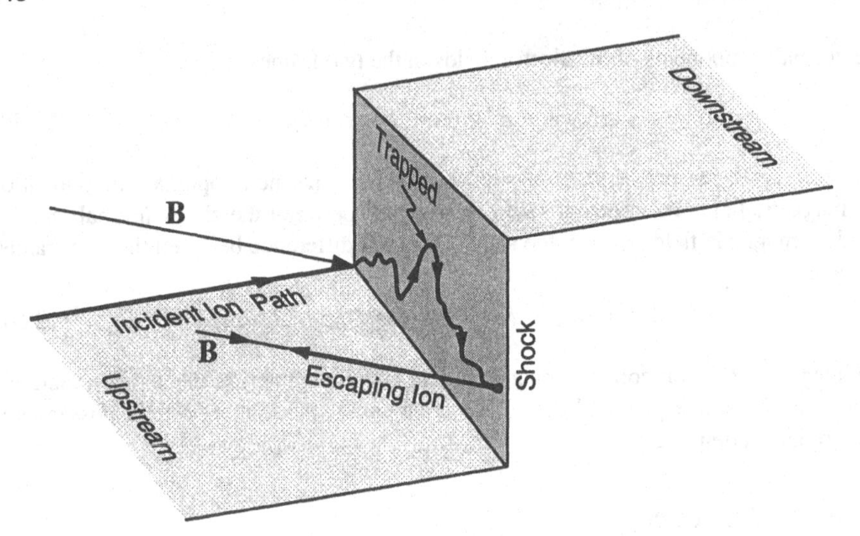

Fig. 12.13. Trapping and acceleration of ion in shock ramp.

specularly reflected ions are captured in the shock reformation process by the newly formed shock and find themselves soon after reflection behind the shock ramp in the downstream region. The upstream escaping particle component is thus composed just of the shock confined ions which are at rest in the de Hoffmann-Teller frame until they become energized sufficiently much and may escape from the shock.

Populating the foreshock region, these ion beams excite a high level of wave intensity, which scatters the ions self-consistently into so-called *kidney beam* distributions. Further down in the foreshock the ion distributions evolve into a nearly isotropic *ring distribution*. This ring distribution is taken up by the solar wind in the foreshock and flows in the direction of the shock. Figure 12.14 shows the evolution of a reflected ion beam in the foreshock into a kidney and a ring around the solar wind beam, when passing from the boundary of the foreshock into the heart of the foreshock.

Upstream Waves and Shock Reformation

The reflected ion beams in the foreshock region are a source of free energy and drive several ion-ion beam instabilities. The two beams involved in the interaction are the dense but cold incident solar wind and the warm reflected ion beam, which is only a little less dense than the solar wind. The dominant ion beam instability in the foreshock is the right-hand resonant instability (see Fig. 5.10). The mode is an Alfvén-cyclotron mode, which scatters the ion beam into the diffuse ion distribution of the deeper foreshock. Part of this diffuse distribution may reach high energies due to addi-

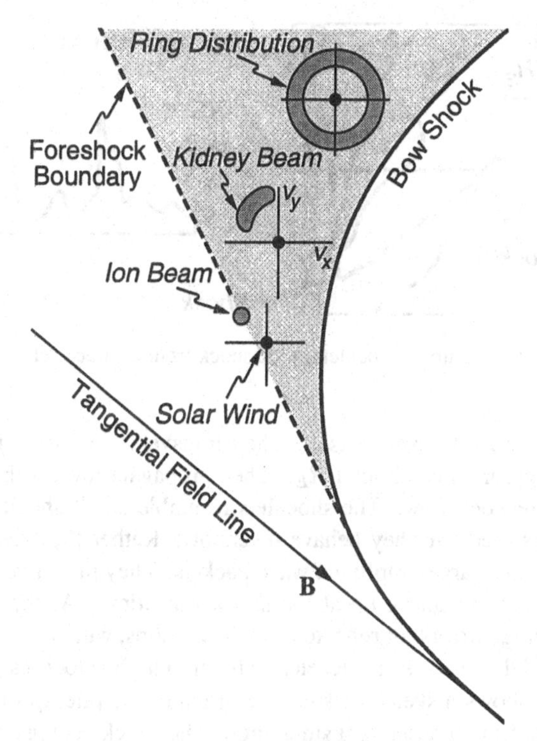

Fig. 12.14. Evolution of reflected ion beam into an isotropic ring distribution.

tional reflection when the ions reach the shock, and this acceleration can be a repeated process. These energetic diffuse ions exert an additional inhomogeneous pressure and may help decelerating and deflecting the solar wind in the foreshock. At the same time, the diffuse ions drive the left-hand resonant instability and generate left-hand polarized wave modes which, when convected downstream, may appear as right-hand modes in a spacecraft frame.

The unstably excited waves are convected downstream by the solar wind. At the same time, they steepen in this strongly driven situation and may undergo parametric decay as well as modulation to form large-amplitude nonlinear structures which are called *slams* or *shocklets*. These structures are also convected downstream towards the shock where they pile up. Such large-amplitude but short-wavelength magnetic pulsations in the quasi-parallel foreshock are a common feature in observations and simulations. The shocklets are important in the shock formation and reformation process. The waves generated by the two kinds of instabilities in the foreshock steepen and reach very large amplitudes, much larger than the initial field.

The large-amplitude shocklets in the quasi-parallel foreshock have steep flanks,

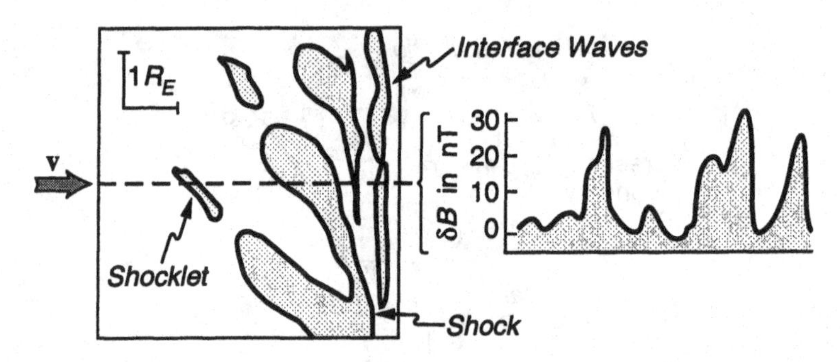

Fig. 12.15. Pile-up of shocklets at the shock front in shock reformation.

comparable to the shock transition itself. Their transverse extension is of the order of the energetic ion gyroradius, about $1\,R_E$. They propagate toward the shock with the super-Alfvénic foreshock flow. The shocklets resemble small shock pieces, but they are not real shocks, because they behave reversibly. Rather they resemble deformed solitary structures, i.e., large-amplitude wave packets. They play a very important role in building the shock by piling up at the shock transition. At the same time, they constitute the moving mirrors for reflected and diffuse ions, which trap the ions between the shock front and themselves and accelerate them to high velocities.

Figure 12.15 shows a sketch of shock reformation by pile-up of large-amplitude magnetic shocklets from a numerical simulation. The shocklets upstream of the shock are generated by the interaction of the solar wind with the diffuse upstream ion component and convected to the right. The shock consists of blobs of merged shocklets, with the shock itself not forming a rigid wall but consisting of merged shocklets. Short wavelength waves are produced in the shock front. These are interface waves. The right-hand side of the figure shows a cut across the figure parallel to the abscissa. The shocklets, the merged wave, and the interface waves appear as steep maxima.

There is a certain periodicity of shock reformation, which obviously has to do with the nonlinear evolution of the solitary structures, their transport to the shock, and the ion dynamics in the foreshock. The quasi-parallel shock is therefore not stationary, but periodically switches between two states. The solitary wave pulses provide the tangential field component which transforms the quasi-parallel shock locally into a quasi-perpendicular shock.

Shock Transition

A particularly complex region is the shock transition, the place of most intense anomalous dissipation. Various kinetic instabilities may arise here, e.g., current-driven ion-acoustic modes, lower-hybrid waves, electron-acoustic modes, and so on. The locations

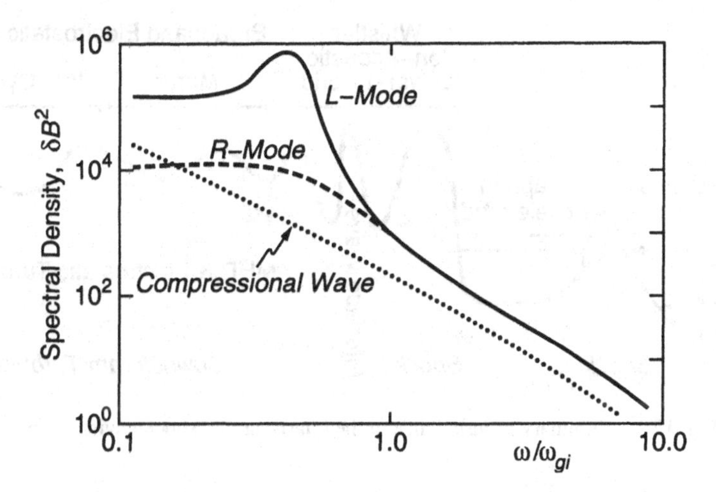

Fig. 12.16. Ion-cyclotron wave spectrum downstream of the shock.

of these instabilities may not coincide because of different threshold conditions for the instabilities. For instance, the lower-hybrid modes are stabilized in the $\beta > 1$ regions, while ion-acoustic modes are stabilized in the high-temperature downstream part of the shock transition. Electron-acoustic modes require the presence of two electron populations and will thus overcome damping in the shock ramp, where the incoming solar wind and the heated electrons mix.

Kinetic waves, including drift wave modes, presumably generate the anomalous resistivities and viscosities needed in the shock ramp to increase entropy and to generate the shock potential and currents. In addition, ion instabilities may be generated in the shock ramp. The dominant instability is a short-wavelength so-called *interface instability*, which arises from the fast dilute solar wind ion stream mixing into the heated dense shock ion distribution within about one ion gyroradius. This mixing of the two ion streams causes short-wavelength magnetic oscillations in the shock, which may be convected downstream into the magnetosheath plasma, where they become damped.

Magnetosheath Turbulence

Another instability is generated by the diffuse hot ion component, which crosses the shock front from the solar wind towards the downstream magnetosheath. These ions contribute to a perpendicular temperature anisotropy just behind the shock and excite ion-cyclotron waves, which scatter these ions over a certain distance until they merge into the downstream magnetosheath plasma. Figure 12.16 gives an example of the magnetic wave spectrum excited by these particles just after the shock. The frequency of the

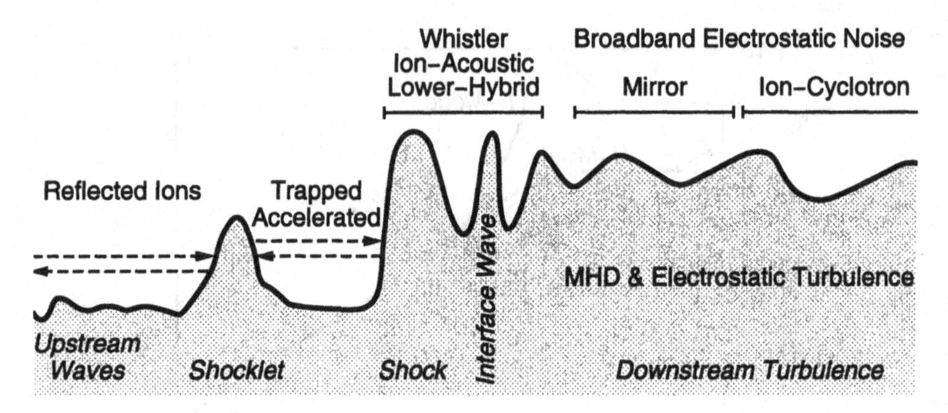

Fig. 12.17. Summary of turbulence in the quasi-parallel shock environment.

left-hand polarized waves is close but below the ion-cyclotron frequency. Hence, these waves are important for downstream heating of the plasma by depleting the remaining free energy of the penetrating ions.

At still larger distances from the shock, in the downstream magnetosheath, one encounters a turbulent plasma exhibiting strong magnetic fluctuations and broadband electric turbulence. The origin of this turbulence is not clear yet. Most of the energy in the spectrum is at the lowest frequencies, signaling the presence of intense magnetohydrodynamic fluid turbulence. A number of ion waves contribute to this turbulence, but most important are ion-cyclotron and mirror mode waves. Both these waves are driven by the transverse temperature anisotropy of the magnetosheath plasma and competing for dominance. Closer to the shock mirror waves may prevail under certain conditions, while ion-cyclotron waves are more important deeper inside the magnetosheath away from the shock. The spectrum of Fig. 12.16 is taken at such a position. Here the compressional waves are suppressed relative to the left-hand mode, but closer to the shock the compressional part is much stronger. In addition, the interface mode may contribute to the turbulence. Numerical simulations show that this mode may propagate deep into the magnetosheath and that it is a relatively short-wavelength low-frequency mode. Also right-hand modes may pass across the shock to contribute to magnetosheath turbulence.

The high-frequency part of the spectrum in the downstream region is mainly electrostatic. It contains Doppler-shifted ion-acoustic waves and drift modes, which may be excited in the density, temperature, and field gradients of the low-frequency turbulence. All these modes have large amplitudes and behave nonlinearly. This leads to the broad power law spectrum typically observed throughout the downstream region of shocks and makes identification of single modes difficult. A summary of the shock wave instability system is given in Fig. 12.17.

12.5. Particle Acceleration

Fast flows and beams of particles are frequently met in space and astrophysics, and their generation poses one of the biggest problems insofar as, from a fundamental point of view, their appearance can hardly been brought into accord with the naive idea that all processes tend towards equilibrium. We have discussed various mechanisms leading to instability, but in all these mechanisms unstable plasma configurations were assumed, very far from equilibrium. Many of them had beams as their sources of free energy. However, such beams must have been injected or excited, leading to the question which mechanisms are responsible for beam injection and, more generally, particle acceleration in a plasma.

The simplest way of accelerating a magnetized plasma is by applying an electric field perpendicular to its internal magnetic field. This is the mechanism exploited in a plasma gun. The plasma which is initially at rest is immediately set into motion with $\mathbf{v}_E = \mathbf{E} \times \mathbf{B}/B^2$ into the direction perpendicular to both, the magnetic and the electric field. The energy gained by the plasma flow is $W = m_i \mathbf{v}_E^2/2$. By increasing either the magnetic or the applied electric field one can reach appreciable energies, only limited by relativistic effects, which appear in very strong electric and magnetic fields.

Particle Acceleration in Reconnection

Most of the fast plasma flows observed in space and astrophysics are believed to be caused by mechanisms similar to a plasma gun. One of these mechanisms is magnetic reconnection, where the applied electric field is the induction field caused in the reconnection process, and the acceleration of the plasma reaches energies of the order of the Alfvén energy. For example, the fast solar wind outflow during solar flares reaches velocities of the order of 1000 km/s, which are the range of fast reconnection speeds in the lower corona. Optical line observations during flares have demonstrated one-sided broadening of the lines, which are successfully interpreted as Doppler broadening due to bulk acceleration in reconnection of the antiparallel magnetic field lines in coronal magnetic arcades. Similarly, fast flows in the magnetotail of the Earth's magnetosphere are interpreted as result of reconnection in the tail current sheet. Hence, although the reconnection process is not yet satisfactorily understood, impulsive bulk acceleration generated by reconnection seems a common process in inhomogeneous magnetized and moving plasmas. This process clearly leads to reorganization of magnetic fields from a regular state into a simpler configuration, relaxation of energy, and increase of entropy.

Current sheets can also lead to particle acceleration. This has been demonstrated by numerical simulations, where test particles are fed into the neutral point region of the reconnecting current sheet. The idea behind such a calculation is based on the special form of particle orbits in a magnetic neutral sheet as shown in Fig. 12.18. The

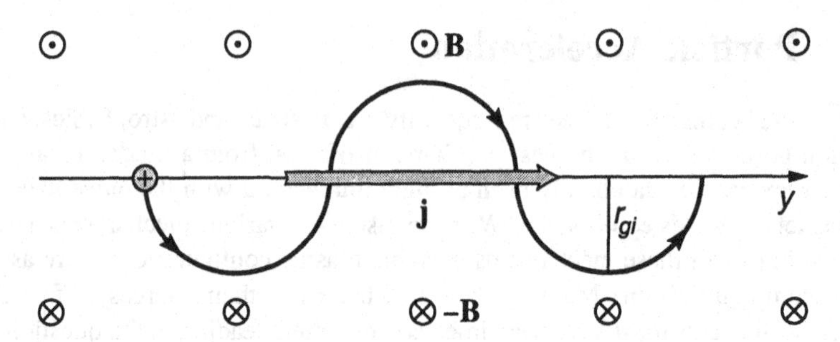

Fig. 12.18. Regular ion orbits in a current neutral sheet.

magnetic field lines are nearly antiparallel outside the sheet and the current flowing in the sheet is assumed to be driven by an electric field parallel to the current. Ions with gyroradii larger than the width of the sheet but not too large perform meandering orbits in the magnetic fields to the both sides of the sheet. At the same time these particles experience the electric field of the sheet and are accelerated along the field direction similar to the acceleration of ions in the foot region of the perpendicular shock. This acceleration takes place until the ions reach sufficiently large energies and are scattered out of the sheet to escape along the magnetic field.

It is believed that this mechanism is responsible for energetic ion beams ejected from the geomagnetic tail into the inner magnetosphere and detected as high-speed flows near the neutral sheet and as fast ion beams in the plasma sheet boundary layer. However, the burstiness of such fast flows and beams and their correlation with geomagnetic activity suggests that reconnection in the neutral sheet is more important than steady current sheets. Magnetohydrodynamic simulations of X-line formation and plasmoid evolution in the tail neutral sheet have been used to investigate the acceleration of ion injected into the X-line. Figure 12.19 shows the idea of such a simulation as well as the schematic velocity distribution and the energy spectrum in the plasma sheet boundary layer. The velocity distribution (left insert) exhibits the nearly stagnant background ion distribution and the beam distribution streaming towards the Earth. The differential energy spectrum (right insert) shows the background power law differential energy flux of energetic ions with the accelerated 60 keV ions popping out as a broad peak. This acceleration is due to the nonadiabatic effects in the reconnection process, the inductive electric fields, and multiple reflection of the ions in the X-line.

Beams of such energetic ions propagate both towards the Earth and along the separatrix of the plasmoid into interplanetary space. Similar ion beam injections, though not directly observable are common during solar flare events. Their indirect signatures are line emissions in gamma radiation detected during solar flares. It is, however, not yet fully established that ion acceleration by reconnection can produce the required energetic beams of GeV energies, which lead to the line emissions via nuclear interactions.

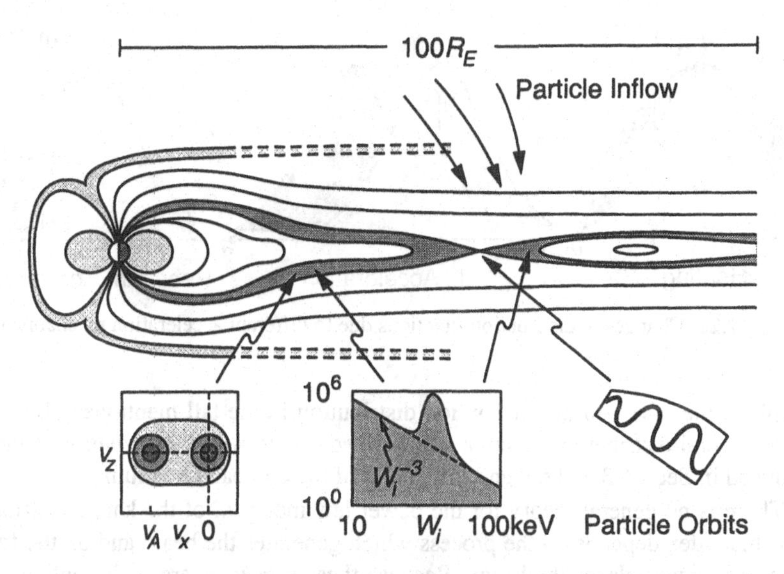

Fig. 12.19. Acceleration of ions during X-line formation in reconnection.

Another place, where reconnection has been found to generate ion beams, is merging at the magnetopause during periods when the magnetic fields in the magnetosheath and the magnetosphere have oppositely directed components. Observations indicate that jetting of plasma occurs in such cases. Sometimes also counterstreaming beams propagating along the magnetic field are observed during magnetic merging.

Accelerated Distributions

The former section dealt with two main types of acceleration, i.e., bulk acceleration and beam generation. There is another type of acceleration which is summarized under the term heating. All these kinds of accelerations lead to changes in the particle distribution function as shown schematically in Fig. 12.20.

Bulk acceleration like the plasma gun process simply shifts the entire distribution function up to the new bulk speed without changing its shape. Heating causes broadening of the distribution which may not necessarily be as symmetric as suggested by the figure. Beam acceleration shifts only one particular (possibly resonant) part of the distribution function to higher speeds as shown in an exaggerated form in Fig. 12.20.

The shape of the right-most distribution function is clearly unstable and will lead to relaxation of the beam into a gradual distribution function. Many observations in space have shown that the final distribution functions which result after such a relaxation process is some kind of power law distribution with negative power law index. An

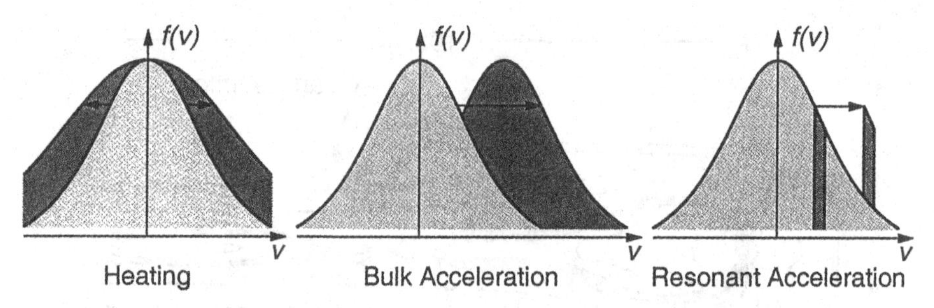

Fig. 12.20. Changes in distribution functions due to different acceleration mechanisms.

example is the $\alpha = -3$ power law ion distribution in the tail mentioned above. But a more general distribution function is described by the *kappa distribution*, which we introduced in Secs. 6.3 and 6.4 and in Fig. 6.8 of the companion volume.

There is no general theory for the power law index, κ, of the kappa distribution, rather this index depends on the process which generates the beam and on the follow-up process which relaxes the beam. Because these processes are both nonlinear, they depend on the properties of the plasma, the conditions for instability and relaxation. It seems, however, that such distributions with different κ are a very general property of plasmas and are encountered in almost every place, where strong interaction between plasma and turbulence is observed.

Auroral Acceleration

In the auroral region magnetic merging and reconnection cannot be made responsible for the local acceleration of both electrons and ions. Observed energetic electron and ion beams thus require a different explanation. Figure 12.21 gives an example of electron distributions measured over active aurora. The left-hand part is a velocity space cross-section through the electron distribution. One recognizes the downward electron beam at large negative parallel and small perpendicular speeds and the backscattered electrons at large positive parallel and small perpendicular speeds as well as the broad plateau of mirrored particles at large transverse and small parallel velocities. The right-hand part shows the differential flux spectrum with the peak caused by the downward electron beam at about 2 keV.

An example of a dynamical electron spectrogram during a spacecraft flight across an active auroral event is given in Fig. 12.22. The signature in energy is that of an *inverted* V, with high fluxes at low parallel energies at the borders of the event and high fluxes at high energies in its center. In the center the trapped fluxes are also enhanced, but exhibit an isotropic distribution, while at the borders one finds the electron beams. Such structures are observed in the lower auroral magnetosphere up to about 3000 km

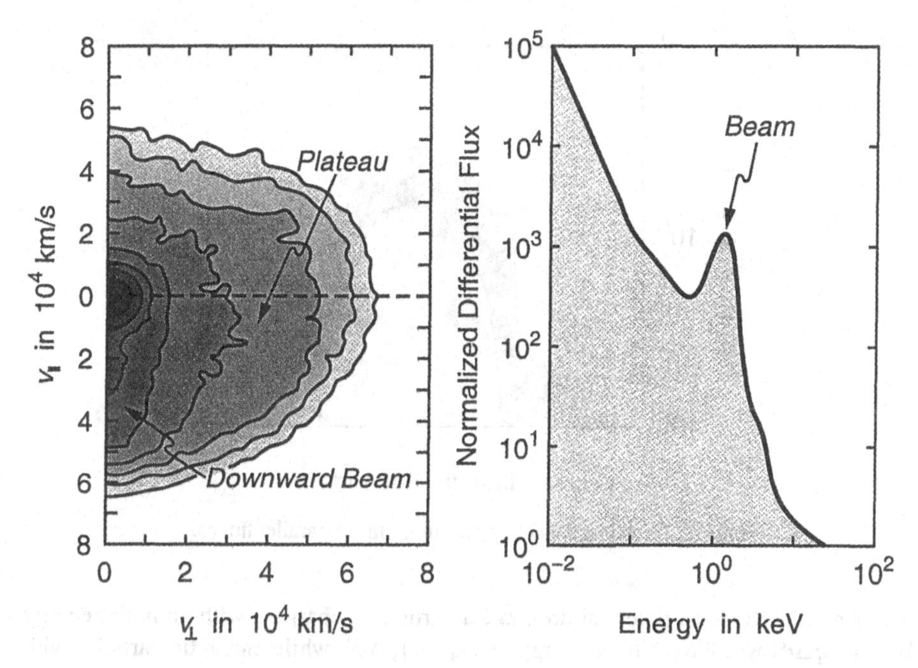

Fig. 12.21. Auroral electron distribution functions.

height and are obviously spatial structures which are being traversed by the spacecraft. Their spatial extension ranges from tens of meters to several 10 km, corresponding to transition times of the order of several 10 ms to several seconds.

Ion acceleration is also observed in the auroral region, but the sites of acceleration do not necessarily coincide for electrons and ions. At altitudes below 2000 km ions are accelerated out of the background hydrogen and oxygen distributions to perpendicular energies of typically 100 eV which, when moving up the field lines, evolve into so-called *ion conics*, conical velocity space distributions resembling a gyrating beam. Two types of such conics have been observed. The first type is a pitch angle distribution which is confined to a narrow range of oblique pitch angles. It is expected that such a distribution results from an acceleration process which itself is confined to a certain altitude. The second type has a broader pitch angle distribution and such conics may be caused by scattering during propagation in wave fields which have a wide extension along the field.

Surely, the easiest way of producing beams is by letting the particles fall through a stationary electric potential confined to a certain altitude range, a *double layer*. When passing across the potential the particle simply picks up the potential drop. The signature of such steady potentials is acceleration of ions and electrons in opposite di-

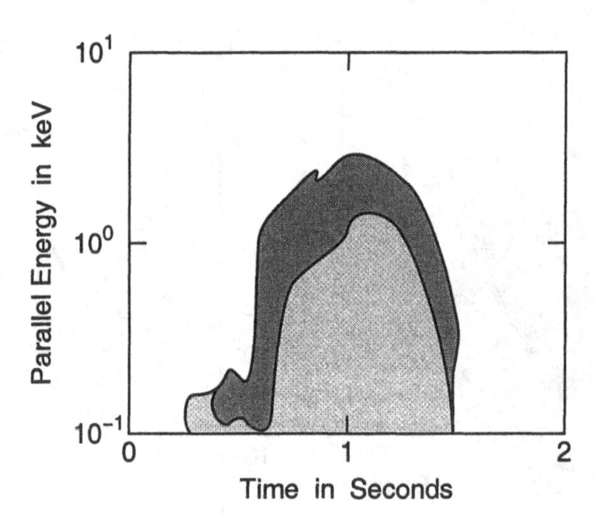

Fig. 12.22. Auroral electron spectrogram for parallel fluxes.

rections. Clearly, for a potential drop $\Delta\phi$ a particle of charge q with an initial energy $W_0 \ll |q\Delta\phi|$ will have a final energy of $W_1 \approx |q\Delta\phi|$, while energetic particles with larger initial energies, $W_0 \gg |q\Delta\phi|$, will remain almost unaffected. Hence the low-energy part of the energy distribution will become shifted to energies of the order of the potential drop. The energy distribution just above the potential drop flattens, and the energy distribution is cut at the potential. This behavior should be observed in both cases, in a non-resistive double layer and in a resistive potential drop, where current instabilities cause anomalous resistivity and Ohm's law generates the potential drop.

An example where a potential drop along the field is generated by differences in the mirror heights of electrons and ions is shown in Fig. 12.23. In this case the mirror force on the particles causes the drop and the parallel current of the precipitating electrons. As a result an U-shaped electric potential structure resembling large-scale double layers is caused. Such a structure is shown in Fig. 12.24. Outside the parallel potential drop the isopotentials are parallel to the field lines. Here the magnetospheric electric field is simply mapped down into the ionosphere. To both sides of the drop this mapping is in opposite directions and the plasma convection flow across the drop is a shear flow. Thus, the observation of shear flows may be an indication of the presence of parallel potentials along the magnetic field. Numerical solutions of the propagation of an Alfvénic pulse from the plasma sheet into the auroral magnetosphere have shown that current instabilities may indeed cause such potential drops at altitudes close to 1 R_E in the auroral region.

If such a pulse starts as a kinetic Alfvén wave somewhere in the plasma sheet, carrying a parallel current down into the ionosphere, it changes character at about 1 R_E al-

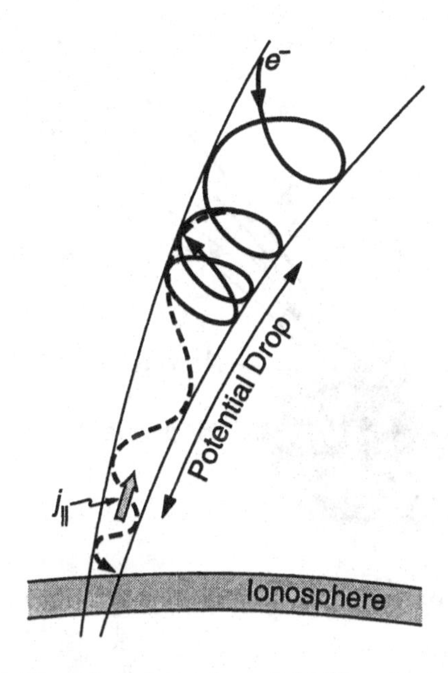

Fig. 12.23. Mirror impedance due to field-aligned potential drop.

titude, where the plasma beta becomes $\beta < m_e/m_i$. The pulse becomes a shear-kinetic Alfvén wave below this altitude, with the current carried by the background plasma. Such field-aligned currents can be unstable to ion-cyclotron and ion-acoustic waves, which give rise to the corresponding anomalous resistivities. Although the ion-acoustic resistivity is theoretically higher, ion-cyclotron resistivity is more likely because of its lower threshold. Reflection of part of the wave pulse at the dense ionosphere helps concentrating the wave energy. The calculations use ion-cyclotron waves and indicate that up to about 50% of the initial kinetic Alfvén wave energy may be transformed into a stationary potential drop along the field and thus used for auroral acceleration.

Ion Holes

The linear threshold for current-driven ion-acoustic waves to become unstable is that the electron current-drift velocity exceeds the ion-acoustic velocity, $v_{de} > c_{ia}$. On this basis it was concluded that ion-cyclotron modes have a lower threshold than ion-acoustic waves. But particle simulations suggest that this might not be true in the non-linear stage, simply because local variations in the in phase space may cause the ion to clump together, leaving behind local accumulations of electrons. This implies density variations which appear as ion-acoustic fluctuations in the field (Fig. 12.25). Numerical simulations have shown that in field-aligned current systems with hot electrons and

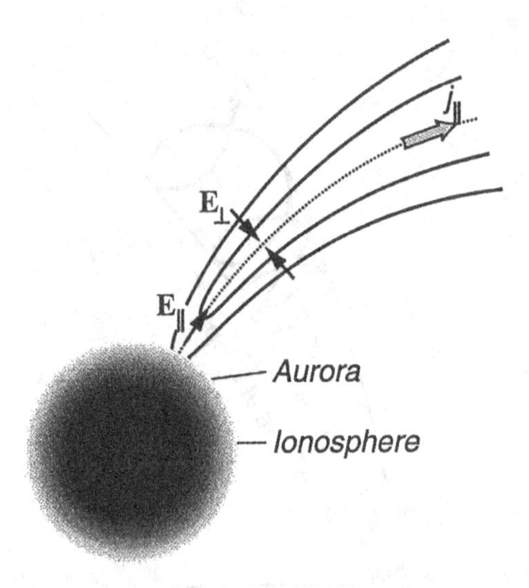

Fig. 12.24. Acceleration of auroral electrons by a U-shaped potential.

warm ions such holes can evolve, grow, and survive for long times, even when linear instability does not arise at low current speeds.

Why such holes grow when being attracted by the bulk of the ion distribution can be understood from the left-hand sketch in Fig. 12.26. The randomly produced

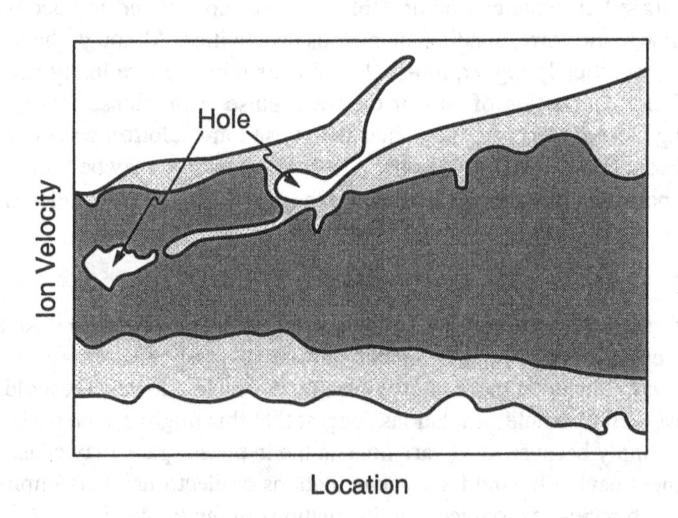

Fig. 12.25. Ion distribution in velocity-configuration space and evolved hole.

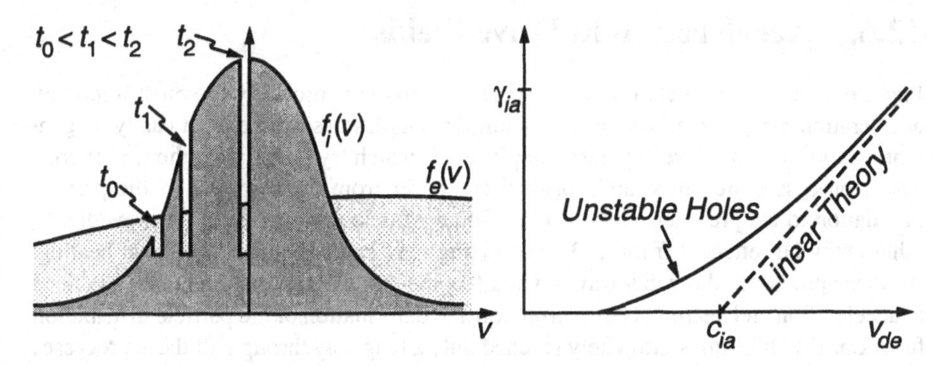

Fig. 12.26. Growth of an ion hole and corresponding ion-acoustic growth rate.

hole appears first at the edge of the ion distribution function. Being a negative charge accumulation, it is attracted by the center of the ion distribution and starts moving without changing its size. But in the denser ion distribution its actual amplitude is much larger than where it started, implying growth of its associated ion-acoustic disturbance. In the right-hand part of Fig. 12.26 the resulting ion-acoustic growth rate is sketched. At small current-drift velocities this growth rate is caused by hole formation, reaching asymptotically the linear ion-acoustic growth rate for speeds exceeding the threshold.

The presence of ion holes causes two effects. First, it enhances the level of ion-acoustic fluctuations and increases the ion-acoustic resistivity. Second, the holes themselves reflect electrons (Fig. 12.27) on both sides, leading to a localized soliton structure. But because the holes move at the ion-acoustic speed, c_{ia}, the solitons are slightly deformed and evolve into small-scale double layers with a non-zero potential drop across them. This potential drop causes electron acceleration. The potentials of many ion holes along a field line will add up to a large potential drop, in which the electrons may be accelerated to appreciable energies.

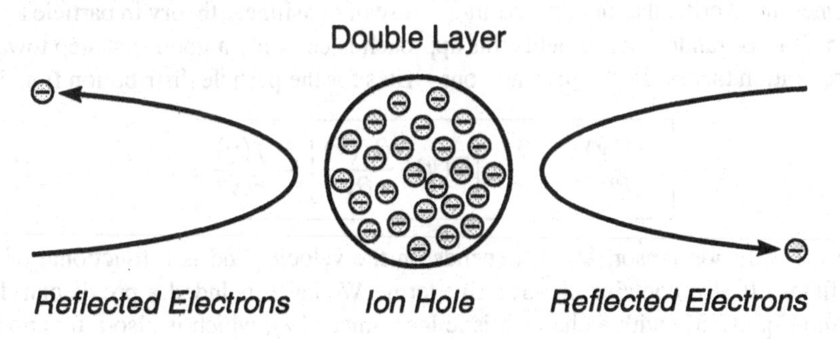

Fig. 12.27. Reflection of electrons from an ion hole.

12.6. Acceleration in Wave Fields

Particle acceleration in various kinds of plasma waves is probably the most important acceleration mechanism. The reason is simply that plasmas evolve most easily by generating a number of waves by unstable processes which try to dissipate the free energy. These processes are not straightforward but suffer from the response of the particle population to the presence of the waves. Some particle are trapped in the wave fields, others may be reflected from it. Phase-mixing may cause disorder and local heating. All these processes show that waves will affect the particle distribution locally and lead to acceleration and heating, beam formation and deformation of the particle distribution function. Equilibrium is ultimately reached only a long way through all these processes which include wave generation and plasma response.

General Formulation

Much effort has been invested looking into the different mechanisms of acceleration in which waves are directly involved. These mechanisms can be divided roughly into the following classes: resonant acceleration in extended wave fields, acceleration in localized wave fields, heating in collapse, ponderomotive force acceleration, shock acceleration, and last not least chaotic acceleration. With the exception of the ponderomotive force and collapse acceleration, all other mechanisms are based on the quasilinear Fokker-Planck diffusion equation in energy or velocity space.

The idea behind this theory is that the particles encounter a random-phased wave field, which scatters them in energy space up to high energies. This scattering is described as energy diffusion, and the problem is reduced to the determination of the diffusion coefficient and solution of the quasilinear equation for the particles. This solution can be found non-selfconsistently or selfconsistently. In the former case the wave field is given and the reaction on the wave field is neglected. In the latter case the damping and amplification of the wave field by the accelerated particle component is taken into account.

One can be critical about the unanimous use of quasilinear theory in particle acceleration. But for random wave fields this approach is certainly a good first step towards an acceleration theory. If it is justified, one writes for the particle distribution function

$$\boxed{\frac{\partial f(\mathbf{v})}{\partial t} = \frac{\partial}{\partial \mathbf{v}} \cdot \left[\mathsf{D}(\mathbf{v}) \cdot \frac{\partial f(\mathbf{v})}{\partial \mathbf{v}}\right] - \frac{f(\mathbf{v})}{\tau(\mathbf{v})}} \tag{12.61}$$

where the diffusion tensor, $\mathsf{D}(\mathbf{v})$, depends on the velocity and is a functional of the wave fields. It also contains an advective term. We have included a provisional loss term into Eq. (12.61) with a characteristic loss-time, $\tau(\mathbf{v})$, which is also a function of velocity.

The quasilinear diffusion coefficient is determined by the wave field, which satisfies the wave kinetic equation. The latter can be written, including wave damping, $\gamma_l(\mathbf{k})$, and sources, $S_w(\mathbf{k})$, as

$$\boxed{\frac{\partial W_w(\mathbf{k},t)}{\partial t} = \frac{\partial}{\partial \mathbf{k}} \cdot \left[\mathsf{D}_w(\mathbf{k}) \frac{\partial W_w(\mathbf{k},t)}{\partial \mathbf{k}} \right] - \gamma_l(\mathbf{k}) W_w(\mathbf{k},t) + S_w(\mathbf{k})} \tag{12.62}$$

The wave diffusion coefficient describes the spread of the wave spectral energy across wavenumber space. With the appropriate expressions for damping rates, source terms, loss coefficients, and particle and wave diffusion coefficients, Eqs. (12.61) and (12.62) describe to lowest order the particle acceleration process in the interaction between particles and waves. Particular models specify these equations for the wave mode under consideration and solve numerically for the distribution function of the particles.

Lower-Hybrid Electron Acceleration

Electron acceleration is one-dimensional because electrons are strongly magnetized, and the acceleration proceeds parallel to the magnetic field. In order to achieve high particle energies, large wave phase velocities are required to satisfy the resonance condition $\omega = kv$. The most important electrostatic wave with high parallel phase velocity is the lower-hybrid wave. Restricting to parallel energy diffusion only, the parallel electron diffusion equation neglecting losses can be written

$$\frac{\partial f_e}{\partial t} = \frac{\partial}{\partial v_\parallel} \left(D_{\parallel\parallel} \frac{\partial f_e}{\partial v_\parallel} \right) \tag{12.63}$$

where $D_{\parallel\parallel}$ is the parallel component of the electron diffusion tensor

$$D_{\parallel\parallel} = \frac{8\pi^2 e^2}{m_e^2} \int d^3k \frac{k_\parallel^2}{k^2} W_{lh}(\mathbf{k}) \delta(\omega - k_\parallel v_\parallel) \tag{12.64}$$

The wave spectral density, $W_{lh}(\mathbf{k})$, of the lower-hybrid waves can, in the simplest model, be taken as given. In this case it is reasonable to approximate it as a product of two functions, each of which depends only on $k_\perp$ or $k_\parallel$, such that $W_{lh}(\mathbf{k}) = \psi_\perp(k_\perp)\psi_\parallel(k_\parallel)$. In addition one may assume that the parallel spectrum is a power law

$$\psi_\parallel(k_\parallel) = A_q |k_\parallel / k_m|^{-q} \tag{12.65}$$

for $|k_\parallel/k_m| > 1$. For $|k_\parallel/k_m| < 1$ the coefficient $A_q = 0$ and thus k_m is the minimum wavenumber where the power law spectrum is cut off at long wavelengths. In addition, because the total wave energy is the integral over k-space of the wave spectral density, we can require that $\int dk_x dk_y \psi_\perp = W$, and $\int dk_\parallel \psi_\parallel = 1$. Moreover, fast electrons have speeds larger than the group velocity of the lower-hybrid wave, $v_{gr\parallel} \ll v_\parallel$. Thus the parallel diffusion coefficient can be approximated as

$$D_{\parallel\parallel} \approx \tilde{A}_q |v_\parallel|^{-3} |k_m v_\parallel/\omega|^q \tag{12.66}$$

where $\tilde{A}_q = 2\pi\omega_{pe}^2 v_0^2 W A_q/m_e n_0$ is another coefficient. We now set formally $\omega = k_\perp v_0$ and introduce normalized parallel velocities and times according to $u = v_\parallel/v_{\rm the}$ and $\tau = t(\tilde{A}_q/v_{\rm the}^5)|k_m v_{\rm the}/\omega|^q$, which allows us to rewrite the parallel electron velocity space diffusion equation into the simple form

$$\frac{\partial f_e(u,\tau)}{\partial \tau} = \frac{\partial}{\partial u}\left(\frac{1}{|u|^{3-q}}\frac{\partial f_e(u,t)}{\partial u}\right) \tag{12.67}$$

This equation must be solved under an appropriate initial condition. For example, one can assume that the initial distribution function at time $\tau = 0$ is a Maxwellian

$$f_0(u) = \pi^{-1/2} n_0 \exp(-u^2) \tag{12.68}$$

It should be noted that the above equation has a trivial solution for $q = 3$, because for the corresponding spectrum the diffusion coefficient is constant, and the solution is simply that of a dispersing heat pulse. However, this solution is of little interest, since it is inappropriate for nearly all conditions encountered in space plasmas. The solution of the diffusion equation for general q can be constructed with the help of the Laplace transform technique. It is represented in the form of an integral

$$f_e(u,\tau) = \int_0^\infty du' g_q(u,u',\tau) f_0(u') \tag{12.69}$$

where the Green's function is determined from the Laplace transform of the above one-dimensional electron diffusion equation. One finds

$$g_q(u,u',\tau) = \frac{(uu')^{2-q/2}}{(5-q)\tau} I_{-\nu}(\eta) \exp\left[-\frac{u^{5-q}+u'^{5-q}}{(5-q)^2\tau}\right] \tag{12.70}$$

with the following abbreviations $\nu = (4-q)/(5-q)$ and $\eta = 2(uu')^{(5-q)/2}/(5-q)^2\tau$, and $I_{-\nu}$ a Bessel function.

For any given parallel power law index, q, this expression shows a complicated dependence on velocity and time, but as is typical for diffusive processes, the entire distribution will decrease with time and will at the same time broaden in u-space with increasing time. This implies that the initial Maxwellian will spread out in u to generate a long extended tail indicating continuous acceleration of electrons to high energies. The long-time behavior for a power law index of $q = 4$, say, is found to evolve like

$$f_e(u,\tau) \approx n_0 \tau^{-1} \exp(-|u|/\tau) \tag{12.71}$$

This function decays much less steeply with velocity than the Gaussian function.

Localized Lower-Hybrid Waves

The theory presented above shows that extended lower-hybrid wave fields may, in continuous interaction with an electron distribution, generate tails on the distribution and accelerate electrons to high energies. But this model is unrealistic, because electrons move rather fast along the magnetic field and will readily leave from the region occupied by the waves. In addition, lower hybrid waves of high intensity may evolve into localized wave packets, with the interaction between electrons and waves limited to the transition time of the electrons across the packets which may be described as cavitons filled with lower-hybrid plasmons. The observation of such lower-hybrid cavitons in correlation with the appearance of auroral electron beams strongly suggests that acceleration in localized wave fields is more important in most cases than acceleration in extended fields.

Localization of the wave field considerably complicates the simple theory of the previous section. Progress can be made by assuming that the electric wave field entering the diffusion coefficient is derived from an electrostatic potential

$$\Phi(\xi, t) = (2\pi)^{-1/2} V(\xi, t) \sum_{\mathbf{k}} Q(k) \exp i[k\xi - \omega(k)t] \tag{12.72}$$

with $V(\xi, t)$ the slowly variable envelope potential, and ξ the spatial coordinate parallel to the direction of the lower-hybrid wave vector. The function $Q(k)$ is the k-spectrum of the waves trapped inside the caviton. If the wave is a single mode, then this spectrum is a delta-function, and the k-dependence is solely given by the transform of the envelope. Since the caviton moves at speed u_0, the coordinate in the moving frame is $x = \xi - u_0 t \cos\theta$, and θ is the angle against the magnetic field. This propagation speed is nearly parallel to the magnetic field.

In order to determine the diffusion coefficient we need the spectral energy density of the waves. This can be calculated from the correlation function knowing the wave electric field

$$E(x, t) = -(2\pi)^{-1/2} V(x) \sum_{\mathbf{k}} R(k) \exp i(kx - \varpi t) \tag{12.73}$$

with $\varpi = \omega - ku_0 \cos\theta$ and $R = ikQ$. For the envelope we assume the caviton function

$$V(x) = A\,\mathrm{sech}(bx) \tag{12.74}$$

But for technical purposes it is simpler to actually use a corresponding Gaussian profile, $V(x) = A\exp(-bx^2)$, which has a similar shape. Calculating the correlation function of the total lower-hybrid wave field

$$C(x, t) = \int\!\!\int d\zeta d\tau E^*(\zeta, \tau) E(x + \zeta, t + \tau) \tag{12.75}$$

one finds that the correlation function can be represented as

$$C(x,t) = C_0 g(\sqrt{2a}, x/2L_\perp) \sum_k |R(k)|^2 \exp i(kx - \varpi t) \tag{12.76}$$

where $C_0 = A^2/8(2a\pi)^{1/2}$, the function g is defined as

$$g(v,w) = \{\mathrm{erf}[v(w+1)] - \mathrm{erf}[v(w-1)]\} \exp(-v^2|w|^2) \tag{12.77}$$

and $a = bL_\perp^2$. The length scale $L_\perp$ is the perpendicular scale of the caviton as obtained from the nonlinear Schrödinger theory. It is related to the amplitude, A, by Eq. (11.58)

$$A^2 L_\perp^2 \sum_k |Q(k)|^2 \approx (12\lambda_{Di} k_B T_e/e)^2 \tag{12.78}$$

The next step is to take the Fourier transform of the correlation function to obtain the spectral energy density as

$$W(\kappa,\Omega) = W_0 \sum_k |R(k)|^2 \delta(\Omega - \varpi) g(a^{1/2}, w) \exp(-a|w|^2) \tag{12.79}$$

and from it to construct the parallel electron velocity space diffusion coefficient

$$D(v) = \frac{W_0 \cos\theta}{2|v|} \sum_k |R(k)|^2 g\left(a^{1/2}, \gamma\right) \exp(-1|\gamma|^2) \tag{12.80}$$

where we used the abbreviations, $W_0 = A^2 L_\perp/8a(2\pi)^{1/2}$, and $w = i(k-\kappa)L_\perp/2a$. In addition, $\gamma = i\alpha k L_\perp/2a$ and $\alpha = [\omega - k(u_0 + v)\cos\theta]/kv\cos\theta$.

This diffusion coefficient is shown in the left-hand part of Fig. 12.28 as a function of the normalized electron velocity, v/v_{the}. It peaks well outside the center of the electron distribution, thus leaving the lowest energy electrons unaffected since they do not come into resonance with the waves trapped in the caviton. Hence, cavitons or localized waves cause different electron dynamics than extended wave fields. Only a relatively narrow range of electron energies will be accelerated by the localized packets, since an electron can spend only finite time in the caviton.

The above diffusion coefficient can be used to solve the diffusion equation for electrons. A Monte-Carlo simulation solution using many electrons initially distributed as a parallel Maxwellian is shown in the right-hand panel of Fig. 12.28. In the long-time limit the distribution function starts exhibiting well expressed maxima outside the main peak. These maxima are at about 16 times the thermal energy of the electrons, in rough agreement with observations. Also, the shape of the spectrum resembles those measured in the auroral magnetosphere.

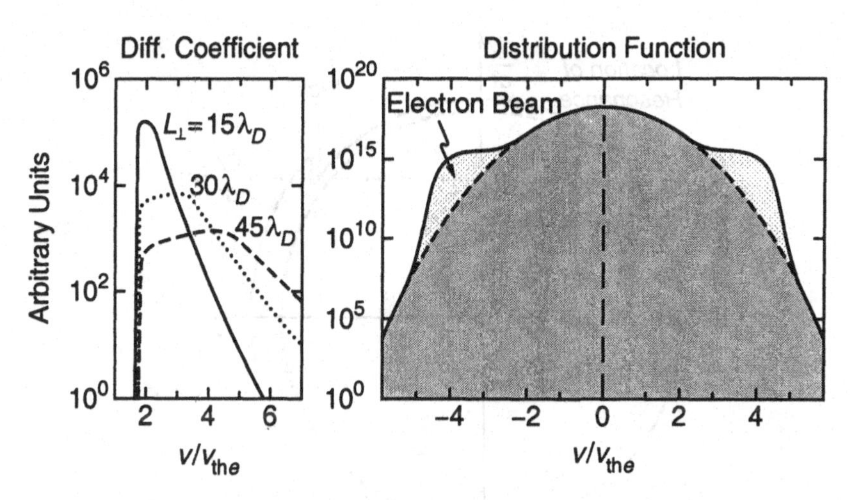

Fig. 12.28. Diffusion coefficient and electron distribution in localized acceleration.

Transverse Ion Heating

Ion heating in lower-hybrid waves leading to ion conics has been observed near 1000 km altitude in the auroral ionosphere. These observations show a good correlation between lower-hybrid cavitons and transverse acceleration of heavy ions like oxygen up to about 10 eV out of the cold background population. A mechanism suggested for this kind of ion acceleration or heating is lower-hybrid caviton collapse, where the transverse wavelength of the trapped lower-hybrid waves shrinks until it is short enough for the waves to resonate with the transverse ion motion. At altitudes above 2000 km ion heating seems to be a result of cyclotron resonance with broadband electrostatic noise. In the intermediate region heating up to about 50–100 eV is uncorrelated with lower-hybrid cavitons and the conics produced are believed to be the result of lower-hybrid heating in the extended wave turbulence. It is predominantly perpendicular because of the nearly perpendicular nature of the waves and the non-magnetized nature of the ions at these frequencies.

The most interesting mechanism of such kind of heating is intrinsic resonance or chaotic heating. The theory of this kind of interaction can be based on an investigation of the Hamiltonian of a single particle in a single lower-hybrid wave of constant amplitude. Let the external magnetic field be $\mathbf{B} = B\hat{\mathbf{e}}_z$, and the wave electric field for purely perpendicular propagation

$$\mathbf{E} = \hat{\mathbf{e}}_y\, E_0 \cos(ky - \omega t) \tag{12.81}$$

where $\mathbf{k} = k\hat{\mathbf{e}}_y$ is the wavenumber, and ω the lower-hybrid wave frequency, which is not exactly equal to ω_{lh}, but much larger than the ion-cyclotron frequency such that

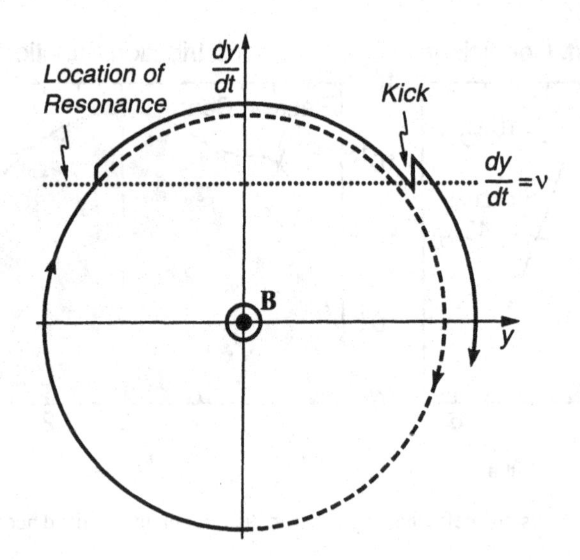

Fig. 12.29. A gyrating ion gets a kick at resonance.

$\nu = \omega/\omega_{gi} \gg 1$. Thus the ions behave nearly unmagnetized at these high frequencies, but they still perform a gyratory motion. At some places on this orbit, where the unmagnetized resonance condition, $\omega = \mathbf{k} \cdot \mathbf{v}$, is satisfied, they experience a kick in their motion extracting energy out of the wave, and become accelerated in the perpendicular direction. Thereby they increase their gyroradius on the expense of the wave amplitude.

We consider a non-selfconsistent problem in which the wave amplitude and energy are kept constant. Using normalized coordinates, with time normalized to $1/\omega_{gi}$, length to $1/k$, and velocity to ω_{gi}/k, the Hamiltonian of the ions becomes

$$\mathcal{H} = \tfrac{1}{2}[(p_x + y)^2 + p_y^2] - \alpha \sin(y - \nu t) \tag{12.82}$$

The coefficient α controls the behavior of the ion motion. It is given by

$$\alpha = kE_0/\omega_{gi}B \tag{12.83}$$

The momentum, p_x, is a constant, because the Hamiltonian does not depend on x.

From this Hamiltonian it is possible to derive the equations of motion of the ion in the lower-hybrid wave field. The first equation is simply

$$dx/dt = y \tag{12.84}$$

The second equation is a driven oscillator equation

$$(d^2y/dt^2) + y = \alpha \cos(y - \nu t) \tag{12.85}$$

Moreover, the above resonance condition for ion acceleration can be written as

$$\nu = dy/dt \tag{12.86}$$

showing that resonance is at the position where the fraction of the frequency equals the velocity. Figure 12.29 shows schematically for a given ion circular orbit around the magnetic field in phase space that a kick is experienced by the gyrating ion whenever its phase space orbit passes the resonance.

The above equations of motion can be solved numerically for different values of α. It turns out that for small α, when the wave field can be neglected, the ion orbits are regular and not disturbed. For large wave amplitudes and sufficiently high frequencies the disturbance of the orbits is large, and the ion gyroradius increases, leading to acceleration in the perpendicular direction. In order to represent this diffusive behavior in energy or phase space one must find an appropriate system of variables. It is most convenient to use cyclic action variables (I_1, w_1) and (I_2, w_2) in which the Hamiltonian can be represented as

$$\mathcal{H} = I_1 - \nu I_2 - \alpha \sin[(2I_1)^{1/2} \sin w_1 - w_2] \tag{12.87}$$

The transformation to the original normalized variables is

$$\begin{aligned} y &= (2I_1)^{1/2} \sin w_1 \\ x &= -I_2 - (2I_1)^{1/2} \cos w_1 \end{aligned} \tag{12.88}$$

and $dw_2/dt = \partial\mathcal{H}/\partial I_2 = \nu$, which gives $w_2 = \nu t$. The equations of motion to be integrated in these variables look

$$\begin{aligned} dw_1/dt &= \partial\mathcal{H}/\partial I_1 = 1 - \alpha(2I_1)^{1/2} \sin w_1 \cos[(2I_1)^{1/2} \sin w_1 - w_2] \\ dw_2/dt &= \partial\mathcal{H}/\partial I_2 = \nu \end{aligned} \tag{12.89}$$

One recognizes that the gyroradius is given by $r_{gi} = (2I_1)^{1/2}$, w_1 is the gyrophase angle, while $w_2 = \nu t$ is the wave phase angle. The best representation of the particle orbit is therefore to consider its crossings of the plane $w_1 = \pi$ in dependence on the gyroradius for different α and different initial start positions.

Figure 12.30 shows three such plots of crossings of the plane $w_1 = \pi$ as function of the gyroradius and of wave phase. The wave frequency is $\omega = 30.23\,\omega_{gi}$ or $\nu = 30.23$, while α assumes the values 1, 2.2, and 4. In the first case the particle orbits are slightly modulated by the presence of the wave, yet behave adiabatically. For larger α the adiabatic motion is destroyed by the large kicks the particles experience near resonance. The closed regular orbits break off creating islands with large regions of stochastic motions between the islands where the particles diffuse across towards larger gyroradii.

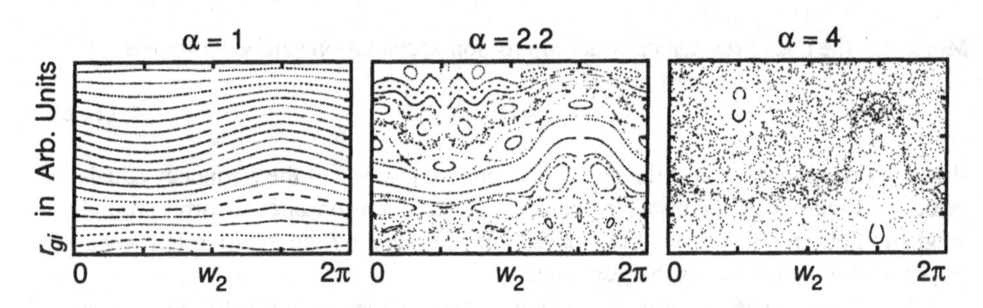

Fig. 12.30. Three different cases of ion orbits in the plane $w_1 = \pi$.

These island disappear for even larger values of α or larger wave amplitudes, indicating strong particle acceleration transverse to the magnetic field.

The above mechanism is nonadiabatic and non-regular. It is based on chaotic resonance of the ions with the wave and leads to chaotic diffusion in phase space and acceleration of ions on the expense of wave energy. Inclusion of parallel motions and parallel wavenumbers slightly changes the picture, but does not change the physics. Actually, ion acceleration due to stochastic motion in wave fields is the most important mechanism of transverse heating and may be responsible for ion conics.

Ponderomotive Force Acceleration

Another mechanism of particle acceleration is based on the ponderomotive force exerted by electromagnetic waves. Since inhomogeneous electromagnetic waves of large temporally varying amplitude exert a radiation pressure onto plasmas, it is immediately understood that the pressure force may act as an accelerating force. Two-fluid theories of ponderomotive forces show that the force is mass-dependent. This property is the reason for mass-selective acceleration in ponderomotive force fields of electromagnetic or plasma waves. The space charge electric field produced by a ponderomotive force on the electrons, $\mathbf{f}_{\mathrm{pme}}$, can be written as

$$\mathbf{E}_{\mathrm{pm}} \approx \mathbf{f}_{\mathrm{pme}}/e \tag{12.90}$$

This field is an ambipolar electric field. Eliminating it from the ion equation of motion, one finds that the ion momentum conservation equation becomes

$$m_i n_0 \frac{d\mathbf{v}_i}{dt} = -\frac{1}{\mu_0}\mathbf{B} \times \nabla \times \mathbf{B} - \nabla p_i + n_0(\mathbf{f}_{\mathrm{pme}} + \mathbf{f}_{\mathrm{pm}i}) \tag{12.91}$$

which shows that the ions can be accelerated by slowly varying electromagnetic wave fields. The ponderomotive forces depend on the model used. Expressions are given in Eq. (11.15). Applying these expressions to an electromagnetic left-circular polarized ion wave, the acceleration parts in the ion equation of motion can be written as

$$\begin{aligned} m_i n_0 \frac{dv_{i\parallel}}{dt} &= -\frac{\omega_{pi}^2}{\omega\, \omega_{gi}} \left[\frac{\omega}{\omega - \omega_{gi}} \nabla_\parallel - \frac{k}{\omega} \left(1 - \frac{\omega_{gi}^2}{(\omega - \omega_{gi})^2} \right) \frac{\partial}{\partial t} \right] W_E \\ m_i n_0 \frac{d\mathbf{v}_{i\perp}}{dt} &= -\frac{\omega_{pe}^2}{\omega_{ge}^2} \left[1 - \frac{m_e}{m_i} \frac{\omega_{ge}^2}{(\omega - \omega_{gi})^2} \right] \nabla_\perp W_E \end{aligned} \tag{12.92}$$

Obviously, there is acceleration of ions in the field of an electromagnetic ion-cyclotron wave with the acceleration acting in both directions. Parallel to the field it will result in acceleration, the transverse effect is heating. The above equations suggest that ion acceleration is significantly enhanced when the frequency is close to the ion-cyclotron frequency. But this conclusion must be taken with care because close to the cyclotron resonance the ponderomotive effect is changed by other kinetic effects.

Diffuse Fermi Acceleration

Very often one observes ion distributions which are not simple beams but diffusely accelerated distributions. We have mentioned such examples in connection with ion reflection and further evolution of the so-called kidney distributions into diffuse ring distributions at the Earth's bow shock wave in the quasi-parallel foreshock region where large-amplitude low-frequency magnetic turbulence evolves into shocklets. The interaction of this turbulence with the non-isotropic ion distribution is believed to lead to strong ion heating and ion acceleration in these cases. This process is usually referred to as *second-order Fermi* acceleration, but in essence it is stochastic acceleration in extended large-amplitude low-frequency electromagnetic waves propagating in the whistler or ion-cyclotron band.

This process is formally described by Eqs. (12.61) and (12.62). The acceleration process is quasilinear. In order to solve these equations, assumptions must be introduced about the damping, growth, and loss rates of particles and waves and about the particle and wave diffusion coefficients. So far these assumptions have been based only on simple considerations, and nonlinear effects have never been taken into account. Usually a spatial diffusion coefficient parallel to the magnetic field, $\kappa_\parallel = \langle(\Delta s)^2\rangle/(2\Delta t)$, is introduced. This coefficient gives the diffusive escape time of the particles as $\tau_d \approx L^2/8\kappa_\parallel$, where $L = 2\Delta s$ is the length of the system along the magnetic field. An approximation for $\kappa_\parallel$ is obtained from pitch angle diffusion theory. The resulting approximative expression for the particle loss-time is

$$\tau_d = \frac{9\pi^2 e^2}{4c^2} \frac{L^2}{m_i^2 v^3} \int_{k_0}^{\infty} \frac{dk_\parallel}{k_\parallel} \left(1 - \frac{k_0^2}{k_\parallel^2} W_w(k_\parallel \right) \tag{12.93}$$

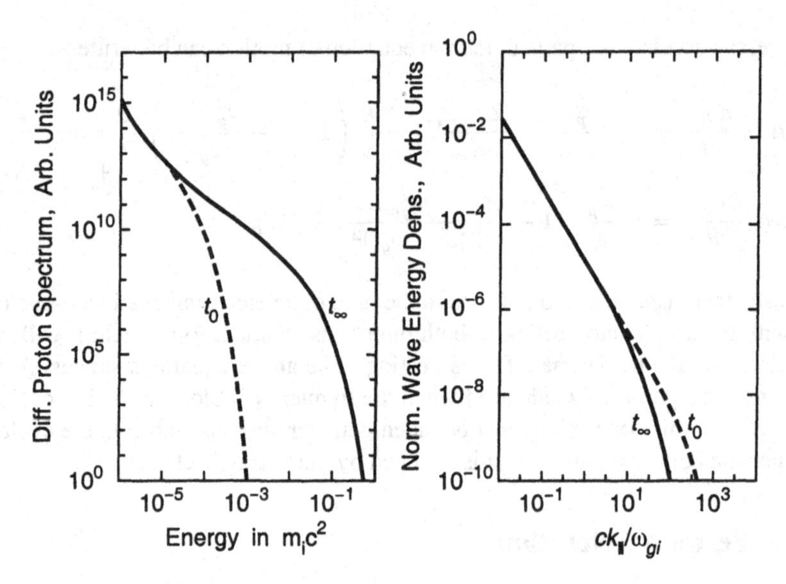

Fig. 12.31. Stochastically accelerated ion energy spectrum in MHD waves.

In a similar way the diffusion coefficient of the ions can be written

$$D(v) = \frac{2\pi^2 e^2}{v} \frac{v_A^2}{c^2} \int_{k_0}^{\infty} \frac{dk_\parallel}{k_\parallel} \left(1 - \frac{k_0^2}{k_\parallel^2} W_w(k_\parallel\right) \tag{12.94}$$

Here k_0 is the smallest wavenumber of the extended wave spectrum, and W_w is the electric wave spectral density. The wave diffusion coefficient is estimated assuming a Kolmogorov law with inertial range turbulence

$$D_w \propto v_A |k_\parallel|^{7/2} \beta_w^{1/2}(k_\parallel) \tag{12.95}$$

where $\beta_w = 2\mu_0 W_w / B^2$ is the wave plasma beta. One needs further assumptions about the wave absorption rate, γ_l, and energy gain in the source and loss terms in Eq. (12.62) to solve the diffusive acceleration equations numerically.

Figure 12.31 shows a numerical example of an ion spectrum resulting from stochastic acceleration in Alfvén waves without any nonlinear evolution of the waves. The wave energy spectrum is given in the right part of the figure as function of the ratio $ck_\parallel/\omega_{gi}$. The main lesson learned from this acceleration process is that with increasing time the evolution of the energy spectrum is towards the generation of a high-energy tail with sharp cut-off at large energies where transit time effects limit the acceleration. Spectra like this indeed resemble measured spectra of particle acceleration in solar flares and in the diffuse foreshock region.

The current theory is semi-selfconsistent. It takes into account the modification of the spectrum and particle fluxes under the interaction, but it introduces severe assumptions about the spatial extent of the waves and about the growth, damping, loss, and diffusion rates. Nevertheless, it explains the generation of very high energy tails as observed in foreshocks, solar flares, and in cosmic rays.

Concluding Remarks

We have excluded many facts and fields from this last chapter, mainly because of lack of space. In particular, we regret not having been able to review slow and intermediate shock theories, which all still are under ongoing discussion. Also shock acceleration in its two variants, first-order Fermi and shock drift acceleration, has been left aside. In spite of their wide application in cosmic ray physics and astrophysics these theories form the simpler part of all acceleration mechanisms.

Further Reading

The general theory of transport coefficients is summarized in [1]. The calculation of anomalous transport coefficients is described in [2] and [4]. The calculation of form factors is given in [3]. Thresholds of ion-acoustic turbulence were determined by Kindel and Kennel, *J. Geophys. Res.* **76** (1971) 3055. Ion-cyclotron collision frequencies are given by Dum and Dupree, *Phys. Fluids* **13** (1970) 2064. Ion-acoustic resistivity is the subject of Dum, *Phys. Fluids*, **21** (1978) 945. Strongly turbulent Langmuir collision frequencies are found in Sagdeev, *Rev. Mod. Phys.*, **51** (1979) 1.

Observational information on shocks is found in [7] and [9]. The older theories of laminar and turbulent shocks are discussed in [8], more recent information is found in [6]. The various critical Mach numbers are discussed by Kennel et al. in [7]. Microinstabilities and their effects on shocks are found in Wu et al., *Space Sci. Rev.*, **37** (1984) 63. Ion distributions in the foreshock are given in Sckopke et al., *J. Geophys. Res.*, **88** (1983) 6121. Measured ion-cyclotron spectra are presented by Sckopke et al., *J. Geophys. Res.*, **95** (1990) 6337. Waves in the magnetosheath are summarized by Lacombe and Belmont in [6]. A summary of electron distribution functions in and near shocks is given by Feldman in [9]. Upstream waves, slams and shocklets and their role in reformation are discussed by Scholer, *J. Geophys. Res.*, **98** (1993) 47.

For the kinetic Alfvén wave mechanism of auroral acceleration we followed Lysak and Dum, *J. Geophys. Res.*, **88** (1983) 365. Ion hole formation is taken from Tetreault, *J. Geophys. Res.*, **96** (1991) 3549 and Gray et al., *Geophys. Res. Lett.*, **18** (1991) 1675. For the acceleration in extended lower-hybrid wave fields see Wu et al., *J. Plasma Phys.*, **25** (1981) 391. The theory of localized electron acceleration follows work by Dubouloz et al., *Geophys. Res. Lett.*, **22** (1995) 2969. Perpendicular ion heating is the subject of Karney, *Phys. Fluids*, **21** (1978) 1584 & **22** (1979) 2188. Ion acceleration in extended wave fields is discussed by Miller and Roberts, *Astrophys. J.*, **452** (1995) 912. Ponderomotive force acceleration was mentioned first by Pottelette et al. in [5].

[1] S. I. Braginskii, in *Reviews of Plasma Physics 1*, ed. M. A. Leontovich (Consultants Bureau, New York, 1965), p. 205.

[2] A. A. Galeev and R. Z. Sagdeev, in *Reviews of Plasma Physics 7*, ed. M. A. Leontovich (Consultants Bureau, New York, 1979), p. 1.

[3] S. Ichimaru, *Basic Principles of Plasma Physics* (W. A. Benjamin, Reading, 1973).

[4] B. B. Kadomtsev, *Plasma Turbulence* (Academic Press, New York, 1965).

[5] R. L. Lysak (ed.), *Auroral Plasma Dynamics* (American Geophysical Union, Washington, 1993).

[6] C. T. Russell (ed.), *Advances of Space Research 15: Physics of Collisionless Shocks* (Pergamon, Oxford, 1995).

[7] R. G. Stone and B. T. Tsurutani (eds.), *Collisionless Shocks in the Heliosphere: A Tutorial Review* (American Geophysical Union, Washington, 1985).

[8] D. A. Tidman and N. A. Krall, *Shock Waves in Collisionless Plasmas* (Wiley Interscience, New York, 1971).

[9] B. T. Tsurutani and R. G. Stone (eds.), *Collisionless Shocks in the Heliosphere: Reviews of Current Research* (American Geophysical Union, Washington, 1985).

[10] V. N. Tsytovich, *Nonlinear Effects in Plasmas* (Plenum Press, New York, 1970).

Epilogue

There would be infinitely more to say about observation and theory of space and astrophysical plasmas. But an introductory text must close at some point. In our two books, *Basic Space Plasma Physics* and *Advanced Space Plasma Physics*, we have tried to give an overview of the current state of the art in space plasma physics on a level which we hope is accessible to the student and to the beginning researcher. Clearly, both volumes should be taken together in order to get a relatively complete picture of space plasma physics, reaching from the elementary level of particle motion in crossed electric and magnetic fields through the state of plasma equilibria up to the more sophisticated level of nonlinear plasma theory.

As far as the available space allowed, we tried to follow the demand of theoretical rigor. However, the reader will find that this intention has been violated at several places, where we skipped the derivation and went on to a verbal description. In all these cases we have put effort into a relatively clear description of the physics involved and a discussion of its consequences. We feel that sometimes such a choice is more valuable than being lost in the rigorous but complicated mathematical jungle. In most of those cases we tried, however, to write down the fundamental mathematical expressions.

As the authors of this text we are left with the unpleasant feeling that we have only touched the problems, skipped a large number of important and interesting effects, which we either felt to go too far beyond an introductory presentation or, worse, we have not been aware of. In the latter case we would be grateful for hints which fields should be included in any possible forthcoming edition. However, the inevitable incompleteness of this course may be compensated by consulting the book edited by M. G. Kivelson and C. T. Russell, *Introduction to Space Physics*. There the reader will find a more phenomenological description of many space plasma phenomena which we have mentioned (or neglected). In a sense the two approaches to space plasma physics given there and here complement each other.

Numerical simulations are the largest field which we have excluded from this introduction to space plasma physics. Contemporary theoretical space plasma physics is to a large part based on numerical simulations. The great advantage of simulations is that they include the nonlinear evolution of the simulated system in a quite natural way. Thus they provide deep insight into the evolution of many phenomena which sometimes cannot even be formulated analytically. There are many examples, especially involving inhomogeneities and nonlinearities, where problems could only be solved by the simulation technique.

Interestingly, numerical simulations create a different view of the phenomena. Numerical simulation can be taken as an experiment done not on real but model plasmas. Space plasma physics has used this possibility in order to investigate those aspects of which real experiments in space can give only sporadic information. This tendency is interesting and challenging, because for a really deep understanding of the various natural and simulated phenomena the combination of observation, simulation, and analytical investigation is required. To provide the basis for the latter was the intention of the writing of this book. But in application to real problems the contemporary researcher will typically enter numerical simulations. A small number of books where an introduction into numerical simulation technique can be found have been mentioned in the last few chapters.

Index